AF787133

Antibacterial, Antifungal, and Antiviral Bioactive Compounds from Natural Products

Antibacterial, Antifungal, and Antiviral Bioactive Compounds from Natural Products

Editors

Xun Song
Chenyang Li
Yifu Guan

MDPI

Basel • Beijing • Wuhan • Barcelona • Belgrade • Novi Sad • Cluj • Manchester

Editors
Xun Song
College of Pharmacy
Shenzhen Technology
University
Shenzhen
China

Chenyang Li
School of Pharmacy
Shenzhen University
Shenzhen
China

Yifu Guan
School of Chemistry and
Chemical Engineering
Guangxi Minzu University
Nanning
China

Editorial Office
MDPI
St. Alban-Anlage 66
4052 Basel, Switzerland

This is a reprint of articles from the Special Issue published online in the open access journal *Molecules* (ISSN 1420-3049) (available at: www.mdpi.com/journal/molecules/special_issues/0X0VNJVF40).

For citation purposes, cite each article independently as indicated on the article page online and as indicated below:

Lastname, A.A.; Lastname, B.B. Article Title. *Journal Name* **Year**, *Volume Number*, Page Range.

ISBN 978-3-7258-0790-1 (Hbk)
ISBN 978-3-7258-0789-5 (PDF)
doi.org/10.3390/books978-3-7258-0789-5

Cover image courtesy of Xun Song

© 2024 by the authors. Articles in this book are Open Access and distributed under the Creative Commons Attribution (CC BY) license. The book as a whole is distributed by MDPI under the terms and conditions of the Creative Commons Attribution-NonCommercial-NoDerivs (CC BY-NC-ND) license.

Contents

About the Editors

Xun Song

Dr. Xun Song is currently an assistant professor at the College of Pharmacy, Shenzhen Technology University. He has more than ten years of research experience in the isolation, identification, structure elucidation, analysis and biological evaluation of small molecules of natural products, leading to the publication of more than 40 peer-reviewed papers. Many of these compounds showed biological activities against cancer, viruses, bacteria and fungi, and some of them have been evaluated in *in vivo* studies. His current research interest is focused on natural product drug discovery and development from natural resources, as well as the development of botanical dietary supplements from herbal medicines. Specifically, he is interested in finding natural lead compounds from the fermentation of mushrooms, and developing dietary supplements from traditional herbal medicines against different disease targets and pathogens such as fungi, bacteria, viruses and cancer chemoprevention.

Chenyang Li

Dr. Chenyang Li is currently the director of the Teaching and Research Section of Traditional Chinese Medicine, a specially appointed associate researcher, a master's supervisor, a back-up high-level leading talent in Shenzhen, a member of the First Committee of Pharmacy and Rehabilitation of the China Rehabilitation Medicine Association, a senior member of the Guangdong Provincial Drug and Food Intelligent Manufacturing Engineering Society, and a member of the Young Scientists Association of Shenzhen University. He is also a young Editorial Board Member of the international journal *Chinese Medicine* and the *Journal of Rehabilitation*, a member of the first committee of the Professional Committee of Pharmaceutical Teachers Development of Guangdong Medical Education Association, a member of the second committee of the Medical Education Development Center of Shenzhen University, and an innovation and entrepreneurship mentor at Shenzhen University.

His current research focuses on natural products from traditional Chinese medicine and microorganisms (including natural product chemistry, biological activity and mechanism of action), as well as innovative tumor drug research based on target design (including drug molecule design, synthesis, biological activity and mechanism of action). He has published more than 20 SCI papers in internationally renowned authoritative journals such as *Theranostics, Journal of Agricultural and Food Chemistry, European Journal of Medicinal Chemistry, Biomedicine & Pharmacotherapy*, and *Phytomedicine*, and has been authorized 4 Chinese national invention patents.

Yifu Guan

Dr. Yifu Guan is currently a professor at the School of Chemistry and Chemical Engineering and the director of the Department of Chemistry at Guangxi Minzu University. He has more than 15 years of research experience in the total synthesis and structural optimization of bioactive natural products, leading to the publication of over 20 peer-reviewed papers and 7 patents. He has achieved the total synthesis of 8 bioactive natural products such as zelkovamycin H, auyuittuqamides E-H and fusarihexins D-E, and more than 200 natural product analogs were obtained in the past 5 years. Many of these analogs showed bioactivities against cancer, bacteria and fungi.

He has undertaken seven research projects as the principal investigator, including two National Natural Science Foundation of China (NSFC) projects, two Science and Technology projects of Guangxi and three other projects. He is an expert reviewer for the NSFC and Science and Technology projects of Guangxi, and a reviewer for several international journals.

Preface

The exploration of natural compounds for their antibacterial, antifungal and antiviral properties represents a compelling avenue of research in contemporary science. This collection delves into the rich diversity of bioactive molecules derived from nature and their potential applications in combating infectious diseases.

The aim of this Special Issue is to report the latest and most promising natural compounds used to combat bacteria, fungi and viruses, with the underlying modes of action outlined. Our motivations for compiling this scientific endeavor are rooted in the need for novel antimicrobial agents amid the escalating global threat of drug-resistant pathogens. Traditional antibiotics and antifungals are facing diminishing effectiveness, necessitating the exploration of alternative strategies. Natural products offer a promising reservoir of bioactive compounds with diverse chemical scaffolds and mechanisms of action, presenting new avenues for drug discovery and development.

This work is intended for a broad audience encompassing researchers, academicians, pharmaceutical scientists and healthcare professionals engaged in microbiology, pharmacology, medicinal chemistry and natural product chemistry. We hope to inspire dialogue and collaboration across disciplines, fostering innovation in the search for effective antimicrobial agents.

The authors contributing to this Special Issue bring together expertise from various scientific backgrounds, including microbiology, pharmacology, organic chemistry and bioinformatics. Their collective efforts have culminated in a comprehensive compilation of research findings, methodologies and perspectives to advance our understanding of natural bioactive compounds and their therapeutic potential.

We express our heartfelt acknowledgment to all those who have contributed to this work, whether through direct assistance, intellectual support or inspiration. In particular, we extend our gratitude to the countless researchers and scientists whose pioneering work has paved the way for discoveries in natural product-based drug development.

We hope that this Special Issue will serve as a timely reference for researchers who are interested in the discovery of useful natural products for the development of novel antimicrobial drugs.

Xun Song, Chenyang Li, and Yifu Guan
Editors

molecules

Editorial

Antibacterial, Antifungal, and Antiviral Bioactive Compounds from Natural Products

Xun Song

College of Pharmacy, Shenzhen Technology University, Shenzhen 518118, China; xsong@szu.edu.cn

1. Introduction

In the relentless pursuit of innovative therapeutic agents, natural products have emerged as a transformative avenue in the battle against infectious diseases. This Special Issue, entitled "Antibacterial, Antifungal, and Antiviral Bioactive Compounds from Natural Products", seeks to unravel the potential within nature's pharmacopeia and its ability to provide a rich source of diverse compounds with remarkable antimicrobial properties [1]. Historically integral to traditional medicine, natural products offer a palette of compounds, including alkaloids, flavonoids, terpenoids, and peptides from various sources, showcasing the versatility of nature's defenses against bacteria, fungi, and viruses [2].

The increasing threat of drug-resistant microorganisms necessitates innovative approaches, and natural compounds present a promising solution [3]. Studies have highlighted their efficacy against a spectrum of bacteria, addressing the multifaceted nature of bacterial infections. Beyond conventional antifungal agents, fungal infections prompt the exploration of natural products from fungi, plants, and bacteria that exhibit a range of antifungal activities. This diversity broadens the scope for discovering new antifungal drugs to combat emerging threats.

In the realm of antiviral research, natural products demonstrate inhibitory effects against various viruses; this is vital in addressing the evolving landscape of viral infections, which is exemplified by global pandemics [4]. Derived compounds offer a foundation for understanding viral interactions and developing targeted interventions.

As we embark on this exploration of bioactive compounds, our aim is to unveil not only their therapeutic potential, but also the intricate mechanisms through which nature defends against microbial adversaries. Additionally, this Special Issue discusses a range of antimicrobial natural products with other biological functions, thus further highlighting the diverse capabilities within nature's pharmacopeia.

2. An Overview of Published Articles

Xiaopeng Hu's article [Contribution 1] discusses the potential inhibitors of the monkeypox virus. Monkeypox outbreaks pose a global health threat, exacerbated by the absence of effective medicines against orthopoxviruses. Molecular modeling, focusing on natural products like traditional Chinese medicine (TCM), has revealed potential inhibitors. Three compounds—rosmarinic acid, myricitrin, quercitrin, and ofloxacin—have demonstrated significant binding to monkeypox DNA topoisomerase I, suggesting promising antiviral effects. Molecular dynamics simulations support their stability. The study by Hu et al. highlights the potential application of these compounds as poxvirus inhibitors, emphasizing the need for further research to assess their therapeutic efficacy.

The review by Mariana Rodríguez-Cisneros et al. [Contribution 2] is focused on the antimicrobial compounds produced by *Burkholderia* Sensu Stricto. The rise in multi-drug-resistant microorganisms is currently demanding the exploration of novel antimicrobial compounds. The World Health Organization has highlighted that *Acinetobacter baumannii*, *Pseudomonas aeruginosa*, and *Enterobacteriaceae* are critical bacteria that require urgent eradication. *Burkholderia*, known for its production of antimicrobials, offers a diverse array

check for updates

Citation: Song, X. Antibacterial, Antifungal, and Antiviral Bioactive Compounds from Natural Products. *Molecules* **2024**, *29*, 825. https://doi.org/10.3390/molecules29040825

Received: 1 February 2024
Accepted: 2 February 2024
Published: 11 February 2024

Copyright: © 2024 by the author. Licensee MDPI, Basel, Switzerland. This article is an open access article distributed under the terms and conditions of the Creative Commons Attribution (CC BY) license (https://creativecommons.org/licenses/by/4.0/).

of compounds, including N-containing heterocycles, polyenes, bacteriocins, and more. These compounds exhibit potential not only as antimicrobials against bacteria and fungi, but also as candidates for anticancer or antitumor agents. The review comprehensively explores Burkholderia's extensive repertoire of antimicrobial compounds, with a focus on those tested in vitro, and includes information on novel compounds discovered through genome-guided approaches.

The review by Meijie Xu et al. [Contribution 3] highlights the 185 antimicrobial natural products that had been extracted from *Fusarium* strains by 2022, showcasing their antibacterial, antifungal, antiviral, and antiparasitic effects. The comprehensive analysis emphasizes the importance of *Fusarium* as a valuable reservoir for new bioactive secondary metabolites. The review concludes by proposing methodologies for the efficient discovery of natural products obtained from *Fusarium* strains.

The fourth article published in this Special Issue is by Ting-Ting Tang et al. [Contribution 4] and focuses on the use of flavonoids obtained from *Scutellaria barbata* D. Don (SB) as inhibitors of HIV-1 and cathepsin L proteases. SB is a medicinal plant rich in flavonoids, and is known for its antitumor, anti-inflammatory, and antiviral properties. The study evaluated SB extracts and identified nine flavonoids that inhibit HIV-1 and SARS-CoV-2 proteases. The results revealed that scutellarein is a lead compound with potent dual inhibitory activity against both proteases. The study emphasized the importance of specific hydroxyl group introductions in flavones regarding the enhancement of anti-protease activities. Notably, luteolin exhibited the potent and selective inhibition of HIV-1 protease. These findings suggest the potential use of SB-derived flavonoids as promising candidates in the development of effective dual protease inhibitors.

In their study, Ben Chung-Lap Chan et al. [Contribution 5] create a vaccine-like supplement against respiratory infection using recombinant *bacillus subtilis* spores expressing SARS-CoV-2 spike protein with natural products. This study explores an alternative approach to COVID-19 vaccination using engineered *Bacillus subtilis* to produce "S spores" that mimic the SARS-CoV-2 spike protein. Administered orally with the natural adjuvants *Astragalus membranaceus* and *Coriolus versicolor*, the S spores induced mild immune responses against COVID-19 without toxicity. Co-administration enhanced the mucosal IgA responses, while prior oral inoculation expedited and strengthened the IgG responses when followed by the commercial vaccine CoronaVac. In vitro studies demonstrated immune activation by AM, CV, and *B. subtilis* spores. This novel combination shows potential in the development of a supplementary vaccine against respiratory infections, addressing concerns regarding needle aversion and side effects.

In the sixth article, Ming-Feng He [Contribution 6] focuses on the antiviral mechanism of the alkaloid coptisine against influenza virus. This study identifies that coptisine, a compound found in Chinese herbs, is a potent inhibitor of the influenza virus, exhibiting an EC_{50} of 10.7 μM in MDCK cells. Coptisine's antiviral effects are attributed to its upregulation of the p21 signaling pathway, leading to the increased expression of p21 and FOXO1. The compound's effectiveness is highlighted in pre-treatment, showing a superior reduction in viral replication compared to co-treatment or post-treatment. Docking analysis suggests that coptisine inhibits MELK activity, forming hydrogen bonds in the catalytic pocket. These findings position coptisine as a promising antiviral agent with the ability to regulate the p21 pathway to combat viral infections.

The research presented by Guangxin Chen [Contribution 7] delves into the antioxidant and anti-inflammatory effects of cepharanthine (CEP). CEP, a biscoclaurine alkaloid extracted from *Stephania cepharantha* Hayata, has been widely utilizes in the treatment of various acute and chronic diseases, including leukopenia, and snake bites. The study explores the therapeutic potential of CEP in countering oxidative stress and inflammation. CEP effectively alleviates weight loss, reduces disease activity, and maintains intestinal integrity in colitis mice. It mitigates malondialdehyde levels, boosts the glutathione content, and inhibits inflammatory responses. Additionally, CEP activates the AMPK-α1/AKT/GSK-3β/NRF2 signaling pathway while suppressing MAPKs and NF-κB p65 pathways. The protective effects extend to leukopenia and snake bites but are compromised in NRF2 knockout mice. This research underscores that CEP is a promising therapeutic agent against oxidative stress and inflammation, emphasizing the potential use of NRF2 as a target for treating inflammatory bowel disease.

Lihan Zhao's study [Contribution 8] focuses on the isolation and anticancer progression of the chemical constituents of *Bridelia balansae* Tutcher. The study highlights the anticancer potential of the dichloromethane extract obtained from *Bridelia balansae* Tutcher roots against HCT116 colorectal cancer cells. Fourteen compounds were identified, including a novel aryltetralin lignan; this was 4′-demethyl-4-deoxypodophyllotoxin (**1**), whose stereochemistry was validated via X-ray crystallography. Compound **1** exhibited significant cytotoxicity (IC_{50} at 20 nM) through the induction of apoptosis, reducing HCT116 cell migration and down-regulating MMP2 and p-Akt, and upregulating p21. This comprehensive analysis positions *B. balansae* as a valuable source of potential anticancer lead compounds, emphasizing the therapeutic promise of compound **1** against the proliferation and metastasis of cancer.

Ruirui Yu's work [Contribution 9] focuses on the isolation and identification of the chemical compounds obtained from *Agaricus blazei* Murrill and their in vitro antifungal activities. This research delves into the antifungal properties of *A. blazei*, an esteemed medicinal and edible fungus. Six isolated compounds from *A. blazei*, including linoleic acid, 1,1′-oxybis(2,4-di-tert-butylbenzene), glycerol monolinoleate, volemolide (17R)-17-methylincisterol, (24s)-ergosta-7-en-3-ol, and dibutyl phthalate, were identified and evaluated for their antifungal activities against *Trichophyton mentagrophology*, *Trichophyton rubrum*, *Candida albicans*, and *Cryptococcus neoformans*. Notably, compound **2** displayed significant inhibition against *T. mentagrophology*, compound **3** against *T. rubrum*, and compound **6** against *C. albicans*. The findings underscore *A. blazei*'s medicinal potential as an antifungal agent, revealing promising avenues for further research.

Jinrong Lin et al. [Contribution 10] performed an evaluation of the anti-cancer activity of lasiokaurin (LAS), a natural antimicrobial diterpenoid, in a triple-negative breast cancer (TNBC) model. TNBC poses challenges due to its aggressiveness and limited treatment options. This study explores LAS as a promising anti-TNBC agent. LAS exhibits significant efficacy in inhibiting TNBC cell growth, with its induction of cell cycle arrest, apoptosis, and DNA damage, and suppression of metastasis. It targets multiple pathways, including PI3K/Akt/mTOR and STAT3, suggesting its potential application in multitarget therapy. Moreover, the application of LAS has been found toe safely reduce tumor growth in a mouse model, suggesting its utilization as a potent candidate for TNBC treatment.

3. Conclusions

In the face of infectious diseases, the urgent need for innovative antimicrobial agents is underscored by antibiotic resistance, persistent fungal infections, and the constant threat of viral outbreaks. The exploration of natural products as reservoirs of antibacterial, antifungal, and antiviral bioactive compounds has garnered significant attention, thus paving the way for potential breakthroughs in the combat of microbial pathogens.

3.1. Antibacterial Bioactive Compounds: A Nature-Inspired Arsenal against Bacterial Infections

Natural products appear to possess a diverse array of antibacterial compounds with therapeutic promise. Compounds such as alkaloids, flavonoids, and terpenoids derived from plants exhibit potent antibacterial activity against a spectrum of strains. Essential oils, exemplified by constituents like thymol and carvacrol, showcase remarkable efficacy against both Gram-positive and Gram-negative bacteria. Additionally, marine organisms have been found contribute unique antibacterial peptides, such as bacteriocins, that exhibit selective toxicity towards bacterial pathogens [5,6].

3.2. Antifungal Bioactive Compounds: Nature's Defense against Fungal Pathogens

Fungal infections pose a persistent threat to global health, particularly in immunocompromised individuals. Natural products offer diverse antifungal agents that target various stages of the fungal life cycle. Polyphenols such as resveratrol and catechins disrupt fungal cell membranes, while secondary metabolites such as mycotoxins exhibit potent antifungal properties [7,8].

3.3. *Antiviral Bioactive Compounds: Harnessing Nature's Defense Mechanisms*

The ongoing battle against viral infections requires the implementation of innovative antiviral strategies. The natural products that are derived from plants, including polyphenols, flavonoids, and alkaloids, interfere with viral processes such as entry, replication, and maturation. Marine organisms also contribute unique antiviral compounds that exhibit promise regarding the inhibition of viral infections, such as sulfated polysaccharides and lectins [9,10].

3.4. *Mechanisms of Action: Unveiling the Versatility of Natural Bioactive Compounds*

The success of natural bioactive compounds against microbial pathogens lies in their diverse mechanisms of action. Antibacterial compounds disrupt cell membranes, inhibit protein synthesis, or interfere with metabolic pathways. Antifungal agents target cell membranes, inhibit ergosterol synthesis, and affect specific cellular processes. Antiviral compounds interfere with the replication, assembly, or entry of the virus into host cells, providing a multi-pronged defense against viral infections.

3.5. *Challenges and Future Directions: Toward Sustainable Antimicrobial Solutions*

While the potential application of natural bioactive compounds is vast, addressing challenges such as bioavailability, standardization, and sustainability is crucial for their successful clinical translation. The development of resistance to natural compounds in microbial pathogens necessitates concerted efforts in research and development.

In conclusion, the exploration of the antibacterial, antifungal, and antiviral bioactive compounds obtained from natural products holds great promise for novel antimicrobial agents. Nature's diverse arsenal, honed over millions of years of evolution, forms a foundation for innovative drug discovery. Their multifaceted mechanisms of action, coupled with recent advancements in extraction techniques and analytical methods, position natural products as invaluable elements in the battle against microbial infections. Interdisciplinary collaborations between chemists, biologists, and clinicians will play a pivotal role in fully unlocking the potential of these natural bioactive compounds, and pave the way for a sustainable and effective approach to global infectious disease management.

Conflicts of Interest: The authors declare no conflict of interest.

List of Contributions:

1. Hu, X.; An, S.; Chu, J.; Liang, B.; Liao, Y.; Jiang, J.; Lin, Y.; Ye, L.; Liang, H. Potential Inhibitors of Monkeypox Virus Revealed by Molecular Modeling Approach to Viral DNA Topoisomerase I. *Molecules* **2023**, *28*, 1444. https://doi.org/10.3390/molecules28031444.
2. Rodríguez-Cisneros, M.; Morales-Ruíz, L.; Salazar-Gómez, A.; Rojas-Rojas, F.; Estrada-de los Santos, P. Compilation of the Antimicrobial Compounds Produced by Burkholderia Sensu Stricto. *Molecules* **2023**, *28*, 1646. https://doi.org/10.3390/molecules28041646.
3. Xu, M.; Huang, Z.; Zhu, W.; Liu, Y.; Bai, X.; Zhang, H. Fusarium-Derived Secondary Metabolites with Antimicrobial Effects. *Molecules* **2023**, *28*, 3424. https://doi.org/10.3390/molecules28083424.
4. Tang, T.; Li, S.; Pan, B.; Xiao, J.; Pang, Y.; Xie, S.; Zhou, Y.; Yang, J.; Wei, Y. Identification of Flavonoids from Scutellaria barbata D. Don as Inhibitors of HIV-1 and Cathepsin L Proteases and Their Structure–Activity Relationships. *Molecules* **2023**, *28*, 4476. https://doi.org/10.3390/molecules28114476.
5. Chan, B.; Li, P.; Tsang, M.; Sung, J.; Kwong, K.; Zheng, T.; Hon, S.; Lau, C.; Cheng, W.; Chen, F.; Lau, C.; Leung, P.; Wong, C. Creating a Vaccine-like Supplement against Respiratory Infection Using Recombinant Bacillus subtilis Spores Expressing SARS-CoV-2 Spike Protein with Natural Products. *Molecules* **2023**, *28*, 4996. https://doi.org/10.3390/molecules28134996.

6. He, M.; Liang, J.; Shen, Y.; Zhang, C.; Yang, K.; Liu, L.; Xie, Q.; Hu, C.; Song, X.; Wang, Y. Coptisine Inhibits Influenza Virus Replication by Upregulating p21. *Molecules* **2023**, *28*, 5398; https://doi.org/10.3390/molecules28145398.
7. Chen, G.; Wen, D.; Shen, L.; Feng, Y.; Xiong, Q.; Li, P.; Zhao, Z. Cepharanthine Exerts Antioxidant and Anti-Inflammatory Effects in Lipopolysaccharide (LPS)-Induced Macrophages and DSS-Induced Colitis Mice. *Molecules* **2023**, *28*, 6070. https://doi.org/10.3390/molecules28166070.
8. Zhao, L.; Xie, W.; Du, Y.; Xia, Y.; Liu, K.; Ku, C.; Ou, Z.; Wang, M.; Zhang, H. Isolation and Anticancer Progression Evaluation of the Chemical Constituents from Bridelia balansae Tutcher. *Molecules* **2023**, *28*, 6165. https://doi.org/10.3390/molecules28166165.
9. Yu, R.; Li, X.; Yi, P.; Wen, P.; Wang, S.; Liao, C.; Song, X.; Wu, H.; He, Z.; Li, C. Isolation and Identification of Chemical Compounds from Agaricus blazei Murrill and Their In Vitro Antifungal Activities. *Molecules* **2023**, 28, 7321. https://doi.org/10.3390/molecules28217321.
10. Lin, J.; Qu, Z.; Pu, H.; Shen, L.; Yi, X.; Lin, Y.; Gong, R.; Chen, G.; Chen, S. In Vitro and In Vivo Anti-Cancer Activity of Lasiokaurin in a Triple-Negative Breast Cancer Model. *Molecules* **2023**, *28*, 7701. https://doi.org/10.3390/molecules28237701.

References

1. Harvey, A.L.; Edrada-Ebel, R.; Quinn, R.J. The re-emergence of natural products for drug discovery in the genomics era. *Nat. Rev. Drug Discov.* **2015**, *14*, 111–129. [CrossRef] [PubMed]
2. Dias, D.A.; Urban, S.; Roessner, U. A historical overview of natural products in drug discovery. *Metabolites* **2012**, *2*, 303–336. [CrossRef] [PubMed]
3. Salam, M.A.; Al-Amin, M.Y.; Salam, M.T.; Pawar, J.S.; Akhter, N.; Rabaan, A.A.; Alqumber, M.A.A. Antimicrobial Resistance: A Growing Serious Threat for Global Public Health. *Healthcare* **2023**, *11*, 1946. [CrossRef] [PubMed]
4. Chakravarti, R.; Singh, R.; Ghosh, A.; Dey, D.; Sharma, P.; Velayutham, R.; Roy, S.; Ghosh, D. A review on potential of natural products in the management of COVID-19. *RSC Adv.* **2021**, *11*, 16711–16735. [CrossRef] [PubMed]
5. Ruiz, G.; Loris, R.; Watzlawick, H. Overview of antimicrobial peptides. *Nat. Rev. Microbiol.* **2019**, *17*, 353–366.
6. Nazzaro, F.; Fratianni, F.; De Martino, L.; Coppola, R.; De Feo, V. Effect of essential oils on pathogenic bacteria. *Pharmaceuticals* **2013**, *6*, 1451–1474. [CrossRef] [PubMed]
7. Silva, A.; Santos, D.; Silva, C.; Amorim, L.; Barreto, E.; Câmara, C.; Souza, J. Antifungal activity and computational approach of (+)-limonene epoxide and its enantiomers. *Front. Microbiol.* **2017**, *8*, 885.
8. Fisher, M.C.; Hawkins, N.J.; Sanglard, D.; Gurr, S.J. Worldwide emergence of resistance to antifungal drugs challenges human health and food security. *Science* **2018**, *360*, 739–742. [CrossRef] [PubMed]
9. Mukhtar, M.; Arshad, M.; Ahmad, M.; Pomerantz, R.J. Antiviral potentials of medicinal plants. *Virus Res.* **2008**, *131*, 111–120. [CrossRef] [PubMed]
10. Carlucci, M.J.; Ciancia, M.; Matulewicz, M.C.; Cerezo, A.S.; Damonte, E.B. Antiherpetic activity and mode of action of natural carrageenans of diverse structural types. *Antivir. Res.* **2012**, *95*, 93–101. [CrossRef] [PubMed]

Disclaimer/Publisher's Note: The statements, opinions and data contained in all publications are solely those of the individual author(s) and contributor(s) and not of MDPI and/or the editor(s). MDPI and/or the editor(s) disclaim responsibility for any injury to people or property resulting from any ideas, methods, instructions or products referred to in the content.

molecules

MDPI

Article

In Vitro and In Vivo Anti-Cancer Activity of Lasiokaurin in a Triple-Negative Breast Cancer Model

Jinrong Lin [1], Zhao Qu [2,*], Huanhuan Pu [1], Li-Sha Shen [3], Xianguo Yi [4], Yu-Shan Lin [5], Rui-Hong Gong [6], Guo-Qing Chen [5,6,7,*] and Sibao Chen [1,5,6,7,*]

1 Institute of Medicinal Plant Development, Chinese Academy of Medical Sciences and Peking Union Medical College, Beijing 100193, China
2 Third-Grade Pharmacological Laboratory on Chinese Medicine Approved by State Administration of Traditional Chinese Medicine, Medical College, China Three Gorges University, Yichang 443002, China
3 Chongqing Academy of Chinese Materia Medica, Chongqing 400065, China
4 College of Animal Science and Technology, Xinyang Agricultural and Forestry University, Xinyang 464000, China; yixianguo@126.com
5 State Key Laboratory of Chinese Medicine and Molecular Pharmacology (Incubation), The Hong Kong Polytechnic University Shenzhen Research Institute, Shenzhen 518057, China
6 Department of Food Science and Nutrition, The Hong Kong Polytechnic University, Hung Hom, Hong Kong 999077, China
7 Research Centre for Chinese Medicine Innovation, The Hong Kong Polytechnic University, Hung Hom, Hong Kong 999077, China
* Correspondence: quzhao@ctgu.edu.cn (Z.Q.); guoqing.chen@polyu.edu.hk (G.-Q.C.); sibao.chen@polyu.edu.hk (S.C.); Tel.: +86-0717-6396558 (Z.Q.); +86-0755-26737401 (G.-Q.C.); +86-0755-26737182 (S.C.); Fax: +86-0755-26972852 (S.C.)

Abstract: Due to its intricate heterogeneity, high invasiveness, and poor prognosis, triple-negative breast cancer (TNBC) stands out as the most formidable subtype of breast cancer. At present, chemotherapy remains the prevailing treatment modality for TNBC, primarily due to its lack of estrogen receptors (ERs), progesterone receptors (PRs), and human epidermal growth receptor 2 (HER2). However, clinical chemotherapy for TNBC is marked by its limited efficacy and a pronounced incidence of adverse effects. Consequently, there is a pressing need for novel drugs to treat TNBC. Given the rich repository of diverse natural compounds in traditional Chinese medicine, identifying potential anti-TNBC agents is a viable strategy. This study investigated lasiokaurin (LAS), a natural diterpenoid abundantly present in Isodon plants, revealing its significant anti-TNBC activity both in vitro and in vivo. Notably, LAS treatment induced cell cycle arrest, apoptosis, and DNA damage in TNBC cells, while concurrently inhibiting cell metastasis. In addition, LAS effectively inhibited the activation of the phosphatidylinositol-3-kinase/protein kinase B/mammalian target of rapamycin (PI3K/Akt/mTOR) pathway and signal transducer and activator of transcription 3 (STAT3), thus establishing its potential for multitarget therapy against TNBC. Furthermore, LAS demonstrated its ability to reduce tumor growth in a xenograft mouse model without exerting detrimental effects on the body weight or vital organs, confirming its safe applicability for TNBC treatment. Overall, this study shows that LAS is a potent candidate for treating TNBC.

Keywords: lasiokaurin; isodon; triple-negative breast cancer; PI3K/Akt/mTOR; STAT3

check for updates

Citation: Lin, J.; Qu, Z.; Pu, H.; Shen, L.-S.; Yi, X.; Lin, Y.-S.; Gong, R.-H.; Chen, G.-Q.; Chen, S. In Vitro and In Vivo Anti-Cancer Activity of Lasiokaurin in a Triple-Negative Breast Cancer Model. *Molecules* **2023**, *28*, 7701. https://doi.org/10.3390/molecules28237701

Academic Editors: RuAngelie Edrada-Ebel, Xun Song, Chenyang Li and Yifu Guan

Received: 12 July 2023
Revised: 17 August 2023
Accepted: 17 August 2023
Published: 22 November 2023

Copyright: © 2023 by the authors. Licensee MDPI, Basel, Switzerland. This article is an open access article distributed under the terms and conditions of the Creative Commons Attribution (CC BY) license (https://creativecommons.org/licenses/by/4.0/).

1. Introduction

Breast cancer stands as the predominant form of solid malignant tumors and is the leading contributor of cancer-related mortality among women worldwide [1]. Triple-negative breast cancer (TNBC), classified as a subtype within the realm of breast cancer, is identified via immunohistochemistry by the lack of estrogen receptors (ERs), progesterone receptors (PRs), and human epidermal growth receptor 2 (HER2). This subtype constitutes approximately 15–20% of the total breast cancer cases [2]. TNBC displays a greater prevalence among young and pre-menopausal women [3,4]. Alongside its elevated risk of relapse and metastasis, TNBC is characterized by a notably brief progression-free survival period and a low overall survival rate. Because of the absence of ERs, PRs, and HER2, TNBC does not respond to endocrine and anti-HER2 therapies. Currently, unresectable TNBC is treated with non-targeted chemotherapeutic agents, such as paclitaxel and anthracyclines. Regrettably, cytotoxic chemotherapy is associated with the emergence of various undesirable side effects [5,6]. Thus, an urgent imperative exists to develop novel drugs with heightened selectivity and reduced side effects for TNBC treatment.

Owing to their substantial abundance and wide-ranging chemical structural diversity, natural products derived from certain traditional medicines have gained widespread recognition for their application in addressing a variety of diseases, including TNBC [7–11]. Belonging to the Lamiaceae family, the genus *Isodon* encompasses a cluster of flowering plants and comprises approximately 100 species, predominantly found across tropical and subtropical regions of Asia [12]. Within the *Isodon* genus, plants harbor an extensive array of diterpenoids with multifaceted biological functions, including anti-cancer, anti-inflammatory, and anti-viral properties [13,14]. Among the diterpenoids in *Isodon* plants, oridonin stands out as a notable example, drawing growing interest for its extensive range of anti-cancer activities [15]. Previous studies demonstrated that oridonin exhibited anti-cancer effects against breast cancer [16–18], colon cancer [19,20], lung cancer [21], nasopharyngeal carcinoma [22], oral cancer [23], pancreatic cancer [24], bladder cancer [25], and neuroblastoma [26]. However, lasiokaurin (LAS; Figure 1A), a significant analogue of oridonin in *Isodon* plants, has not yet attracted much attention from the research community. Only a limited number of earlier studies have noted LAS's anti-microbial and anti-tumor activities [27,28]. Nevertheless, the exploration of LAS's potential as an anti-TNBC agent remains largely uncharted territory. In light of this gap, the present study is dedicated to investigating LAS's anti-cancer effects on TNBC, along with the underlying mechanisms.

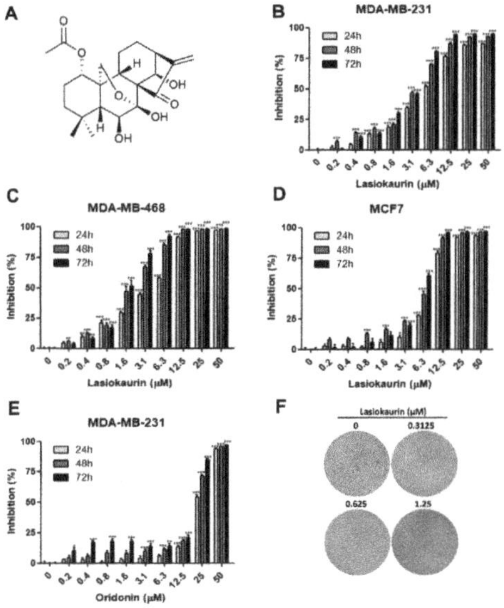

Figure 1. LAS inhibited breast cancer cell proliferation. (**A**) Chemical structure of LAS. (**B–D**) Cell viability of MDA-MB-231, MDA-MB-468, and MCF7 was separately measured by MTT assay after LAS treatment. (**E**) Cell viability of MDA-MB-231 was measured by MTT assay after oridonin treatment. (**F**) Colony formation ability of MDA-MB-231 cells treated with LAS for 13 days. Data are presented as means ± SEM from three independent experiments. * $p < 0.05$, ** $p < 0.01$, *** $p < 0.001$, compared to control.

2. Results

2.1. LAS Inhibits the Proliferation of TNBC Cells

To evaluate the in vitro anti-cancer activity of LAS in breast cancer, two TNBC cell lines, namely MDA-MB-231 and MDA-MB-468, along with the ER and PR positive cell line MCF7, were employed. Additionally, the MCF-10A cell line, serving as a model for normal human breast cells, was utilized to analyze the cytotoxicity of LAS. The MTT assay results demonstrated a notable dose- and time-dependent reduction in cell viability with increasing concentrations of LAS (0.2–50 μM), as depicted in Figure 1B–D. Notably, LAS exhibited an inhibitory effect on both breast cancer cells and normal breast cells, although the impact on normal cells was comparatively weaker (Figure S1). Furthermore, we explored the impact of oridonin on MDA-MB-231 cell viability (Figure 1E), and the IC_{50} values for LAS and oridonin were separately presented in Table 1. These findings collectively highlight that LAS was more potent in diminishing TNBC cell viability compared to oridonin. Additionally, LAS demonstrated a relatively low level of toxicity to normal cells. To assess the extended inhibitory impact of LAS on TNBC cells, colony formation assays were conducted. As expected, even at a low and non-toxic concentration of LAS (0.3125 μM), the colony formation ability of MDA-MB-231 cells was notably suppressed (Figure 1F). These findings collectively indicated that LAS exhibited the potential to inhibit TNBC cell proliferation and survival in vitro.

Table 1. IC_{50} values (μM) of LAS in human breast cancer cell lines.

Compound	Cell Line	24 h	48 h	72 h
LAS	MDA-MB-231	5.43	3.37	2.9
	MDA-MB-468	3.42	1.84	1.6
	MCF7	8.35	5.69	5.16
	MCF-10A	25.84	6.69	5.95
Oridonin	MDA-MB-231	23.38	18.96	16.79

2.2. LAS Modulates Cell Cycle Progression in TNBC Cells

The progression of the cell cycle plays a crucial role in determining cell proliferation outcomes [29]. Therefore, the current study focused on the regulation of cell cycle progression to uncover the potential underlying mechanisms of LAS. Cell cycle distribution in the LAS-treated MDA-MB-231 and MDA-MB-468 cells was evaluated by PI staining using flow cytometric analysis. Figure 2A–D shows the effect of the rising LAS concentration (1.25–20 μM) on the MDA-MB-231 cell cycle. Remarkably, LAS led to a dose- and time-dependent induction of cell cycle arrest specifically at the G2/M phase. Interestingly, this arrest effect saw a decline in cells treated with LAS concentrations exceeding 5 μM. Parallel observations were made in the LAS-treated MDA-MB-468 cells (Figure S2A–D). The outcomes of the aforementioned experiments collectively indicated that G2/M phase cell cycle arrest in TNBC cells was notably induced only with low LAS concentrations, and other mechanisms could be involved in LAS concentrations exceeding 5 μM.

2.3. LAS Induces Apoptosis and DNA Damage in TNBC Cells

In addition to cell cycle regulation, cell death activation is also considered as an alternative strategy for cancer therapy [30]. Apoptosis is a major form of programmed cell death, while several natural compounds have been identified to promote apoptosis in cancer cells [31]. Herein, to explore the effect of LAS on inducing cell apoptosis, Annexin V-FITC/PI staining was performed via flow cytometry. Figure 3A–C reveals that lower LAS concentrations (1.25 μM and 2.5 μM) did not induce cell apoptosis in MDA-MB-231 cells. Interestingly, the landscape shifted when TNBC cells were exposed to 5–20 μM LAS for 24 and 48 h, resulting in a substantial rise in the proportion of apoptotic cells. Remarkably, the consistent results were mirrored in the LAS-treated MDA-MB-468 cells (Figure S3A–C).

The collective evidence underscored that LAS concentrations exceeding 5 μM exerted a marked induction of cell apoptosis.

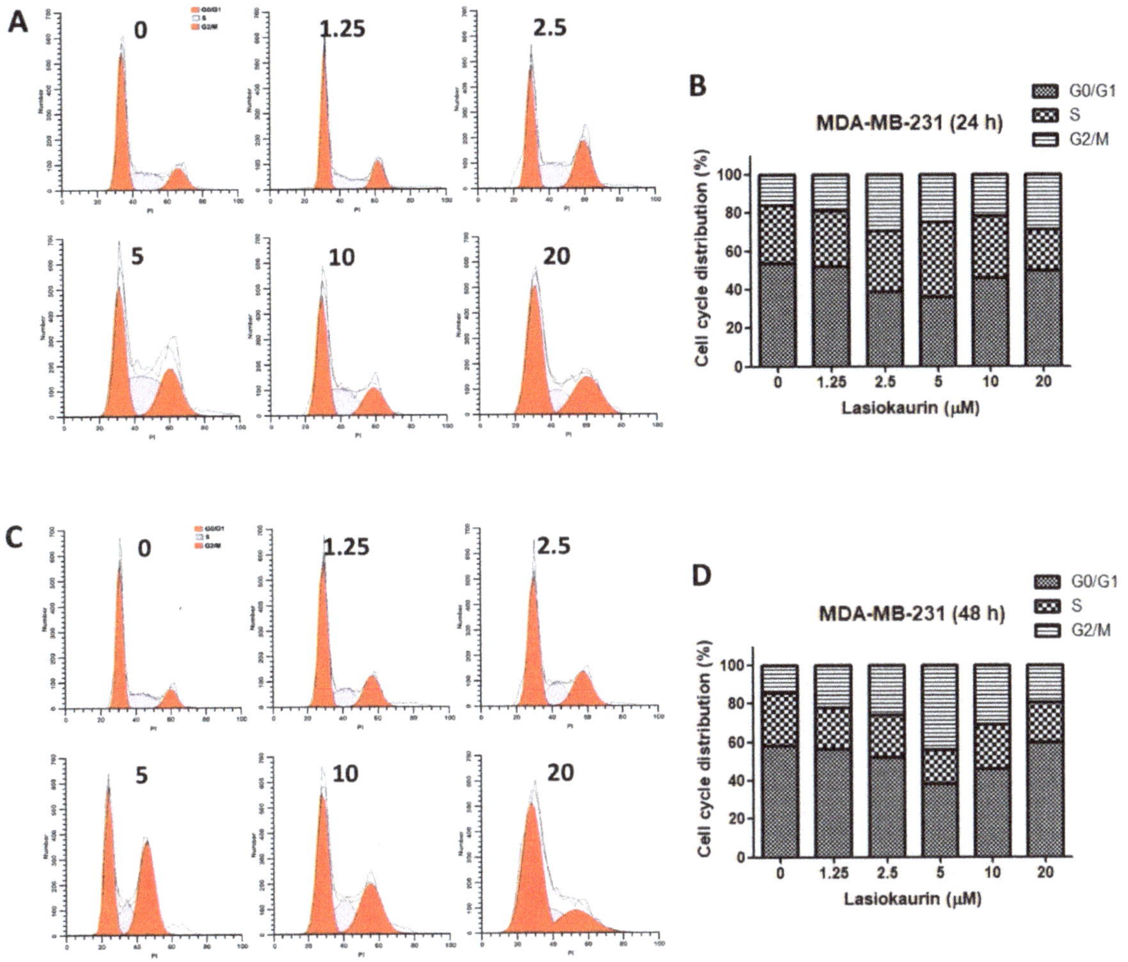

Figure 2. LAS induced cell cycle arrest in MDA-MB-231 cells. MDA-MB-231 cells were stained with PI after LAS treatment and the cell cycle analyzed by flow cytometry. Representative DNA fluorescence histograms of cell cycle distribution after 24 h (**A**) and 48 h (**C**) treatment were presented. Bar charts showed the percentage of different phases after 24 h (**B**) and 48 h (**D**) treatment.

DNA damage is considered as a significant strategy for killing cancer cells. Therefore, the expression levels of PARP, which is involved in repairing DNA damage [32], were detected. Figure 3D reveals that LAS significantly suppressed PARP expression in MDA-MB-231 cells, suggesting LAS's potential to induce DNA damage in TNBC cells. Correspondingly, consistent outcomes emerged in the LAS-treated MDA-MB-468 cells (Figure S3D). The above results demonstrated that treating TNBC cells with lower LAS concentrations predominantly triggered cell cycle arrest, whereas higher LAS concentrations induced both cell apoptosis and DNA damage.

2.4. LAS Inhibits the Migration and Invasion of TNBC Cells

As the most aggressive form of breast cancer, TNBC exhibits heightened metastasis rates, inevitably leading to elevated mortality rates [33]. Therefore, an anti-TNBC compound with good potential should have anti-metastatic properties [34]. To determine whether LAS could affect the migration and invasion abilities of TNBC cells, wound-healing and transwell invasion assays were performed. As shown in Figure 4A, cell treatment with LAS for 24 h post-wounding remarkably dampened the migratory capacity of MDA-MB-231 cells in a dose-dependent manner. Correspondingly, LAS exerted a dose-dependent inhibition on the invasive potential of MDA-MB-231 cells (Figure 4B). Similarly, LAS's impact on restraining migration and invasion was echoed in MDA-MB-468 cells (Figure S4A,B). Collectively, these findings suggested that LAS could inhibit TNBC cell metastasis in vitro.

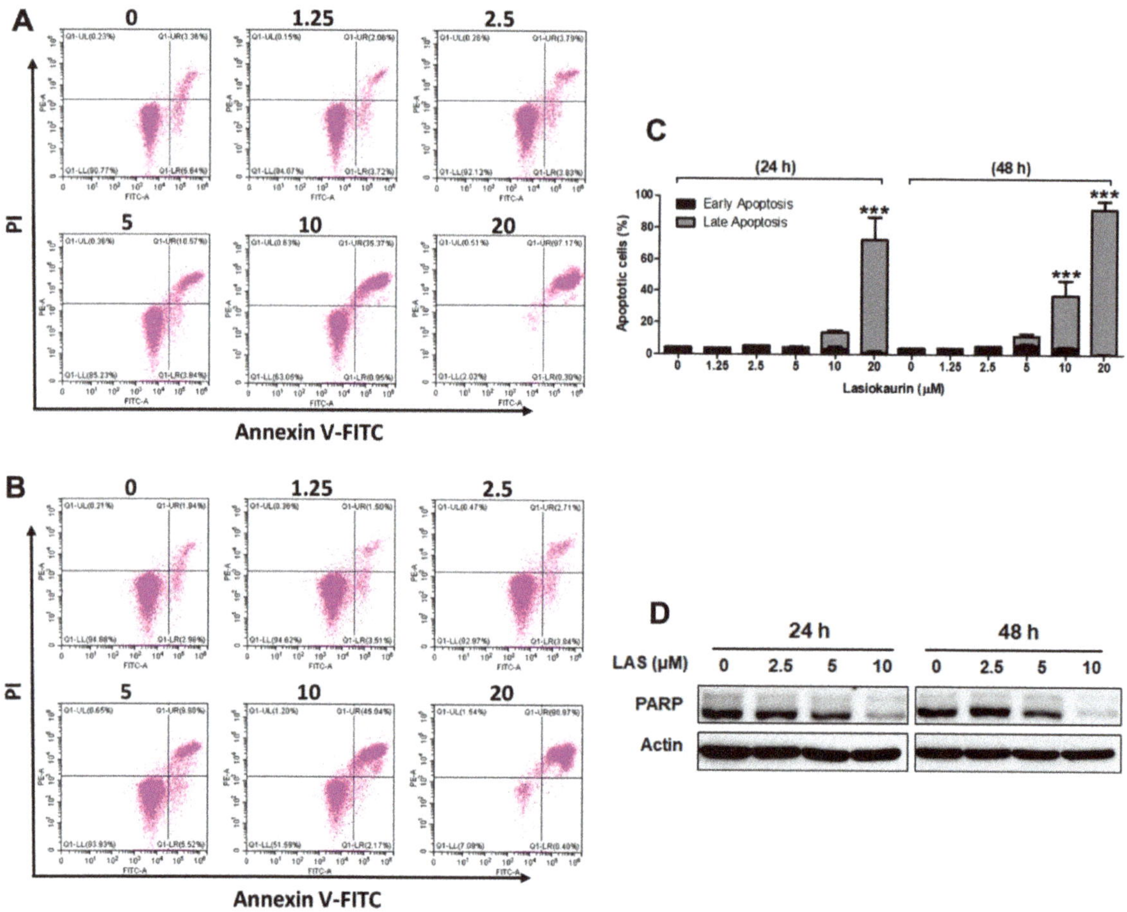

Figure 3. LAS induced apoptosis and DNA damage in MDA-MB-231 cells. MDA-MB-231 cells were treated with LAS for 24 h (**A**) and 48 h (**B**), stained with Annexin V-FITC/PI, and cell apoptosis was analyzed by flow cytometry. (**C**) Representative flow cytometry Annexin V/PI data. *** $p < 0.001$, compared to control. (**D**) Cell extracts were prepared from MDA-MB-231 cells and immunoblotted with the indicated antibodies. β-Actin was used as an internal control.

2.5. LAS Inhibits PI3K/Akt/mTOR and STAT3 Signalling in TNBC Cells

The PI3K/Akt/mTOR pathway is one of the most significant and active pathways which are involved in TNBC development [35]. This pathway is known to play a pivotal role in regulating various cellular processes, including cell growth, proliferation, and metastasis [36]. Given its significance, inhibiting the PI3K/Akt/mTOR pathway holds promise as a therapeutic strategy for TNBC [37]. In this context, we investigated whether LAS could inhibit this pathway. Figure 5 shows a significant reduction in the phosphorylation levels of PI3K, Akt, and mTOR upon LAS treatment in MDA-MB-231 cells. Additionally, it is widely known that mTOR acts via multiprotein complexes, such as mTORC1 and mTORC2 [38], which interplay with the PI3K/Akt pathway. To discern the complex responsible for LAS's inhibitory effect on TNBC, we evaluated the expression of Rictor, Raptor, and Gβ1, three partners that can bind to mTOR [39]. The results demonstrated decreased levels of Rictor, Raptor, and Gβ1 in the LAS-treated MDA-MB-231 and MDA-MB-468 cells, thus affirming the inhibition of both mTORC1 and mTORC2 activities (Figures 5 and S5).

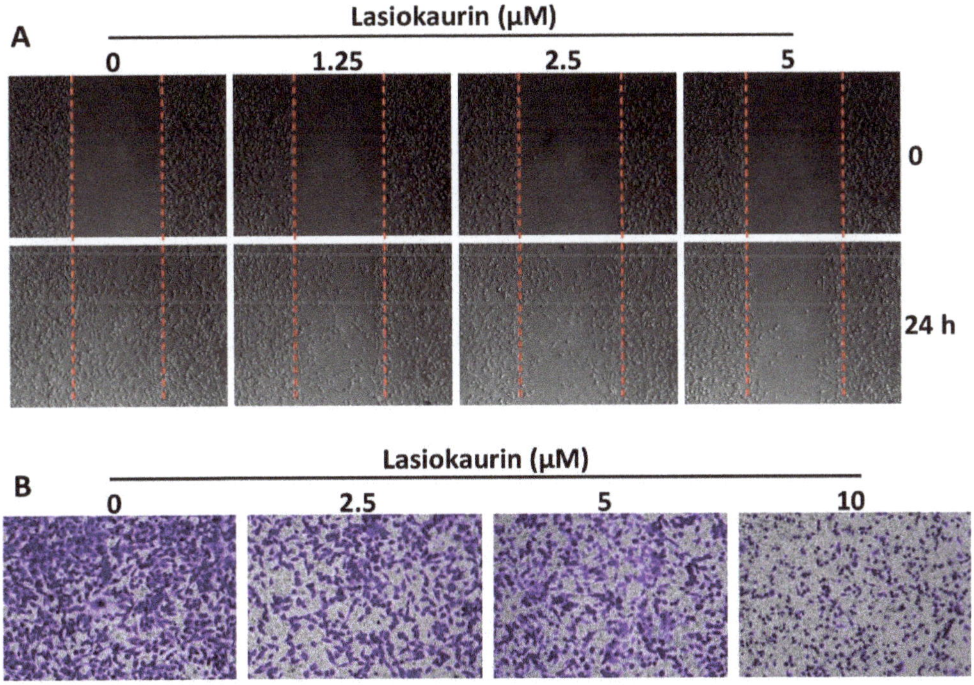

Figure 4. LAS inhibited the migration and invasion of MDA-MB-231 cells. (**A**) Cell migration was measured by wound-healing assay. (**B**) Cell invasion ability was assessed by transwell invasion assay.

Additionally, it has been reported that the transcription factor STAT3 is upregulated and constitutively activated in TNBC [40]. Prior research further demonstrated that STAT3 activation plays a crucial role in cancer cell proliferation, invasion, and migration [41], suggesting its potential as a therapeutic target for TNBC treatment. Consequently, we investigated the impact of LAS on STAT3 within TNBC cells. Our findings revealed that the expression levels of both STAT3 and p-STAT3 significantly decreased in a dose-dependent manner following LAS treatments at 24 h and 48 h (Figures 5 and S5).

2.6. LAS Inhibits Tumor Growth in a Xenograft Nude Mouse Model

To investigate the in vivo anti-cancer effect of LAS, a mouse xenograft model was established via inoculating MDA-MB-231 cells into mammary fat pads. Following tumor inoculation, the mice received 5 mg/kg (LD group) or 10 mg/kg (HD group) LAS intraperitoneally daily for 20 consecutive days. Mice in the vehicle and positive control groups were also intraperitoneally injected with saline and docetaxel (10 mg/kg), respectively. After 20 days, the mice were sacrificed and the xenograft tumors and body organs were resected for evaluation. As shown in Figure 6A–D, the tumor volume and weight were significantly reduced in the LAS-LD and LAS-HD groups, thus indicating that LAS retarded tumor growth. While LAS-LD exhibited slightly lower efficacy than docetaxel, LAS-HD at the same dose showed comparable efficacy to docetaxel. Moreover, LAS had no discernible impact on the body weight, signifying its safety for application. In order to validate the aforementioned findings, we assessed the weight and histopathological attributes of the hearts, lungs, livers, spleens, and kidneys in mice from each group. As shown in Figures 6E–J and 7, in comparison to the vehicle group, there were no noteworthy alterations in the organ weight or histopathological characteristics observed in either the LAS-treated groups or the vehicle group. Collectively, these findings indicated that LAS could suppress tumor growth in vivo without inducing discernible toxicity.

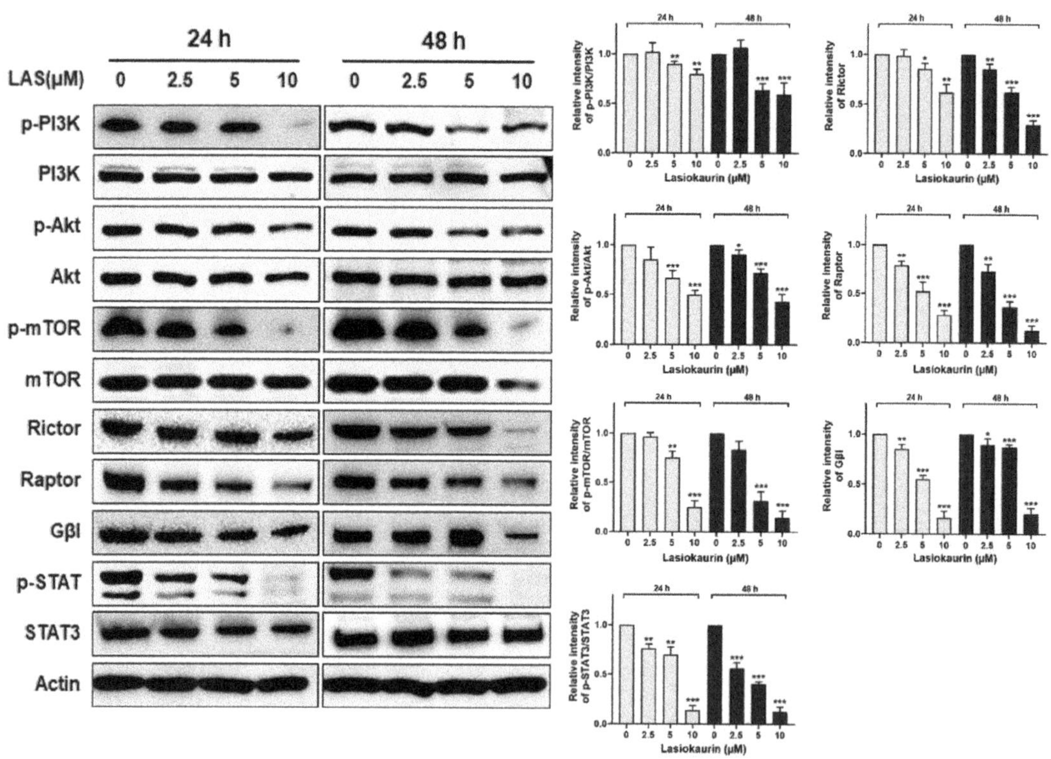

Figure 5. LAS inhibited PI3K/Akt/mTOR pathway and STAT3 in MDA-MB-231 cells. MDA-MB-231 cells were treated with LAS at concentrations of 2.5, 5, 10 μM for 24 or 48 h. Cell pellets collected and immunoblotted with the indicated antibodies. β-Actin was used as an internal control. Quantitative analysis of protein expression was in the right panel. * $p < 0.05$, ** $p < 0.01$, *** $p < 0.001$, compared to control.

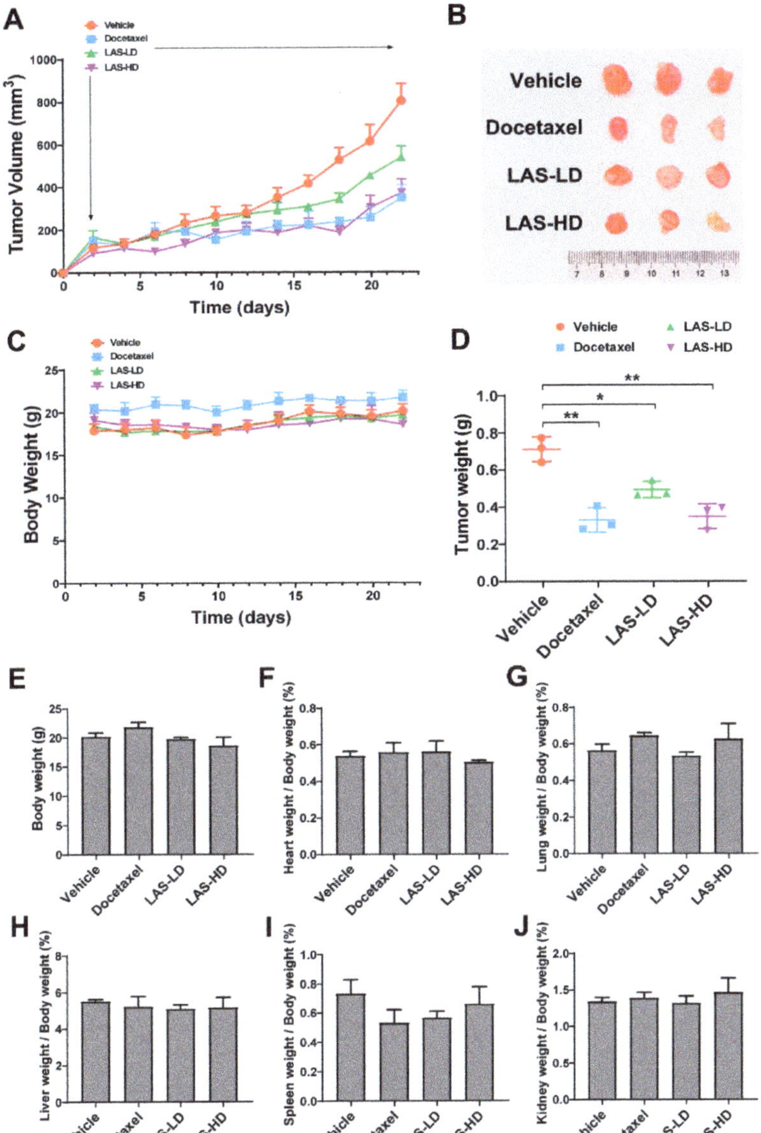

Figure 6. LAS inhibited in vivo MDA-MB-231 xenograft tumor growth. A xenograft model was established by subcutaneous inoculation of MDA-MB-231 cells into BALB/c nude mice mammary fat pads. When the average tumor volumes reached 120 mm3, mice were randomly divided into four groups and administrated with vehicle (5% of Cremophor EL, 5% of ethanol in saline), LAS-LD (5 mg/kg), or LAS-HD (10 mg/kg) daily, or docetaxel (10 mg/kg) via i.p. injection. The treatment period lasted for 20 days and all mice were sacrificed. (**A**) Tumor volumes were measured throughout the experimental period. (**B**) Images of tumors at the end of experiment. (**C**) Mouse body weights throughout the experimental period. (**D**) Tumor weights at experimental endpoint. (**E–J**) Organ weights normalized to body weights and expressed at the percentage of body weight. Data are expressed as means ± SEM.* $p < 0.05$, ** $p < 0.01$, compared to the vehicle group.

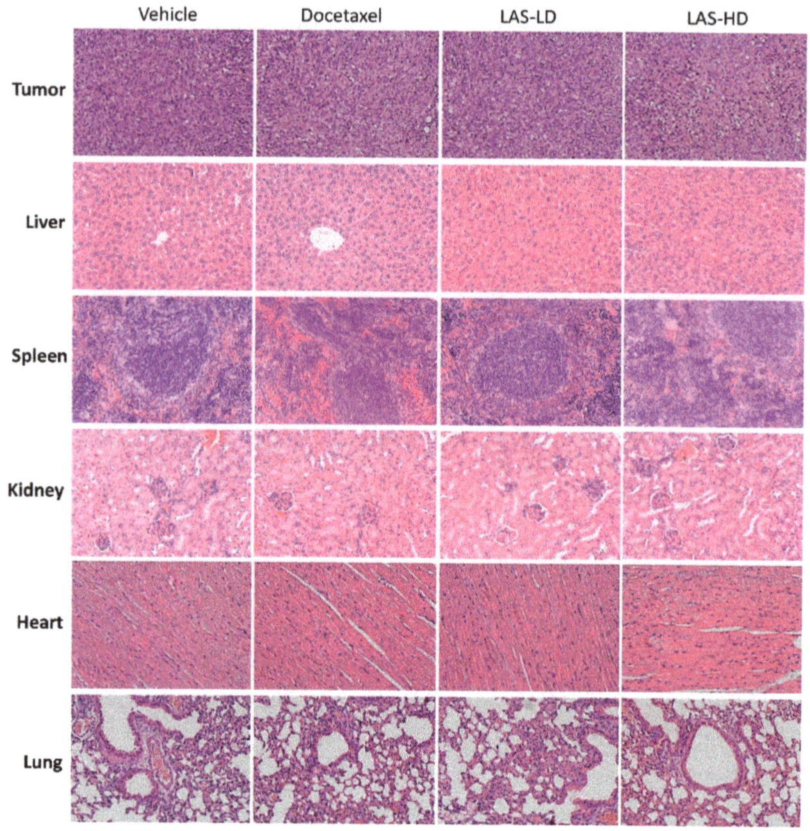

Figure 7. Photomicrograph of H&E histology of tissues (tumor, liver, spleen, kidney, heart, lung) after LAS treatment. Representative images are shown (20×).

3. Discussion

Clinically, TNBC stands as the most aggressive subtype of breast cancer, characterized by high recurrence and metastasis rates. Owing to the absence of relevant receptor markers, patients with TNBC derive limited benefits from conventional targeted chemotherapies. Consequently, despite the adversities of its side effects, non-specific chemotherapy remains the established treatment approach for TNBC patients. Therefore, the imperative pursuit of novel, effective TNBC treatments devoid of side effects assumes paramount significance.

In recent years, a range of innovative treatment strategies has emerged, encompassing groundbreaking concepts like artificial intelligence (AI). Despite these advancements, natural compounds are still the most significant sources of novel drugs. Several anti-cancer drugs, such as paclitaxel, vinblastine, and camptothecin, were born out of the screening of natural compounds and have been widely used in clinical practice.

In our present study, we investigated the anti-TNBC activity of LAS, a natural diterpenoid discovered within the *Isodon* genus, serving as an analogue to oridonin. We initiated our exploration with MTT and colony formation assays, revealing the pronounced inhibitory effects of LAS on the proliferation of TNBC cells. To elucidate the potential anti-cancer mechanism of LAS in TNBC, we investigated the distribution of the cell cycle. The results showed that LAS, at lower doses, significantly induced cell cycle arrest in the G2/M phase. Interestingly, as the LAS concentrations increased, its efficacy in inducing cell cycle arrest diminished. This observation suggests that the primary mechanism driving the inhibitory impact of low LAS concentrations on TNBC is cell cycle arrest. Furthermore, additional mechanisms might come into play as LAS concentrations rise. Subsequently, we assessed LAS-induced apoptosis in TNBC cells using Annexin V-FITC/PI staining with flow cytometry. The findings demonstrated a dose-dependent increase in the proportion of apoptotic cells with rising LAS concentrations. Given the effectiveness of DNA damage response in treating TNBC, we further investigated whether LAS could induce DNA damage in TNBC cells [42]. The results indicated that LAS could trigger DNA damage in TNBC cells by inhibiting PARP expression, an enzyme crucial for DNA damage repair [43]. Collectively, these outcomes confirm the hypothesis: low LAS concentrations facilitate TNBC cell cycle arrest in the G2/M phase, while higher concentrations induce TNBC cell apoptosis and DNA damage.

In clinical practice, a potential therapeutic strategy for TNBC involves inhibiting the migration and invasion abilities of TNBC cells [44]. Consequently, an effective anti-TNBC compound should possess anti-metastatic properties as well. With this context in mind, we proceeded to investigate the impact of LAS on inhibiting the migration and invasion capabilities of TNBC cells. In addition to its observed effect in suppressing cell migration during the wound-healing assay, LAS also demonstrated the ability to curtail the invasive potential of TNBC cells in transwell invasion assays. This combined evidence suggests that LAS holds promise in restraining TNBC metastasis.

It has been reported that the PI3K/Akt/mTOR pathway is overactivated in over 60% of patients with TNBC, eventually contributing to cancer cell proliferation, metastasis, and survival [45]. Previous research demonstrated a correlation between the expression of p-mTOR and unfavorable prognosis in early-stage TNBC patients [46]. In recent years, various inhibitors targeting the PI3K/Akt/mTOR pathway have undergone clinical trials. Examples include PQR309, a dual PI3K/mTOR inhibitor, as well as Ipatasertib and AZD5363, both Akt inhibitors, and temsirolimus, an mTOR inhibitor [37]. However, these endeavors have yielded less-than-optimal outcomes, underscoring the complexity of treating multifaceted conditions like TNBC with single-target chemotherapy. This scenario has prompted a shift toward multitarget therapeutics for TNBC treatment, a strategy that has exhibited enhanced effectiveness [47]. For example, promising results have emerged from studies targeting both mTOR and STAT3 in TNBC treatment [48]. Furthermore, it is worth noting that the transcription factor STAT3 is frequently overexpressed and constitutively active in TNBC, thus holding a crucial role in anti-TNBC strategies [40]. The present study showed that LAS not only inhibited the activation of the PI3K/Akt/mTOR pathway but also dampened that of STAT3. This observation indicates LAS can be a potential multitarget therapeutic candidate for TNBC.

The inhibitory effects of LAS on tumor growth were evaluated in a subcutaneous TNBC xenograft mouse model. Remarkably, even at low doses, LAS significantly curtailed tumor growth in vivo. Furthermore, the escalating doses of LAS exhibited a similar inhibitory effect to that of docetaxel, one of the most efficacious chemotherapy drugs for TNBC. Importantly, LAS administration did not result in weight loss, abnormal behavior, or histopathological changes in critical organs. These observations underscore the safety of applying LAS in TNBC treatment.

4. Materials and Methods

4.1. Compounds

LAS ($C_{22}H_{30}O_7$; cat. no. CAS28957-08-6) was purchased from Jiangsu Yongjian Pharmaceutical Technology Co., Ltd. (Yangzhou, China), while oridonin ($C_{20}H_{28}O_6$; cat. no. CAS28957-04-2) was purchased from Shanghai Aladdin Biochemical Technology Co., Ltd. (Shanghai, China) The purity of both compounds was >98% as analyzed by high-performance liquid chromatography.

4.2. Cell Culture

The human TNBC cell lines, MDA-MB-231 and MDA-MB-468, and the non-TNBC cell line, MCF7, were purchased from the American Type Culture Collection (ATCC,

Manassas, VA, USA). All cell lines were maintained in DMEM supplemented with 10% heat-inactivated FBS and 1% penicillin/streptomycin solution at 37 °C in a humidified incubator with 5% CO_2. The cell lines were used within two months after resuscitation and mycoplasma contamination was assessed utilizing the PCR Mycoplasma Detection Kit (Beijing Transgen Biotech Co., Ltd., Beijing, China).

4.3. Cell Viability Assay

A 3-(4, 5-dimethylthiazol-2-yl)-2,5-diphenyltetrazolium bromide (MTT) assay was used to investigate the effects of LAS and oridonin on breast cancer cell viability according to the report [49]. Briefly, following incubation for 24 h, cells were seeded in 96-well plates at a density of 5 × 103 cells/well (MDA-MB-231 and MCF7) or 1 × 10^4 cells/well (MDA-MB-468) and were then treated with different concentrations of LAS or oridonin for an additional 24, 48, or 72 h. Untreated cells served as the control group. Subsequently, cells were treated with 0.5 mg/mL MTT (MilliporeSigma, Burlington, MA, USA) and incubated for 4 h at 37 °C. After discarding culture media, the wells were supplemented with DMSO to dissolve the formed formazan. The optical density (OD) at a wavelength of 570 nm was measured using a Biotek Synergy H1 microplate reader (BioTek Instruments, Inc., Winooski, VT, USA). The experiments were performed in six parallel wells and repeated for three times. The half-maximal inhibitory concentration (IC_{50}) values were calculated using Graphpad Prism 5 software (GraphPad Software Inc., La Jolla, CA, USA).

4.4. Colony Formation Assay

According to a method in the publication [50], MDA-MB-231 cells were seeded in triplicate in 6-well plates at a density of 600 cells/well and were then treated with different concentrations (0, 0.3125, 0.625, and 1.25 μM) of LAS. Following culturing for 13 days, when the colonies were visible, the cell culture was terminated. Subsequently, cells were fixed with anhydrous methanol for 5 min, dried, and stained with 0.1% crystal violet solution for 10 min at room temperature. The excess dyes were washed away with MilliQ water, the formed colonies were dried, and images were then captured.

4.5. Cell Cycle Assay

Cell cycle distribution was assessed using the Cell Cycle and Apoptosis Analysis kit (Beyotime Biotechnology, Pudong, China). In accordance with the report [51], MDA-MB-231 and MDA-MB-468 cells were seeded into 6-well plates and treated with LAS for 24 or 48 h. Subsequently, cells were harvested, fixed with ice-cold 70% ethanol at 4 °C overnight, rinsed in PBS, and were then incubated with propidium iodide (PI) and RNase for 30 min at 37 °C in the dark. Flow cytometric analysis was performed using the BD AccuriC6 flow cytometry system (Becton Dickson Immunocytometry-Systems, San Diego, CA, USA) and the cell cycle distribution was analyzed using ModFit LT 5.0 (Verity Software House, Topsham, ME, USA).

4.6. Annexin V-FITC Apoptosis Assay

Cell apoptosis assay was carried out using the Annexin V-FITC Apoptosis Detection kit (Beyotime Biotechnology). Using a method in the report [52], MDA-MB-231 and MDA-MB-468 cells were seeded into 6-well plates and were then treated with LAS for 24 or 48 h. Following digestion with 0.25% trypsin (without EDTA), cells were harvested and incubated with 195 μL binding buffer supplemented with 5 μL FITC-labelled Annexin V and 10 μL PI for 20 min in the dark at room temperature. The fluorescence of cells was immediately quantified on the CytoFLEX flow cytometer (Beckman Coulter, Inc., Brea, CA, USA).

4.7. Wound-Healing Assay

The migration ability of TNBC was evaluated by wound-healing assays, as previously described [53]. Briefly, MDA-MB-231 cells at a density of 4 × 10^4 cells/well were seeded into both chambers of the culture insert (Ibidi GmbH, Gräfelfing, Germany). After allowing cells to attach overnight, the inserts were removed to create a wound. The cells were then washed with serum-free medium to remove non-adherent cells, followed by treatment with various concentrations of LAS in 2% FBS-containing medium. Images of the migrated cells were captured under a microscope at 0 and 24 h in the same three randomly selected fields.

4.8. Transwell Invasion Assay

Transwell invasion assay was performed as previously reported using Corning transwell insert chambers with a pore size of 8 μm [50]. MDA-MB-231 and MDA-MB-468 cells were seeded at a density of 5×10^4 cells/chamber in 200 μL serum-free DMEM in the upper Matrigel-coated chamber of the transwell insert. The lower chamber was supplemented with 10% FBS medium (500 μL) as a chemo-attractant. Following incubation for 24 h at 37 °C, cells on the upper surface of the membrane were carefully removed with a cotton swab. Cells that had invaded to the lower surface of the membrane were fixed with 4% polyformaldehyde for 15 min and stained with crystal violet for 10 min. Following washing with MilliQ water, the membrane was air-dried and cells were counted under a light microscope (magnification, 10×).

4.9. Western Blot Analysis

Following cell treatment with different concentrations of LAS for 24 and 48 h, cells were harvested for western blot analysis. Cell pellets were firstly lysed in RIPA buffer and the protein concentration was measured using the Pierce™ BCA Protein Assay (Thermo Fisher Scientific, Inc., Waltham, MA, USA). Equal amounts of protein extracts were separated by SDS-PAGE and were then transferred onto PVDF membranes (Bio-Rad Laboratories, Inc., Hercules, CA, USA). The membranes were blocked with 5% non-fat milk followed by incubation with primary antibodies in 5% BSA at 4 °C overnight. Following washing with TBS-Tween-20 buffer, blots were incubated with the corresponding secondary antibodies. The primary antibodies used were the following: anti-poly (ADP ribose) polymerase (PARP; cat. no. ab191217), anti-phosphorylated (p)-phosphatidylinositol-3-kinase (PI3K; cat. no. ab182651), anti-p-mammalian target of rapamycin (mTOR; cat. no. ab137133), anti-protein kinase B (Akt; cat. no. ab179463), anti-Raptor (cat. no. ab26264), anti-Rictor (cat. no. ab70374), anti-G protein beta subunit like (Gbl; cat. no. ab228832), anti-STAT3 (cat. no. ab68153), anti-p-STAT3 (cat. no. ab76315; all from Abcam, 1:1000 dilution), PI3K (cat. no. 3011), mTOR (cat. no. 9964T; both from Cell Signaling Technology, Inc., 1:1000 dilution), p-Akt (cat. no. 66444-1; ProteinTech Group, Inc. (Rosemont, IL, USA), 1:1000 dilution), and β-actin (ZSGB-Bio, 1:5000 dilution).

4.10. Establishment of Tumor Xenograft Model

Female BALB/c nude mice were obtained from ZhuHai Bestest Biotechnology Co., Ltd. (Zhuhai, China). Mice were housed at room temperature (23 ± 2 °C) with a 12 h light/dark cycle and were given ad libitum access to food and water. All animal experiments were carried out at the Hong Kong Polytechnic University, according to the protocol approved by the Animal Subjects Ethics Sub-committee. Mice were inoculated with 5×10^6 MDA-MB-231 cells suspended in 0.1 mL DMEM without FBS or penicillin/streptomycin, at the fourth mammary fat pad. When the average tumor volume reached 120 mm^3, mice were randomly divided into four groups (n = 3 mice/group). Groups were balanced for mean tumor size. The treatment groups were as follows: Mice in the vehicle/negative group were injected intraperitoneally daily with an equivalent amount of solvent (5% Cremophor EL and 5% ethanol in saline; 10 mL/kg). Mice in the positive control group were treated with 10 mg/kg docetaxel, while those in the LAS group with 5 [LAS-Low Dose (LD)] or 10 mg/kg [LAS-High Dose (HD)] LAS. The body weight and tumor size of the mice were measured every other day. Tumor volume was calculated using the following formula: volume = (length × width^2)/2. When the average tumor volume in the vehicle group reached ~800 mm^3 (the maximum tumor volume reached 1094 mm^3) on day 20 post-treatment, the experiment was terminated. The mice were euthanized by CO_2 inhalation at a displacement rate of 50% cage volume per minute in a cage with 10 L for 2 min. After confirming the death of the mice, their tumors and vital organs, including heart, liver, spleen, lung, and kidney, were harvested and weighed.

4.11. Haematoxylin and Eosin (H&E) Staining

Tumors and vital organs were fixed in 10% neutral buffered formalin for three days and the tissues were then embedded in paraffin. Subsequently, the paraffin-embedded tissues were cut into 5 μm sections, followed by staining with H&E. Images of the H&E-stained sections were captured under a light microscope (magnification, 20×).

4.12. Statistical Analysis

All data were expressed as the mean ± SEM. The results were analyzed using SPSS 20.0 statistical software. The differences among multiple groups were compared by one-way ANOVA, followed by LSD post hoc or Dunnett's tests. $p < 0.05$ was considered to indicate a statistically significant difference.

5. Conclusions

At present, there exists a scarcity of clinically efficacious pharmaceuticals employed in the treatment of TNBC. Despite the utilization of conventional cytotoxic agents like paclitaxel and anthracyclines, their efficacy in effecting a curative response for TNBC remains suboptimal, concurrently giving rise to pronounced adverse effects. In contrast, LAS exhibits good inhibitory activity against TNBC both in vivo and in vitro. Additionally, LAS shows no toxicity when applied in vivo. Furthermore, LAS inhibits the PI3K/Akt/mTOR and STAT3 pathways, which are two pivotal signaling pathways associated with TNBC development. Although deeper mechanisms and prognostic roles of LAS in TNBC need to be explored in the future, our findings provide a preliminary basis for evaluating it as a promising therapeutic candidate for TNBC.

Supplementary Materials: The following supporting information can be downloaded at: https://www.mdpi.com/article/10.3390/molecules28237701/s1, Figure S1: cell viability of MCF-10A was measured by MTT assay after LAS treatment; Figure S2: LAS induced cell cycle arrest in MDA-MB-468 cells; Figure S3: LAS induced apoptosis and DNA damage in MDA-MB-468 cells; Figure S4: LAS inhibited the migration and invasion of MDA-MB-468 cells; Figure S5: LAS inhibited PI3K/Akt/mTOR pathway and STAT3 in MDA-MB-468 cells.

Author Contributions: Conceptualization: S.C. and G.-Q.C. Data curation and formal analysis: J.L., Z.Q., H.P., L.-S.S., X.Y., Y.-S.L. and R.-H.G. Funding acquisition: L.-S.S., S.C. and G.-Q.C. Writing: J.L., Z.Q. and H.P. Review and editing: S.C. and G.-Q.C. All authors contributed to the article and approved the submitted version. All authors have read and agreed to the published version of the manuscript.

Funding: This research was funded by of the Research Centre for Chinese Medicine Innovation of the Hong Kong Polytechnic University (E-ABCT-BBBB-3), the Hong Kong Polytechnic University Start-up Fund (P0036741 and P0038596), Chongqing Science and Technology Commission (JA21026), and Shenzhen Science and Technology Innovation Commission (JCYJ20220531090802006).

Institutional Review Board Statement: All animal experimental procedures were performed according to the Institutional Guidelines and Animal Ordinance of the Department of Health, and approved by the Hong Kong Polytechnic University Animal Subjects Ethics Sub-committee (20-21/184-ABCT-R-GRF, 23 March 2021).

Informed Consent Statement: Not applicable.

Data Availability Statement: The data presented in this study are available from the corresponding author upon reasonable request.

Conflicts of Interest: The authors declare no conflict of interest.

References

1. Sung, H.; Ferlay, J.; Siegel, R.L.; Laversanne, M.; Soerjomataram, I.; Jemal, A.; Bray, F. Global Cancer Statistics 2020: GLOBOCAN Estimates of Incidence and Mortality Worldwide for 36 Cancers in 185 Countries. *CA Cancer J. Clin.* **2021**, *71*, 209–249. [CrossRef] [PubMed]
2. Garrido-Castro, A.C.; Lin, N.U.; Polyak, K. Insights into Molecular Classifications of Triple-Negative Breast Cancer: Improving Patient Selection for Treatment. *Cancer Discov.* **2019**, *9*, 176–198. [CrossRef] [PubMed]
3. Lee, K.L.; Kuo, Y.C.; Ho, Y.S.; Huang, Y.H. Triple-Negative Breast Cancer: Current Understanding and Future Therapeutic Breakthrough Targeting Cancer Stemness. *Cancers* **2019**, *11*, 1334. [CrossRef] [PubMed]
4. Bianchini, G.; Balko, J.M.; Mayer, I.A.; Sanders, M.E.; Gianni, L. Triple-negative breast cancer: Challenges and opportunities of a heterogeneous disease. *Nat. Rev. Clin. Oncol.* **2016**, *13*, 674–690. [CrossRef]
5. Shastry, M.; Yardley, D.A. Updates in the treatment of basal/triple-negative breast cancer. *Curr. Opin. Obstet. Gynecol.* **2013**, *25*, 40–48. [CrossRef]
6. Loibl, S.; Poortmans, P.; Morrow, M.; Denkert, C.; Curigliano, G. Breast cancer. *Lancet* **2021**, *397*, 1750–1769. [CrossRef]
7. Shi, C.; Yang, E.J.; Tao, S.; Ren, G.; Mou, P.K.; Shim, J.S. Natural products targeting cancer cell dependency. *J. Antibiot.* **2021**, *74*, 677–686. [CrossRef]
8. Gairola, K.; Gururani, S.; Bahuguna, A.; Garia, V.; Pujari, R.; Dubey, S.K. Natural products targeting cancer stem cells: Implications for cancer chemoprevention and therapeutics. *J. Food Biochem.* **2021**, *45*, e13772. [CrossRef]
9. Chen, H.; Yang, J.; Yang, Y.; Zhang, J.; Xu, Y.; Lu, X. The Natural Products and Extracts: Anti-Triple-Negative Breast Cancer in Vitro. *Chem. Biodivers.* **2021**, *18*, e2001047. [CrossRef]
10. Yap, K.M.; Sekar, M.; Seow, L.J.; Gan, S.H.; Bonam, S.R.; Mat Rani, N.N.I.; Lum, P.T.; Subramaniyan, V.; Wu, Y.S.; Fuloria, N.K.; et al. *Mangifera indica* (Mango): A Promising Medicinal Plant for Breast Cancer Therapy and Understanding Its Potential Mechanisms of Action. *Breast Cancer* **2021**, *13*, 471–503. [CrossRef]
11. Yap, K.M.; Sekar, M.; Wu, Y.S.; Gan, S.H.; Rani, N.; Seow, L.J.; Subramaniyan, V.; Fuloria, N.K.; Fuloria, S.; Lum, P.T. Hesperidin and its aglycone hesperetin in breast cancer therapy: A review of recent developments and future prospects. *Saudi J. Biol. Sci.* **2021**, *28*, 6730–6747. [CrossRef] [PubMed]
12. Harley, R.M.; Atkins, S.; Budantsev, A.L.; Cantino, P.D.; Conn, B.J.; Grayer, R.; Harley, M.M.; de Kok, R.; Krestovskaja, T.; Morales, R.; et al. Labiatae. In *Flowering Plants Dicotyledons: Lamiales (Except Acanthaceae Including Avicenniaceae)*; Kadereit, J.W., Ed.; Springer: Berlin/Heidelberg, Germany, 2004; pp. 167–275.
13. Sun, H.D.; Huang, S.X.; Han, Q.B. Diterpenoids from Isodon species and their biological activities. *Nat. Prod. Rep.* **2006**, *23*, 673–698. [CrossRef] [PubMed]
14. Liu, M.; Wang, W.G.; Sun, H.D.; Pu, J.X. Diterpenoids from Isodon species: An update. *Nat. Prod. Rep.* **2017**, *34*, 1090–1140. [CrossRef] [PubMed]
15. Li, X.; Zhang, C.T.; Ma, W.; Xie, X.; Huang, Q. Oridonin: A Review of Its Pharmacology, Pharmacokinetics and Toxicity. *Front. Pharmacol.* **2021**, *12*, 645824. [CrossRef] [PubMed]
16. Guo, J.; Chen, T.; Ma, Z.; Qiao, C.; Yuan, F.; Guo, X.; Liu, J.; Shen, Y.; Yu, L.; Xiang, A. Oridonin inhibits 4T1 tumor growth by suppressing Treg differentiation via TGF-beta receptor. *Int. Immunopharmacol.* **2020**, *88*, 106831. [CrossRef]
17. Sun, B.; Wang, G.; Liu, H.; Liu, P.; Twal, W.O.; Cheung, H.; Carroll, S.L.; Ethier, S.P.; Mevers, E.E.; Clardy, J.; et al. Oridonin inhibits aberrant AKT activation in breast cancer. *Oncotarget* **2018**, *9*, 23878–23889. [CrossRef]
18. Li, C.; Wang, Q.; Shen, S.; Wei, X.; Li, G. Oridonin inhibits VEGF-A-associated angiogenesis and epithelial-mesenchymal transition of breast cancer in vitro and in vivo. *Oncol. Lett.* **2018**, *16*, 2289–2298. [CrossRef]
19. Bu, H.Q.; Shen, F.; Cui, J. The inhibitory effect of oridonin on colon cancer was mediated by deactivation of TGF-beta1/Smads-PAI-1 signaling pathway in vitro and vivo. *OncoTargets Ther.* **2019**, *12*, 7467–7476. [CrossRef]
20. Zhou, J.; Li, Y.; Shi, X.; Hao, S.; Zhang, F.; Guo, Z.; Gao, Y.; Guo, H.; Liu, L. Oridonin inhibits tumor angiogenesis and induces vessel normalization in experimental colon cancer. *J. Cancer* **2021**, *12*, 3257–3264. [CrossRef]
21. Xu, L.; Bi, Y.; Xu, Y.; Zhang, Z.; Xu, W.; Zhang, S.; Chen, J. Oridonin inhibits the migration and epithelial-to-mesenchymal transition of small cell lung cancer cells by suppressing FAK-ERK1/2 signalling pathway. *J. Cell. Mol. Med.* **2020**, *24*, 4480–4493. [CrossRef]
22. Liu, W.; Huang, G.; Yang, Y.; Gao, R.; Zhang, S.; Kou, B. Oridonin inhibits epithelial-mesenchymal transition of human nasopharyngeal carcinoma cells by negatively regulating AKT/STAT3 signaling pathway. *Int. J. Med. Sci.* **2021**, *18*, 81–87. [CrossRef] [PubMed]
23. Yang, J.; Ren, X.; Zhang, L.; Li, Y.; Cheng, B.; Xia, J. Oridonin inhibits oral cancer growth and PI3K/Akt signaling pathway. *Biomed. Pharmacother.* **2018**, *100*, 226–232. [CrossRef] [PubMed]
24. Liu, Q.Q.; Chen, K.; Ye, Q.; Jiang, X.H.; Sun, Y.W. Oridonin inhibits pancreatic cancer cell migration and epithelial-mesenchymal transition by suppressing Wnt/beta-catenin signaling pathway. *Cancer Cell Int.* **2016**, *16*, 57. [CrossRef] [PubMed]
25. Che, X.; Zhan, J.; Zhao, F.; Zhong, Z.; Chen, M.; Han, R.; Wang, Y. Oridonin Promotes Apoptosis and Restrains the Viability and Migration of Bladder Cancer by Impeding TRPM7 Expression via the ERK and AKT Signaling Pathways. *Biomed. Res. Int.* **2021**, *2021*, 4340950. [CrossRef]

26. Zhu, H.Q.; Zhang, C.; Guo, Z.Y.; Yang, J.M.; Guo, J.H.; Chen, C.; Yao, Q.H.; Liu, F.; Zhang, Q.W.; Gao, F.H. Oridonin induces Mdm2-p60 to promote p53-mediated apoptosis and cell cycle arrest in neuroblastoma. *Cancer Med.* **2019**, *8*, 5313–5326. [CrossRef]
27. Li, D.H.; Hu, P.; Xu, S.T.; Fang, C.Y.; Tang, S.; Wang, X.Y.; Sun, X.Y.; Li, H.; Xu, Y.; Gu, X.K.; et al. Lasiokaurin derivatives: Synthesis, antimicrobial and antitumor biological evaluation, and apoptosis-inducing effects. *Arch. Pharm. Res.* **2017**, *40*, 796–806. [CrossRef]
28. Fujita, E.; Nagao, Y.; Node, M.; Kaneko, K.; Nakazawa, S.; Kuroda, H. Antitumor activity of the Isodon diterpenoids: Structural requirements for the activity. *Experientia* **1976**, *32*, 203–206. [CrossRef]
29. Nik Nabil, W.N.; Xi, Z.; Liu, M.; Li, Y.; Yao, M.; Liu, T.; Dong, Q.; Xu, H. Advances in therapeutic agents targeting quiescent cancer cells. *Acta Mater. Medica* **2022**, *1*, 56–71. [CrossRef]
30. Gong, R.H.; Yang, D.J.; Kwan, H.Y.; Lyu, A.P.; Chen, G.Q.; Bian, Z.X. Cell death mechanisms induced by synergistic effects of halofuginone and artemisinin in colorectal cancer cells. *Int. J. Med. Sci.* **2022**, *19*, 175–185. [CrossRef]
31. Kim, C.; Kim, B. Anti-Cancer Natural Products and Their Bioactive Compounds Inducing ER Stress-Mediated Apoptosis: A Review. *Nutrients* **2018**, *10*, 1021. [CrossRef]
32. Yap, T.A.; Plummer, R.; Azad, N.S.; Helleday, T. The DNA Damaging Revolution: PARP Inhibitors and Beyond. *Am. Soc. Clin. Oncol. Educ. Book* **2019**, *39*, 185–195. [CrossRef] [PubMed]
33. Yin, L.; Duan, J.J.; Bian, X.W.; Yu, S.C. Triple-negative breast cancer molecular subtyping and treatment progress. *Breast Cancer Res.* **2020**, *22*, 61. [CrossRef] [PubMed]
34. Zeichner, S.B.; Terawaki, H.; Gogineni, K. A Review of Systemic Treatment in Metastatic Triple-Negative Breast Cancer. *Breast Cancer* **2016**, *10*, 25–36. [CrossRef] [PubMed]
35. Costa, R.L.B.; Han, H.S.; Gradishar, W.J. Targeting the PI3K/AKT/mTOR pathway in triple-negative breast cancer: A review. *Breast Cancer Res. Treat.* **2018**, *169*, 397–406. [CrossRef]
36. Xu, F.; Na, L.; Li, Y.; Chen, L. Retraction Note to: Roles of the PI3K/AKT/mTOR signalling pathways in neurodegenerative diseases and tumours. *Cell Biosci.* **2021**, *11*, 157. [CrossRef] [PubMed]
37. Khan, M.A.; Jain, V.K.; Rizwanullah, M.; Ahmad, J.; Jain, K. PI3K/AKT/mTOR pathway inhibitors in triple-negative breast cancer: A review on drug discovery and future challenges. *Drug Discov. Today* **2019**, *24*, 2181–2191. [CrossRef]
38. Sarbassov, D.D.; Ali, S.M.; Sabatini, D.M. Growing roles for the mTOR pathway. *Curr. Opin. Cell Biol.* **2005**, *17*, 596–603. [CrossRef]
39. Jhanwar-Uniyal, M.; Wainwright, J.V.; Mohan, A.L.; Tobias, M.E.; Murali, R.; Gandhi, C.D.; Schmidt, M.H. Diverse signaling mechanisms of mTOR complexes: mTORC1 and mTORC2 in forming a formidable relationship. *Adv. Biol. Regul.* **2019**, *72*, 51–62. [CrossRef]
40. Qin, J.J.; Yan, L.; Zhang, J.; Zhang, W.D. STAT3 as a potential therapeutic target in triple negative breast cancer: A systematic review. *J. Exp. Clin. Cancer Res.* **2019**, *38*, 195. [CrossRef]
41. Siersbaek, R.; Scabia, V.; Nagarajan, S.; Chernukhin, I.; Papachristou, E.K.; Broome, R.; Johnston, S.J.; Joosten, S.E.P.; Green, A.R.; Kumar, S.; et al. IL6/STAT3 Signaling Hijacks Estrogen Receptor alpha Enhancers to Drive Breast Cancer Metastasis. *Cancer Cell* **2020**, *38*, 412–423.e9. [CrossRef]
42. Williams, K.; Sobol, R.W. Mutation research/fundamental and molecular mechanisms of mutagenesis: Special issue: DNA repair and genetic instability. *Mutat. Res.* **2013**, *743*, 1–3. [CrossRef] [PubMed]
43. Geenen, J.J.J.; Linn, S.C.; Beijnen, J.H.; Schellens, J.H.M. PARP Inhibitors in the Treatment of Triple-Negative Breast Cancer. *Clin. Pharmacokinet.* **2018**, *57*, 427–437. [CrossRef]
44. Singh, D.D.; Yadav, D.K. TNBC: Potential Targeting of Multiple Receptors for a Therapeutic Breakthrough, Nanomedicine, and Immunotherapy. *Biomedicines* **2021**, *9*, 876. [CrossRef] [PubMed]
45. Kalimutho, M.; Parsons, K.; Mittal, D.; Lopez, J.A.; Srihari, S.; Khanna, K.K. Targeted Therapies for Triple-Negative Breast Cancer: Combating a Stubborn Disease. *Trends Pharmacol. Sci.* **2015**, *36*, 822–846. [CrossRef] [PubMed]
46. Ueng, S.H.; Chen, S.C.; Chang, Y.S.; Hsueh, S.; Lin, Y.C.; Chien, H.P.; Lo, Y.F.; Shen, S.C.; Hsueh, C. Phosphorylated mTOR expression correlates with poor outcome in early-stage triple negative breast carcinomas. *Int. J. Clin. Exp. Pathol.* **2012**, *5*, 806–813.
47. Vitali, F.; Cohen, L.D.; Demartini, A.; Amato, A.; Eterno, V.; Zambelli, A.; Bellazzi, R. A Network-Based Data Integration Approach to Support Drug Repurposing and Multi-Target Therapies in Triple Negative Breast Cancer. *PLoS ONE* **2016**, *11*, e0162407. [CrossRef]
48. Vundavilli, H.; Datta, A.; Sima, C.; Hua, J.; Lopes, R.; Bittner, M. Bayesian Inference Identifies Combination Therapeutic Targets in Breast Cancer. *IEEE Trans. Biomed. Eng.* **2019**, *66*, 2684–2692. [CrossRef]
49. Tada, H.; Shiho, O.; Kuroshima, K.; Koyama, M.; Tsukamoto, K. An improved colorimetric assay for interleukin 2. *J. Immunol. Methods* **1986**, *93*, 157–165. [CrossRef]
50. Qu, Z.; Lin, Y.; Mok, D.K.; Bian, Q.; Tai, W.C.; Chen, S. Arnicolide D Inhibits Triple Negative Breast Cancer Cell Proliferation by Suppression of Akt/mTOR and STAT3 Signaling Pathways. *Int. J. Med. Sci.* **2020**, *17*, 1482–1490. [CrossRef]
51. Chen, G.; Gong, R.; Shi, X.; Yang, D.; Zhang, G.; Lu, A.; Yue, J.; Bian, Z. Halofuginone and artemisinin synergistically arrest cancer cells at the G1/G0 phase by upregulating p21Cip1 and p27Kip1. *Oncotarget* **2016**, *7*, 50302–50314. [CrossRef]

52. Qu, Z.; Lin, Y.; Mok, D.K.; Bian, Q.; Tai, W.C.; Chen, S. Brevilin A, a Natural Sesquiterpene Lactone Inhibited the Growth of Triple-Negative Breast Cancer Cells via Akt/mTOR and STAT3 Signaling Pathways. *OncoTargets Ther.* **2020**, *13*, 5363–5373. [CrossRef] [PubMed]
53. Saran, U.; Chandrasekaran, B.; Tyagi, A.; Shukla, V.; Singh, A.; Sharma, A.K.; Damodaran, C. A small molecule inhibitor of Notch1 modulates stemness and suppresses breast cancer cell growth. *Front. Pharmacol.* **2023**, *14*, 1150774. [CrossRef] [PubMed]

Disclaimer/Publisher's Note: The statements, opinions and data contained in all publications are solely those of the individual author(s) and contributor(s) and not of MDPI and/or the editor(s). MDPI and/or the editor(s) disclaim responsibility for any injury to people or property resulting from any ideas, methods, instructions or products referred to in the content.

molecules

MDPI

Article

Isolation and Identification of Chemical Compounds from *Agaricus blazei* Murrill and Their In Vitro Antifungal Activities

Ruirui Yu [1], Xiaojian Li [1], Peng Yi [1,2], Ping Wen [3], Shuhong Wang [3], Chenghui Liao [4], Xun Song [2], Haiqiang Wu [1], Zhendan He [1,2,*] and Chenyang Li [1,*]

1 School of Pharmacy, Shenzhen University Medical School, Shenzhen University, Shenzhen 518055, China; 15220201716@163.com (R.Y.); 2210245016@email.szu.edu.cn (X.L.); 15115433780@163.com (P.Y.); wuhq@szu.edu.cn (H.W.)

2 College of Pharmacy, Shenzhen Technology University, Shenzhen 518118, China; xsong@szu.edu.cn

3 Shenzhen Institute for Drug Control, Shenzhen 518057, China; lanhair@126.com (P.W.); wangshuhongszyj@163.com (S.W.)

4 School of Pharmacy, Harbin Medical University, Harbin 150081, China; liao.chenghui66@gmail.com

* Correspondence: hezhendan@szu.edu.cn (Z.H.); lcy@szu.edu.cn (C.L.)

Abstract: This study explores the antifungal properties of *Agaricus blazei* Murrill, a valuable medicinal and edible fungus. Six compounds (**1–6**) were first isolated from *A. blazei* using various isolation techniques and identified using spectroscopic methods. These compounds include linoleic acid, 1,1′-oxybis(2,4-di-tert-butylbenzene), glycerol monolinoleate, volemolide (17R)-17-methylincisterol, (24s)-ergosta-7-en-3-ol, and dibutyl phthalate. This study also assesses the antifungal activities of these compounds against *Trichophyton mentagrophology*, *Trichophyton rubrum*, *Candida albicans*, and *Cryptococcus neoformans*. The results demonstrate varied sensitivities against these pathogenic fungi, with compound **2** showing significant inhibition against *T. mentagrophology*, compound **3** showing significant inhibition against *T. rubrum*, and compound **6** showing significant inhibition against *C. albicans*. This study underscores the medicinal potential of *A. blazei* as an antifungal agent and sheds light on its valuable research implications.

Keywords: *Agaricus blazei* Murrill; fungus; structure identification; antifungal activity

check for updates

Citation: Yu, R.; Li, X.; Yi, P.; Wen, P.; Wang, S.; Liao, C.; Song, X.; Wu, H.; He, Z.; Li, C. Isolation and Identification of Chemical Compounds from *Agaricus blazei* Murrill and Their In Vitro Antifungal Activities. *Molecules* **2023**, *28*, 7321. https://doi.org/10.3390/molecules28217321

Academic Editor: Armando Zarrelli

Received: 28 August 2023
Revised: 23 October 2023
Accepted: 27 October 2023
Published: 28 October 2023

Copyright: © 2023 by the authors. Licensee MDPI, Basel, Switzerland. This article is an open access article distributed under the terms and conditions of the Creative Commons Attribution (CC BY) license (https://creativecommons.org/licenses/by/4.0/).

1. Introduction

Fungi stand out as exceptional creators of a broad spectrum of natural products, contributing to biodiversity and offering a treasure trove of compounds imbued with a wide range of biological functions [1,2]. Notably, mushrooms, a type of fungi [3], assume a pivotal role in traditional agriculture, cuisine, and medicinal practices [4–6]. The fungi within these mushrooms produce unique bioactive compounds [7–9], with derived compounds also fulfilling nutritional roles [10]. These compounds not only serve as food sources, but also hold potential as biopharmaceuticals [11], earning them the designation of "drug alternatives".

As we know it, there is a large number and a wide variety of fungal diseases in the world that have a great range of influence. There are several classes of antifungal drugs, including azoles, echinocandins, polyenes, and allylamines. These drugs work through various mechanisms to inhibit fungal growth and replication. Azoles, for example, interfere with fungal cell membrane formation, while echinocandins target cell wall synthesis [12,13]. Although many drugs are used as current treatment of fungal diseases, there are still many defects and deficiencies. Therefore, the development of antifungal drugs and the search for antifungal active ingredients are still of great significance. Many reports have shown that fungal families have a variety of biological activities. Therefore, the focus of this study was on the extraction, isolation, and identification of antifungal activities of ABM.

A. blazei (ABM), also known as the Brazilian mushroom, is native to the southern regions of North America, Brazil, Peru, and various parts of China (Figure 1). In China, it primarily thrives in high-altitude forests of northeastern regions, as well as Yunnan, Guizhou, and the Sichuan border areas. *Shennong Baicao Classic*, an ancient Chinese herbal text, documents the notable effects of *A. blazei*, such as "harmonizing the spleen and stomach" and "stabilizing the mind." It is especially recommended for individuals with low immunity and malnutrition. Its nutritional value is substantial as it is rich in various components, including polysaccharides (such as oligosaccharides, mannan, xylose, and α- and β-glucans) [14–16], proteins [17], terpenes [18], fats, saponins, sterols [19,20], trace elements [21], and essential amino acids. This abundance of nutritional content underscores its significance in both cuisine and medicine [22]. Its range of biological activities is diverse, and its impact is particularly notable in various domains; for example, its anti-tumor effects [23–25], antiviral properties [26,27], antioxidation capabilities [16], hypoglycemic effects [28], hepatoprotective benefits [29,30], immune regulation [31], and other aspects are particularly remarkable [32]. It is considered a precious dual-purpose fungus for drugs and food [33], and it is also gaining reputation for its excellent medical value and biochemical properties. It was included in the "Redlist of China's Biodiversity—Macrofungi" in 2018.

Figure 1. The fruit body of *Agaricus blazei* Murrill.

Professor Wu Yiyuan of the Chinese Institute of Medical Cancer showed that the tumor-inhibitory effect of agaricum Antellae, which is a close relative of *Ganoderma lucidum*, can be about five times higher than *G. lucidum*, ranking first among the various medicinal fungi; this fungus is recognized as one of the 15 effective anti-cancer medicinal fungi, and is also known as "the last food for cancer patients on Earth". In short, ABM not only can be used as a valuable edible fungus with a beautiful taste and benefits for health care, but it also has immeasurable development potential in its scientific research and application value. However, as far as the current literature reports are concerned, most studies on ABM have focused on the anti-tumor and immune-regulating effects of its extracts [34,35] and polysaccharides [36], and there are no reports on the chemical components of non-polysaccharides and their antifungal activities. In this study, the non-polysaccharide components of ABM Antler were extracted and separated, the structure of these compounds was characterized, and their antifungal activities were studied.

Many scholars have tried to extract the chemical components of ABM, among which the effective methods [37] are acid extraction, hot water extraction, enzyme extraction, chromatography, and ethanol precipitation [38]. In the present study, a large scale extraction of *A. blazei* was carried out with reference to the above mentioned methods, from which the compounds were isolated, their structures were resolved using spectroscopic methods and their antifungal activities were evaluated. In a word, the main objective of the present work was to determine the antifungal activity of *A. blazei*.

2. Results

2.1. Identification of Compounds 1–6

Linoleic acid (1): Compound **1** was extracted as a yellow oil, and its molecular formula is $C_{18}H_{32}O_2$. This compound shows the following spectroscopic properties: ^{1}H NMR and ^{13}C NMR data (Figure S1) showed that the structural type of compound **1** is fatty acid. ^{13}C NMR showed that there is an obvious carbonyl carbon signal (δ_C 173.27), four olefinic carbons (δ_C 130.24, 130.02, 128.09, 127.91) and one methyl carbon (δ_H 14.08). The remaining signals are methylene carbon signals. All the obtained data of compound **1** are identical to rhapontigenin [39]. Hence, compound **1** was elucidated as linoleic acid (Figure 2).

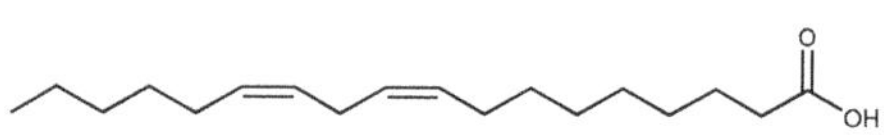

1

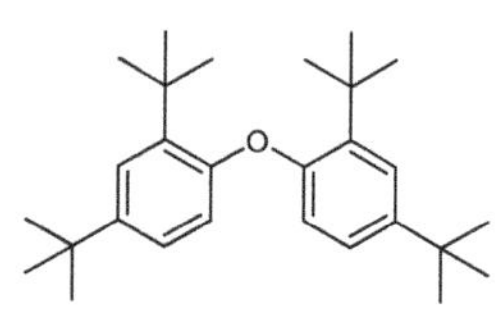

2

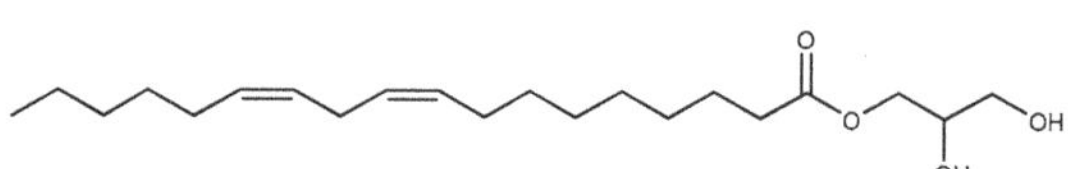

3

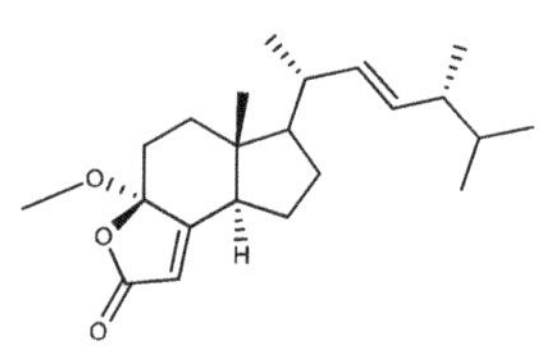

4

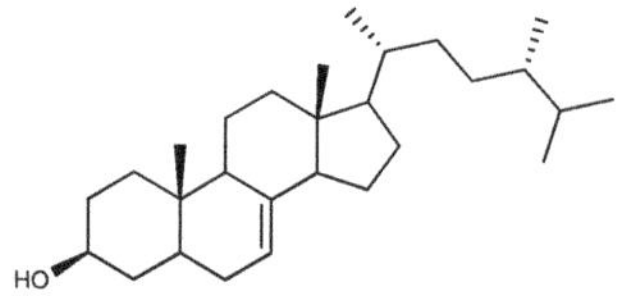

5

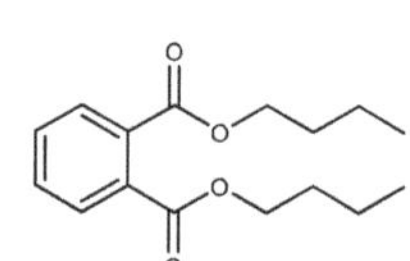

6

Figure 2. The structure of compounds **1–6**.

1,1′-Oxybis(2,4-di-tert-butylbenzene) (**2**): Compound **2** was extracted as a yellowish oily matter, and its molecular formula is $C_{28}H_{42}O$. This compound shows the following spectroscopic properties: The ^{1}H NMR spectrum of 2 in $CDCl_3$ showed signals attributable to an ABX-type aromatic ring at δ_H 7.56 (d, J = 8.6 Hz, 2H), 7.38 (t, J = 2.2 Hz, 2H), 7.15 (dd, J = 8.6, 2.5 Hz, 2H), which indicated the presence of a 1,2,4-trisubstituted benzene ring. In addition, the locations of δ_H 1.36 (s, 18H) and δ_H 1.31 (s, 18H) suggest that there are two sets of non-equivalent hydrogen signals, each containing three methyl groups. ^{13}C NMR data show that compound **2** has 12 carbon signals, including eight olefins or aromatic hydrocarbons (δ_C 147.80, 147.76, 147.23, 138.66, 138.60, 124.61, 124.13, 119.24), two quaternary carbons (δ_C 35.02, 34.67) and two obvious methyl groups (δ_C 31.59, 30.33). After identification, the above data of ^{1}H-NMR and ^{13}C-NMR (Figure S1) are basically consistent with those reported in [40]. Thus, compound **2** was considered to be 1,1′-Oxybis(2,4-di-tert-butylbenzene) (Figure 2).

Glycerol monolinoleate (**3**): Compound **3** was extracted as a colorless oil, and its molecular formula is $C_{21}H_{38}O_4$. This compound shows the following spectroscopic properties: The structural type of compound **3** was determined to be fatty acid from ^{1}H NMR and ^{13}C NMR data. ^{13}C NMR data can clearly observe that compound **3** has a carbonyl (δ_C 174.35) and two four olefin carbon signals (δ_C 130.25, 130.02, 128.09, 127.91). It is inferred that δ_C 70.27 is connected with carbonyl carbon and olefin carbon, and the residual carbon signals are mostly concentrated in the low chemical shift in the high field region, suggesting the existence of fat chains. The above NMR data (Figure S1) are basically consistent with those reported in the literature [41]. Thus, compound **3** was elucidated as glycerol monolinoleate (Figure 2).

Volemolide (17R)-17-methylincisterol (**4**): Compound **4** was extracted as a colorless oil, and its molecular formula is $C_{22}H_{34}O_3$. This compound shows the following spectroscopic properties: ^{13}C NMR data showed that compound **4** contains an obvious carbonyl carbon (δ_C 170.70) and four olefinic carbons (δ_C 169.11, 134.65, 132.87, 114.21). The combination of characteristic carbon signal and residual carbon signal shows that there is a tricyclic plus side chain in the structure of compound **4**, which is similar to that of sterols. The above NMR data (Figure S1) are basically consistent with those reported in the literature [42]. Hence, compound **4** was identified as volemolide (17R)-17-methylincisterol (Figure 2).

(24S)-ergosta-7-en-3-ol (**5**): Compound **5** was extracted as a colorless acicular crystal (dichloromethane), and its molecular formula is $C_{28}H_{48}O$. This compound shows the following spectroscopic properties: ^{13}C NMR data show that compound **5** contains 28 carbon signals, in which two olefinic carbons at δ_C 139.64, 117.43, hydroxyl carbon at δ_C 71.09 and five methyl carbon at δ_C 19.04, 17.61, 15.45, 13.05, 11.86 can be obviously observed. combined with all the ^{1}H NMR and ^{13}C NMR data, it can be inferred that the structural type of compound **5** is sterol. The above NMR data (Figure S1) are basically consistent with those of the literature [43]. Hence, compound **5** was identified as (24S)-ergosta-7-en-3-ol (Figure 2).

Dibutyl phthalate (**6**): Compound **6** was extracted as a colorless transparent oily liquid, and its molecular formula is $C_{16}H_{22}O_4$. This compound shows the following spectroscopic properties: ^{1}H NMR data showed that compound **6** is a typical ortho-substitution of benzene ring [δ_H 7.71 (dd, J = 5.7,3.3 Hz, 2H), 7.52 (dd, J = 5.8,3.3 Hz, 2H)] and symmetrical [δ_H 4.30 (t, J = 6.7 Hz, 4H), 1.75–1.68 (m, 4H), 1.48–1.40 (m, 4H), 0.96 (t, J = 7.4 Hz, 6H)]. ^{13}C NMR data show that compound **6** has carbonyl (δ_C 167.76), quaternary carbon (δ_C 132.31), alkene carbon (δ_C 130.94, 128.85), dioxycarbon (δ_C 65.59), methylene carbon (δ_C 30.57,19.19) and methyl carbon (δ_C 13.74). The above NMR data (Figure S1) are basically consistent with those reported in the literature [10]. Thus, compound **6** was identified as dibutyl phthalate (Figure 2).

2.2. Antifungal Activity of Compounds ***1–6***

T. mentagrophytes and *T. rubrum* are illustrative examples of non-yeast fungi, whereas *C. albicans* and *C. neoformans* belong to the yeast category. Consequently, an assessment of antifungal activity was conducted on these four types of fungi (see Figure 3). Notably, at a concentration of 25 µg/mL, compounds **1**, **3**, and **6** displayed inhibitory rates surpassing 50% against *T. mentagrophytes*, with compound **3** exhibiting the most potent activity. For *T. rubrum*, compounds **2** and **6** exhibited inhibitory rates exceeding 50%, with compound **6** demonstrating the most robust activity. The inhibitory rates of compounds **1**, **2**, **3**, and **4** against *C. albicans* ranged between 20% and 40%, with compound **4** exhibiting the most pronounced effect. However, all of these compounds exhibited relatively modest antibacterial activity against *C. neoformans*.

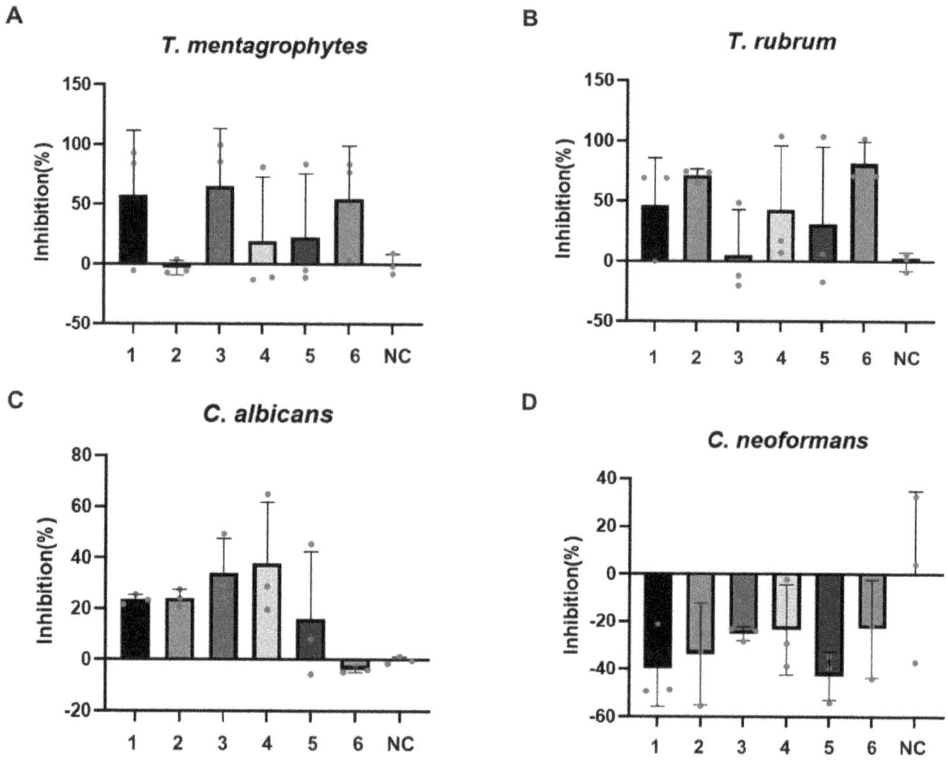

Figure 3. Effects of compounds **1–6** on four fungi. (**A**) *T. mentagroph*, (**B**) *T. rubrum*, (**C**) *C. albicans*, (**D**) *C. neoformans*.

2.3. Discussion

A. blazei, a rare and valuable edible fungus that serves as both a nourishing food and a medicinal resource, is renowned for its remarkable nutritional benefits in everyday consumption. While its abundant nutritional value is commonly harnessed for dietary supplementation, comprehensive investigations into its bioactive constituents have been limited among scholars. Furthermore, there is a dearth of literature detailing the extraction and isolation of its chemical compounds, as well as a lack of in-depth exploration of its biological functionalities. Recent advancements in scientific and medical research have brought to light the potent and distinctive therapeutic attributes of naturally derived compounds. These encompass a spectrum of effects including antibacterial, anti-allergic, anti-cancer, immunosuppressive, and anti-inflammatory properties. Notably, the antifungal potential of the chemical constituents within *A. blazei* has remained largely unexplored until now.

In this study, compounds **1–6** were successfully isolated from *A. blazei*'s fruit bodies for the first time. After spectroscopic methods we can know that these six compounds were identified as linoleic acid (**1**), 1,1′-oxybis(2,4-di-tert-butylbenzene) (**2**), glycerol monolinoleate (**3**), volemolide (17R)-17-methylincisterol (**4**), (24s)-ergosta-7-en-3-ol (**5**), and dibutyl phthalate (**6**). The previous literature reports on *A. blazei* are mostly focused on anti-tumor, anti-virus, enhancing immune function and so on, but there are no related reports on antifungal activity. Therefore, in order to better understand and apply the resources of *A. blazei* and clarify its chemical composition and pharmacodynamic basis, this study was carried out. It is worth noting that all of these purified compounds were first found in *A. blazei*. The antifungal activities of these compounds were studied for the first time in vitro. The results showed that compounds **1–6** showed different sensitivities to different fungi. Particularly noteworthy was the exceptional inhibitory effect of compound **2** on *T. mentagrophology*, while compound **3** demonstrated a notable inhibitory effect against *T. rubrum*, and compound **6** displayed a high level of inhibitory activity against *C. albicans*. This comprehensive dataset underscores the medicinal significance of *A. Blazei* as a potent antifungal agent against pathogenic fungi. Moreover, it has the potential to reshape current perceptions surrounding *A. blazei* mushrooms and their analogous counterparts, which may dispel preconceived notions. The findings of this study are expected to stimulate further investigation from researchers, encouraging more profound explorations. Generally speaking, the above findings are helpful to promote the research progress of chemical constituents and antifungal activity of *A. blazei*. Nevertheless, it is pertinent to acknowledge that the present examination of individual compounds from *A. blazei* remains at a relatively preliminary stage, and this study has certain limitations. A future avenue of research necessitates the isolation of additional compounds, thereby facilitating more intricate analyses into the underlying mechanisms of action of *A. blazei* within an in vitro context.

However, it is essential to acknowledge that the present examination of individual compounds from *A. blazei* remains at a relatively preliminary stage. While this study marks a partial progress in our understanding of *A. blazei*'s medicinal properties, it also has certain limitations. To fully unlock the therapeutic potential of *A. blazei*, future research endeavors should focus on the isolation and characterization of additional compounds, facilitating more intricate analyses into the underlying mechanisms of action within an in vitro context. One promising avenue for further exploration is the elucidation of the synergistic effects of these compounds when combined. It is possible that the unique chemical composition of *A. blazei* contributes to its antifungal properties, and a more comprehensive study of these compounds in combination could provide valuable insights into their collective efficacy. In addition, research on the mechanism of antimicrobial action is crucial to help modify its structure, reduce toxicity and increase efficacy, and translate the research results into practical medical applications.

3. Materials and Methods

3.1. Reagents and Instruments

The *A. blazei* samples were sourced from an *A. blazei* cultivation base situated in Shiming Village, Si Qian Town, Jiangmen City, Guangdong Province. These samples were provided in the form of dried mushrooms. The extraction process utilized solvents of analytical grade, which were procured from Tianjin Yongda Chemical Reagent Co., Ltd. (Tianjin, China). These solvents encompassed methanol, petroleum ether, dichloromethane, ethyl acetate, and ethanol, among others. For chromatography, the essential reagents included methanol and ultrapure water, which were acquired from Fisher/Annergi Chemistry of Thermo Fisher Scientific (Shanghai, China).

Electric heating sleeve: Model 98-1-B from Tianjin Tester Instrument Co., Ltd. (Tianjin, China); electronic analytical balance: Maximum of 220 g, from Sedolis Co., Ltd.; semi-prepared liquid chromatography: Model LC-52 from Cypress (Beijing) Technology Co., Ltd. (Beijing, China); high-performance liquid chromatography: Agilent 1200 from Agilent Technologies; rotary evaporator: Model R-260PRO from BUCHI Labortechnik AG; rapid preparation liquid chromatography: Model Biotage Isolera One from Shanghai Musen Biotechnology (Shanghai, China); nuclear magnetic resonance spectrometer: Model AVANCE III 500 MHz/600 MHz from Bruker.

3.2. Separation and Purification

A total of 7.5 kg of dried *A. blazei* mushrooms was meticulously ground using a grinder and then packed into non-woven traditional Chinese medicine bags. A portion of the dried ABM mushroom powder was carefully placed into a sizeable flask and submerged in methanol. The flask, outfitted with an electric jacket and connected to a condenser, underwent reflux for roughly 2 h. Following this, the mixture was filtered, and a fresh solvent was introduced to the system, initiating a subsequent 2 h reflux. This iterative process was repeated until the complete extraction of the ABM dried mushroom samples was accomplished. The amassed extracts were pooled, and methanol was subsequently recuperated using a large-scale rotary evaporator to yield a concentrated ABM extract. The concentrated extract was then introduced into the rotating vessel of a sizable rotary evaporator, while ensuring that the quantity of the extract added at each instance did not surpass half of the capacity of the rotary bottle. Manipulating the water bath's temperature and vacuum conditions, along with adjusting the rotation speed, facilitated the gradual addition of the ABM extract as the solvent within the rotating vessel as water evaporated. This cycle was repeated until all extracts were effectively concentrated, culminating in the recovery of the methanol solvent and the acquisition of the preliminary fraction of the crude methanol extract of *A. blazei*.

For column chromatography with silica gel, the ratio of ABM concentrate/mixed silica gel (80–100 mesh)/chromatographic silica gel (200–300 mesh) = 1:1:8. Petroleum ether extract (ABM-A), dichloromethane extract (ABM-DCM), ethyl acetate extract (ABM-B), acetone extract (ABM-BT), and methanol extract (ABM-E) were obtained via gradient elution with petroleum ether, dichloromethane, ethyl acetate, acetone, and methanol, respectively. Finally, these extracts were concentrated to obtain individual solvent concentrates, which were stored until use. The preliminary fractionation activity screening (secondary fraction) was carried out, and the same method was also used for the silica gel column chromatography.

For the ODS column chromatography, after analysis using TLC, HPLC, ELSD and other methods, the sample, mixed sample ODS, and chromatographic ODS were mixed at a ratio of 1:1:20 for each fraction, and then loaded into the appropriate column. Rough cut: the proportion of methanol to water was subjected to gradient elution at 30%, 50%, 70%, and 100%, respectively, and then collected into different fractions. Then, the TLC point plate, HPLC liquid phase, and evaporative light-scattering detector (ELSD) were used again for spectrum analysis and quality analysis.

Acetone crude extract and dichloromethane crude extract were separated and purified using a combination of silica gel column chromatography (sample/mixed sample silica gel/chromatography silica gel ≈ 1:1:8), ODS column chromatography (sample/mixed sample ODS/ODS = 1:1:20), and Biotage HPLC (methanol/water). The obtained compounds are listed below.

ABM-DCM (20 g) was eluted using silica gel column chromatography (PE/EA) to obtain 14 segments of small-fraction ABM-DCM-A~N. ABM -DCM-F (50 mg) was directly synthesized using HPLC (96% Me, flow rate of 3 mL/min) to obtain compound **2** (16.8 mg). ABM-DCM-L (ABM-DCM was eluted using silica gel column chromatography, with PE/EA = 100: 15 cm 100: 50, 3.15 g) was eluted with methanol/water using ODS chromatography to obtain nine segments, of which ABM-DCM-L-6 (312 mg) was semi-prepared using HPLC (Cyprus, Phenomenex semi-preparation column, 82% Me, flow rate of 5 mL/min) to obtain compound **3** (22.7 mg) and compound **4** (4.3 mg). Compound **5** (104.2 mg) was prepared from small-fraction ABM-DCM-J-93%-4 (≈220 mg) via conditional touch and HPLC (Cypress, phenomenex semi-preparation column, 92% Me, flow rate of 5 mL/min). Compound **6** (220 mg) was directly obtained via ODS chromatography with methanol/water (25–100%).

3.3. Antibacterial Activity Determination

In this study, four fungal strains, *T. mentagroph*, *T. rubrum*, *C. albicans*, and *C. neoformans*, were cultured in vitro.

Culture preparation: for the aseptic operation, 2.5 mL of the medium was added to a 15 mL fast cap culture tube, a single colony was carefully removed from the Petri dish using a sterile rod, and the tip of the rod was rotated in the culture medium (while being careful not to touch the end of the rod at anything, except the colony and the medium); the culture tube was placed in a 37 °C incubator, at a 250 rpm speed oscillation, for culturing to the logarithmic phase.

Culture plate preparation: compounds **1–6** at 0.2 mg each were dissolved in 200 μL of DMSO, and 5 μL of the sample solution and DMSO as a control were added to a 96-well plate; then, the culture plate was put into reserve.

Diluted pathogen culture: 5 mL of aseptic medium and 500 μL of culture were added to a 15 mL conical tube and mixed, and then the medium or culture was used as diluent to dilute OD600 to 0.030–0.060. After mixing the diluted culture with the culture medium at a proportion of 1:10, 195 μL of inoculation medium and 200 μL of aseptic culture medium were added to a 96-well plate as the blank group, and then the plate was placed in a sealed bag. The plate was fixed in a vibrating screen with a plate bracket or tape and cultured at 37 °C for 16 h.

3.4. Statistical Analysis

All in vitro assays were performed in triplicate. The data were analyzed and processed using GraphPad Prism 9.0 software. The data were expressed as mean ± SD. Differences between the groups were compared using a *t*-test. p-value < 0.05 was considered as representing a statistically significant difference.

4. Conclusions

In recent years, the exploration of *A. blazei*, a rare and valuable edible fungus, has garnered increased attention due to its dual role as a nourishing food and a potential medicinal resource. Its reputation for remarkable nutritional benefits in everyday consumption has made it a sought-after dietary supplement. However, despite its extensive use, comprehensive investigations into its bioactive constituents and their potential medicinal properties have remained limited among scholars. Furthermore, there is a notable dearth of literature detailing the extraction and isolation of its chemical compounds, as well as a lack of in-depth exploration of its biological functionalities.

In this study, compounds **1–6** have been successfully isolated from the non-polysaccharide constituents of *A. blazei* via processes like column chromatography and semi-preparative HPLC. These compounds underwent meticulous characterization via techniques such as ^{1}H-NMR and ^{13}C-NMR spectroscopy. These six compounds have been identified as linoleic acid (**1**), 1,1′-oxybis(2,4-di-tert-butylbenzene) (**2**), glycerol monolinoleate (**3**), volemolide (17R)-17-methylincisterol (**4**), (24s)-ergosta-7-en-3-ol (**5**), and dibutyl phthalate (**6**). Of particular significance is the fact that all these purified compounds were previously unidentified in *A. blazei*. Additionally, a pioneering assessment of the in vitro antifungal activities of these compounds against *T. mentagrophytes*, *T. rubrum*, *C. albicans*, and *C. neoformans* was carried out for the first time. The results of this assessment unveiled varying degrees of sensitivity against the four fungi, with specific compounds exhibiting distinct inhibitory effects. Compound **2** emerged as particularly effective against *T. mentagrophytes*, while compound **3** demonstrated a notable inhibitory effect against *T. rubrum*. Remarkably, compound **6** displayed a high level of inhibitory activity against *C. albicans*. Different types of compounds in *A. blazei* have been shown to have different antifungal activities against different fungi. The identified compounds have simple structures, including fatty acid glycerides and aromatic compounds, and have the potential to be developed into novel antifungal agents. In a word, *A. blazei*, as a representative macrofungus, demonstrates inherent antifungal activities. This study informs for further research and advancements related to *A. blazei*, positioning it as a potential source of natural antifungal products. Moreover, it contributes to expanding the knowledge on the structural diversity of natural compounds and facilitating the elucidation of *A. blazei*'s effective material basis.

However, it is essential to acknowledge that the present examination of individual compounds from *A. blazei* remains at a relatively preliminary stage. While this study represents a significant leap forward in our understanding of *A. blazei*'s medicinal properties, it also has certain limitations. To fully unlock the therapeutic potential of *A. blazei*, future research endeavors should focus on the isolation and characterization of additional compounds, facilitating more intricate analyses into the underlying mechanisms of action within an in vitro context. One promising avenue for further exploration is the elucidation of the synergistic effects of these compounds when combined. It is possible that the unique chemical composition of *A. blazei* contributes to its antifungal properties, and a more comprehensive study of these compounds in combination could provide valuable insights into their collective efficacy. Moreover, in vivo studies are essential to validate the in vitro findings and to assess the safety and efficacy of *A. blazei* as a potential antifungal therapy in living organisms. Clinical trials involving *A. blazei*-based treatments may be the next step in translating these promising findings into practical medical applications.

In conclusion, the discovery of compounds with antifungal properties within *A. blazei* marks further progress in our understanding of this remarkable fungus's medicinal potential. While there are still many questions to be answered and challenges to be overcome, it deserves further study.

Supplementary Materials: The following supporting information can be downloaded at https://www.mdpi.com/article/10.3390/molecules28217321/s1. Figure S1: The ^{1}H-NMR and ^{13}C-NMR spectra of the six identified compounds. The spectra are included in the Supplementary Materials. Figure S2: 13C NMR spectrum of compound **1** in Methanod-d4. Figure S3: 1H NMR spectrum of compound **2** in CDCl3. Figure S4: 13C NMR spectrum of compound **2** in CDCl3. Figure S5: 1H NMR spectrum of compound **3** in CDCl3. Figure S6: 13C NMR spectrum of compound **3** in CDCl3. Figure S7: 1H NMR spectrum of compound **4** in CDCl3. Figure S8: 13C NMR spectrum of compound **4** in CDCl3. Figure S9, 1H NMR spectrum of compound **5** in CDCl3. Figure S10: 13C NMR spectrum of compound **5** in CDCl3. Figure S11: 1H NMR spectrum of compound **6** in CDCl3. Figure S12: 13C NMR spectrum of compound **5** in CDCl3.

Author Contributions: Conceptualization, C.L. (Chenyang Li), Z.H., X.S. and H.W.; methodology, C.L. (Chenyang Li); formal analysis, R.Y. and X.L.; software, R.Y., C.L. (Chenghui Liao) and P.Y.; investigation, C.L. (Chenyang Li) and R.Y.; data curation, R.Y., P.W. and S.W.; writing, R.Y. and X.L.; supervision, C.L. (Chenyang Li) and Z.H.; funding acquisition, C.L. (Chenyang Li) and Z.H. All authors have read and agreed to the published version of the manuscript.

Funding: This research was funded by the Shenzhen Natural Science Fund (the Stable Support Plan Program), grant no. 20220811110339002; the Shenzhen Science and Technology Innovation Committee, grant no. JCYJ20190808122213241; the Guangdong Basic and Applied Basic Research Foundation, grant no. 2019B1515120029; and the National Natural Science Foundation of China, grant no. 81973293.

Institutional Review Board Statement: Not applicable.

Informed Consent Statement: Not applicable.

Data Availability Statement: Not applicable.

Acknowledgments: We thank Yi-bao Jin from the Shenzhen Institute for Drug Control for the support in performing the samples' NMR test.

Conflicts of Interest: The authors declare no conflict of interest.

Sample Availability: Samples of the identified compounds are available from the authors.

References

1. Pérez-Gómez, L.; Pérez-Martínez, A.T.; Matheeussen, A.; Pieters, L.; Mendez, D.; Quirós-Molina, Y.; Trujillo, R.; Tuenter, E.; Cos, P. Phytochemical characterization and antifungal potential of leaf extracts of *Mosiera bullata*. *Nat. Prod. Res.* **2023**, 1–10. [CrossRef] [PubMed]
2. Phan, N.K.N.; Huynh, T.K.C.; Nguyen, H.P.; Le, Q.T.; Nguyen, T.C.T.; Ngo, K.K.H.; Nguyen, T.H.A.; Ton, K.A.; Thai, K.M.; Hoang, T.K.D. Exploration of Remarkably Potential Multitarget-Directed N-Alkylated-2-(substituted phenyl)-1-benzimidazole Derivatives as Antiproliferative, Antifungal, and Antibacterial Agents. *ACS Omega* **2023**, *8*, 28733–28748. [CrossRef] [PubMed]
3. Huang, K.; El-Seedi, H.; Xu, B. Critical review on chemical compositions and health-promoting effects of mushroom *Agaricus blazei* Murill. *Curr. Res. Food Sci.* **2022**, *5*, 2190–2203. [CrossRef]

4. Pan, Z.C.; Zhang, Y.Z.; Liang, Z.Q.; Wang, Y.; Zeng, N.K. Extraction, Characterization, and In Vitro Hypoglycemic Activity of a Neutral Polysaccharide from the New Medicinal Mushroom *Cantharellus yunnanensis* (Agaricomycetes). *Int. J. Med. Mushrooms* **2023**, *25*, 19–31. [CrossRef]
5. Peng, B.; Huang, W.-H.; Zhao, J.; Liu, H.-G. Chemical constituents from *Ilex hainanensis*. *Zhong Yao Cai* **2012**, *35*, 1251–1254. [PubMed]
6. Fujimiya, Y.; Suzuki, Y.; Oshiman, K.I.; Kobori, H.; Moriguchi, K.; Nakashima, H.; Matumoto, Y.; Takahara, S.; Ebina, T.; Katakura, R. Selective tumoricidal effect of soluble proteoglucan extracted from the basidiomycete, *Agaricus blazei* Murill, mediated via natural killer cell activation and apoptosis. *Cancer Immunol. Immunother.* **1998**, *46*, 147–159. [CrossRef]
7. Dong, Y.; Wang, T.; Zhao, J.; Gan, B.; Feng, R.; Miao, R. Polysaccharides Derived from Mushrooms in Immune and Antitumor Activity: A Review. *Int. J. Med. Mushrooms* **2023**, *25*, 1–17. [CrossRef]
8. Dinçer, E.; Işık, H.; Hepokur, C.; Tutar, U.; Çelik, C. Cytotoxic, Antioxidant, Antibiofilm, and Antimicrobial Activities of Mushroom Species from Turkey. *Int. J. Med. Mushrooms* **2023**, *25*, 75–86. [CrossRef]
9. Arunachalam, K.; Sasidharan, S.P.; Yang, X. A concise review of mushrooms antiviral and immunomodulatory properties that may combat against COVID-19. *Food Chem. Adv.* **2022**, *1*, 100023. [CrossRef]
10. Qu, X.Y.; Gu, Q.Q.; Cui, C.B.; Fang, Y.C.; Liu, H.B.; Zhu, T.J.; Zhu, W.M. Structural identification and antitumor activity of secondary metabolites of marine-derived actinomycete 3295. *Chin. J. Mar. Drugs* **2004**, *23*, 1–4.
11. Rong, P.X.; He, X.Q.; Ayyash, M.; Liu, Y.; Wu, D.T.; Geng, F.; Li, H.B.; Ng, B.S.; Liu, H.Y.; Gan, R.Y. Untargeted metabolomics analysis of non-volatile metabolites and dynamic changes of antioxidant capacity in Douchi with edible mushroom by-products. *Food Chem.* **2023**, *431*, 137066. [CrossRef] [PubMed]
12. Marzi, M.; Farjam, M.; Kazeminejad, Z.; Shiroudi, A.; Kouhpayeh, A.; Zarenezhad, E. A Recent Overview of 1,2,3-Triazole-Containing Hybrids as Novel Antifungal Agents: Focusing on Synthesis, Mechanism of Action, and Structure-Activity Relationship (SAR). *J. Chem.* **2022**, *2022*, 7884316. [CrossRef]
13. Demirpolat, A.; Akman, F.; Kazachenko, A.S. An Experimental and Theoretical Study on Essential Oil of *Aethionema sancakense*: Characterization, Molecular Properties and RDG Analysis. *Molecules* **2022**, *27*, 6129. [CrossRef]
14. Mizuno, T. Medicinal Properties and Clinical Effects of Culinary-Medicinal Mushroom *Agaricus blazei* Murrill (Agaricomycetideae) (Review). *Int. J. Med. Mushrooms* **2002**, *4*, 14. [CrossRef]
15. Carneiro, A.A.; Ferreira, I.C.; Dueñas, M.; Barros, L.; Da Silva, R.; Gomes, E.; Santos-Buelga, C. Chemical composition and antioxidant activity of dried powder formulations of *Agaricus blazei* and *Lentinus edodes*. *Food Chem.* **2013**, *138*, 2168–2173. [CrossRef] [PubMed]
16. Wu, S.; Li, F.; Jia, S.; Ren, H.; Gong, G.; Wang, Y.; Lv, Z.; Liu, Y. Drying effects on the antioxidant properties of polysaccharides obtained from *Agaricus blazei* Murrill. *Carbohydr. Polym.* **2014**, *103*, 414–417. [CrossRef]
17. Chen, T.; Li, K.; Xu, J.; He, X.; Chen, F.; Jiang, Z. Nutritional composition of Brazilian mushroom from Fujian. *J. Edible Mushrooms* **1999**, *06*, 55–58.
18. Sun, P.; Wei, H.; Yang, K.; Wu, X.; He, G. Isolation, purification and physicochemical properties of *Agaricus blazei* polysaccharides. *Chin. Tradit. Herb. Drug* **2006**, *37*, 190–192.
19. Gao, H.; Gu, W.Y. Quantitative determination of ergosterol in *Agaricus brasiliensis* by triple-wavelength spectrophotometry. *Chin. J. Anal. Chem.* **2007**, *35*, 586–588.
20. Takaku, T.; Kimura, Y.; Okuda, H. Isolation of an antitumor compound from *Agaricus blazei* Murill and its mechanism of action. *J. Nutr.* **2001**, *131*, 1409–1413. [CrossRef]
21. Tsai, S.Y.; Tsai, H.L.; Mau, J.L. Non-volatile taste components of *Agaricus blazei*, *Agrocybe cylindracea* and *Boletus edulis*. *Food Chem.* **2008**, *107*, 977–983. [CrossRef]
22. Schepetkin Igor, A.; Quinn Mark, T. Botanical polysaccharides: Macrophage immunomodulation and therapeutic potential. *Int. Immunopharm.* **2006**, *6*, 317–333. [CrossRef] [PubMed]
23. Yeh, M.Y.; Shang, H.S.; Lu, H.F.; Chou, J.; Yeh, C.; Chang, J.B.; Hung, H.F.; Kuo, W.L.; Wu, L.Y.; Chung, J.G. Chitosan oligosaccharides in combination with *Agaricus blazei* Murill extract reduces hepatoma formation in mice with severe combined immunodeficiency. *Mol. Med. Rep.* **2015**, *12*, 133–140. [CrossRef]
24. Ohno, N.; Furukawa, M.; Miura, N.N.; Adachi, Y.; Motoi, M.; Yadomae, T. Antitumor b-Glucan from the Cultured Fruit Body of *Agaricus blazei*. *Biol. Pharm. Bull.* **2001**, *24*, 820–828. [CrossRef]
25. Bertollo, A.G.; Mingoti, M.E.D.; Plissari, M.E.; Betti, G.; Junior, W.A.R.; Luzardo, A.R.; Ignácio, Z.M. *Agaricus blazei* Murrill mushroom: A review on the prevention and treatment of cancer. *Pharmacol. Res. Mod. Chin. Med.* **2022**, *2*, 100032. [CrossRef]
26. Sorimachi, K.; Ikehara, Y.; Maezato, G.; Okubo, A.; Yamazaki, S.; Akimoto, K.; Niwa, A. Inhibition by *Agaricus blazei* Murill fractions of cytopathic effect induced by western equine encephalitis (WEE) virus on VERO cells in vitro. *Biosci. Biotechnol. Biochem.* **2001**, *65*, 1645–1647. [CrossRef]
27. Zhao, Y.; Tian, N.; Wang, H.; Yan, H. Chemically Sulfated Polysaccharides from *Agaricus blazei* Murill: Synthesis, Characterization and Anti-HIV Activity. *Chem. Biodivers.* **2021**, *18*, e2100338. [CrossRef]
28. Wei, Q.; Zhan, Y.; Chen, B.; Xie, B.; Fang, T.; Ravishankar, S.; Jiang, Y. Assessment of antioxidant and antidiabetic properties of *Agaricus blazei* Murill extracts. *Food Sci. Nutr.* **2019**, *8*, 332–339. [CrossRef]
29. Al-Dbass, A.M.; Al-Daihan, S.K.; Bhat, R.S. *Agaricus blazei* Murill as an efficient hepatoprotective and antioxidant agent against CCl4-induced liver injury in rats. *Saudi J. Biol. Sci.* **2012**, *19*, 303–309. [CrossRef]

30. Yu, L.; Yang, S.; Sun, L.; Jiang, Y.F.; Zhu, L.Y. Effects of selenium-enriched *Agaricus blazei* Murill on liver metabolic dysfunction in mice, a comparison with selenium-deficient *Agaricus blazei* Murill and sodium selenite. *Biol. Trace Elem. Res.* **2014**, *160*, 79–84. [CrossRef]
31. Wang, P.; Li, X.T.; Sun, L.; Shen, L. Anti-Inflammatory Activity of Water-Soluble Polysaccharide of *Agaricus blazei* Murill on Ovariectomized Osteopenic Rats. *Evid.-Based Complement. Altern. Med.* **2013**, *2013*, 164817. [CrossRef] [PubMed]
32. Hetland, G.; Tangen, J.-M.; Mahmood, F.; Mirlashari, M.R.; Nissen-Meyer, L.S.H.; Nentwich, I.; Therkelsen, S.P.; Tjønnfjord, G.E.; Johnson, E. Antitumor, Anti-inflammatory and Antiallergic Effects of *Agaricus blazei* Mushroom Extract and the Related Medicinal Basidiomycetes Mushrooms, *Hericium erinaceus* and *Grifola frondosa*: A Review of Preclinical and Clinical Studies. *Nutrients* **2020**, *12*, 1339. [CrossRef] [PubMed]
33. Zhou, Y.; Xu, B. New insights into molecular mechanisms of "Cold or Hot" nature of food: When East meets West. *Food Res. Int.* **2021**, *144*, 110361. [CrossRef]
34. Cui, L.; Sun, Y.; Xu, H.; Xu, H.; Cong, H.; Liu, J. A polysaccharide isolated from *Agaricus blazei* Murill (ABP--AW1) as a potential Th1 immunity-stimulating adjuvant. *Oncol. Lett.* **2013**, *6*, 1039–1044. [CrossRef]
35. Tangen, J.M.; Holien, T.; Mirlashari, M.R.; Misund, K.; Hetland, G. Cytotoxic Effect on Human Myeloma Cells and Leukemic Cells by the *Agaricus blazei* Murill Based Mushroom Extract, Andosan™. *Biomed. Res. Int.* **2017**, *2017*, 2059825. [CrossRef] [PubMed]
36. Chung, I.M.; Park, H.Y.; Chun, S.C.; Kim, J.J.; Ahmad, A. New glycosidic and other constituents from hulls of *Oryza sativa*. *Chem. Nat. Compd.* **2007**, *43*, 417–421. [CrossRef]
37. Otgonsugar, P.; Buyankhishig, B.; Undrakhbayar, T.; Bilguun, B.; Sasaki, K.; Davaapurev, B.O.; Batkhuu, J.; Byambajav, T.; Murata, T. Phytochemical investigation of aerial parts of *Woodsia ilvensis* and its plasmin-inhibitory activity in vitro. *Phytochemistry* **2023**, *215*, 113826. [CrossRef]
38. Chebaro, Z.; Abdallah, R.; Badran, A.; Hamade, K.; Hijazi, A.; Maresca, M.; Mesmar, J.E.; Baydoun, E. Study of the antioxidant and anti-pancreatic cancer activities of *Anchusa strigosa* aqueous extracts obtained by maceration and ultrasonic extraction techniques. *Front. Pharmacol.* **2023**, *14*, 1201969. [CrossRef]
39. Wheeler, J.J.; Domenichiello, A.F.; Jensen, J.R.; Keyes, G.S.; Maiden, K.M.; Davis, J.M.; Ramsden, C.E.; Mishra, S.K. Endogenous Derivatives of Linoleic Acid and their Stable Analogs Are Potential Pain Mediators. *JID Innov.* **2023**, *3*, 100177. [CrossRef]
40. Jing, S.; Qu, Z.; Zhao, C.; Li, X.; Guo, L.; Liu, Z.; Zheng, Y.; Gao, W. Dihydroisocoumarins and Dihydroisoflavones from the Rhizomes of *Dioscorea collettii* with Cytotoxic Activity and Structural Revision of 2,2′-Oxybis(1,4-di-tert-butylbenzene). *Molecules* **2021**, *26*, 5381. [CrossRef]
41. Okuyama, E.; Hasegawa, T.; Matsushita, T.; Fujimoto, H.; Ishibashi, M.; Yamazaki, M. Analgesic Components of Saposhnikovia Root (*Saposhnikovia divaricata*). *Chem. Pharm. Bull.* **2001**, *49*, 154–160. [CrossRef] [PubMed]
42. Kobata, K.; Wada, T.; Hayashi, Y.; Shibata, H. Volemolide, a Novel Norsterol from the Fungus *Lactarius volemus*. *Biosci. Biotechnol. Biochem.* **2014**, *58*, 1542–1544. [CrossRef]
43. Gao, J.; Shen, J.; Yang, X.; Liu, J. The constituents of *Russula ochroleuca* Basidiomycetes. *Acta Bot. Yunnanicam* **2001**, *23*, 385–393.

Disclaimer/Publisher's Note: The statements, opinions and data contained in all publications are solely those of the individual author(s) and contributor(s) and not of MDPI and/or the editor(s). MDPI and/or the editor(s) disclaim responsibility for any injury to people or property resulting from any ideas, methods, instructions or products referred to in the content.

molecules

Article

Isolation and Anticancer Progression Evaluation of the Chemical Constituents from *Bridelia balansae* Tutcher

Lihan Zhao [1,2,†], Wen-Jian Xie [1,†], Yin-Xiao Du [1,†], Yi-Xuan Xia [1], Kang-Lun Liu [1], Chuen Fai Ku [1], Zihao Ou [3], Ming-Zhong Wang [4,*] and Hong-Jie Zhang [1,*]

1 School of Chinese Medicine, Hong Kong Baptist University, Kowloon, Hong Kong SAR, China; liukanglun@hkbu.edu.hk (K.-L.L.)
2 Department of Biology, Georgia State University, Atlanta, GA 30303, USA
3 Department of Materials Science and Engineering, Stanford University, Stanford, CA 94305, USA
4 College of Pharmacy, Shenzhen Technology University, Shenzhen 518118, China
* Correspondence: wangmingzhong@sztu.edu.cn (M.-Z.W.); zhanghj@hkbu.edu.hk (H.-J.Z.)
† These authors contributed equally to this work.

Abstract: The dichloromethane extract of the roots of *Bridelia balansae* Tutcher (Phyllanthaceae) was found to show potential anticancer activity against HCT116 colorectal cancer cell. Our bioassay-guided phytochemical investigation of the roots of *B. balansae* led to the identification of 14 compounds including seven lignans (**1–7**), three phenylbenzene derivatives (**8–10**), two flavanone (**11–12**), and two triterpenoids (**13–14**). Among them, 4′-demethyl-4-deoxypodophyllotoxin (**1**) is the first aryl-tetralin lignan compound identified from this plant species. In addition, the stereochemistry of **1** was validated by X-ray crystallography for the first time, and its distinguished cytotoxic effect on HCT116 cells with an IC_{50} value at 20 nM was induced via an apoptosis induction mechanism. Compound **1** could also significantly decrease the migration rate of HCT116 cells, indicating its potential application against cancer metastasis. The western blot analysis showed that **1** has the potential to inhibit cell proliferation and metastasis. Treatment of **1** resulted in the downregulation of matrix metalloproteinases 2 (MMP2) and p-Akt, while p21 was upregulated. Collectively, the present study on the phytochemical and biological profile of *B. balansae* has determined the plant as a useful source to produce promising anticancer lead compounds.

Keywords: *Bridelia balansae*; lignan compounds; colorectal cancer; cytotoxicity; apoptosis

check for updates

Citation: Zhao, L.; Xie, W.-J.; Du, Y.-X.; Xia, Y.-X.; Liu, K.-L.; Ku, C.F.; Ou, Z.; Wang, M.-Z.; Zhang, H.-J. Isolation and Anticancer Progression Evaluation of the Chemical Constituents from *Bridelia balansae* Tutcher. *Molecules* **2023**, *28*, 6165. https://doi.org/10.3390/molecules28166165

Academic Editor: René Csuk

Received: 31 July 2023
Revised: 16 August 2023
Accepted: 17 August 2023
Published: 21 August 2023

Copyright: © 2023 by the authors. Licensee MDPI, Basel, Switzerland. This article is an open access article distributed under the terms and conditions of the Creative Commons Attribution (CC BY) license (https://creativecommons.org/licenses/by/4.0/).

1. Introduction

Natural products, with sources from plants, marine organisms, and microorganisms, have been applied for the treatment of a range of types of diseases in folk medicine since ancient times and played a key role in drug discovery historically [1,2]. With classical natural product chemistry methodologies, a variety of bioactive compounds from natural product resources have been discovered as current drug candidates [1,3]. These therapeutic areas include cardiovascular diseases [4] and multiple sclerosis [5], but cancer and infectious diseases remain the therapeutic areas of which natural products made the major contributions [6,7]. Specifically, over 60% of currently used anticancer agents are derived from natural products, and intensive studies by a random selection screening program funded by the United States National Cancer Institute (NCI) led to the discovery and development of important chemotherapeutics such as the vinca alkaloids, vinblastine, vincristine, and the isolation of the cytotoxic podophyllotoxins [8,9].

The plant genus *Bridelia*, mainly distributed in tropical and subtropical regions of the Africa and Asia, belongs to the Phyllanthaceae family (formerly Euphorbiaceae) with approximately 60 species [10]. Among them, a dozen species of *Bridelia* genus have been used as traditional folk and ethnic medicinal plants in Africa and southeast Asia countries, for the treatment of infective diseases caused by parasitic, bacterial, malarial, viral, oral, or sexual pathogens, glucometabolic and bowel disorders, arthritis [10], and hypertension [11]. They can also be used as analgesics, antianemic purposes, anticonvulsants [10], poison antidotes [12], and antioxidatives [13], or for the purposes of wound healing [12], relieving itching [13], and resolving cough, fever, and jaundice [14]. Consistent with their traditional use, the crude extracts from this plant genus have been reported to have a wide range of bioactivities including potential anticancer activities [15,16]. However, the study of the substance basis of pharmacological effects are rarely involved.

The dichloromethane fraction of the roots of *Bridelia balansae* Tutcher was found to show potential anticancer activity, with an IC_{50} value of 0.79 μg/mL against human colorectal carcinoma cells HCT116. According to APG III System, *B. balansae* belongs to the *Phyllanthaceae* family, and could be found mainly in Lingnan area of China, including Fujian, Guangdong, Guangxi, Guizhou, Hainan, Sichuan, Taiwan, and Yunnan Province, together with other countries in Asia such as Japan (Ryukyu Islands), Laos, and Vietnam. It has been used as a folk medicine to treat gastropathy and glometulonephropathy in China [17], but there is no prior report of the bioactivity of the chemical constituents from this plant. Thus, we are encouraged to carry out further phytochemical studies to discover bioactive compounds from this plant based on the potent bioactivity of the plant extract. Herein, we describe the isolation, structural identification, and biological activity evaluation of the compounds obtained from *B. balansae*.

2. Results

2.1. Compound Isolation and Structure Identification

Compounds **1–14** were identified as 4′-demethyl-4-deoxypodophyllotoxin (**1**) (Figures S1–S7) [18,19], deoxypodophyllotoxin (**2**) [20], polygamain (**3**) [21], yatein (**4**) [22], pinoresinol (**5**) [23], medioresinol (**6**) [24], (+)-Syringaresinol (**7**) [25], gallic acid (**8**) [26], gallicin (**9**) [27], methyl-3,5-dimethoxy-4-hydroxybenzoate(**10**) [28], gallocatechin (**11**) [29], and epigallocatechin gallate (**12**) [30], epifriedelanol (**13**) [31], and friedelin (**14**) [31] (Figure 1) by comparing the NMR and HRMS data reported from previous studies. The prior absolute stereochemistry of **1** was only established by using NMR method since it was first discovered in 1984, and no X-ray crystallographic data have been reported so far to validate the absolute configuration [18]. Considering the chirality diversity of the carbon skeleton of aryltetralin lignans, we nurtured the fine crystals of **1** from acetone to obtain high-quality X-ray crystallographic data with CuKα radiation. The resulted Flack parameter of 0.13 (17) from the refinement of the X-ray data, together with the comparison of the reported optical rotation data of podophyllotoxin [32], confirmed the absolute stereochemistry of **1** (Figure 2). In addition, we measured the melting point (248.5–249.0 °C) and the optical rotation ($[\alpha]^{20}_{D} = -96°$, $c = 0.3$, $CHCl_3$) of **1**, which are consistent with the literature report [19]. The isolated and purified compounds are stored at −80 °C and then dissolved into DMSO for treatment. Dissolved compounds were stored at −20 °C. Through a bioassay-guided phytochemical investigation of the roots of *B. balansae*, we were able to identify seven lignans, three phenylbenzene derivatives, two flavanone, and two triterpenoids. The newly isolated compounds were tested for their anticancer activity potential against HCT116 colorectal cancer cells. We subsequently investigated the mechanism investigation of **1**.

2.2. Cytotoxic Effect of Isolated Compounds against HCT116 Cancer Cells

The in vitro cytotoxicity of each isolated compound at 20 μg/mL was evaluated against the human colorectal carcinoma HCT116 cells (Table 1). Among the tested compounds, **1**, **8**, and **9** were the most active ones with an inhibition rate larger than 50% at 20 μg/mL, and thus further tested for their IC_{50} values (Table 2). Among these three compounds, only **1** showed potent cytotoxicity against HCT116 cells with an IC_{50} value of 0.02 μM, and thus was chosen for further mechanism studies.

Table 1. The inhibition rate of isolated compounds against HCT 116 cells at 20 μg/mL.

Compound	Inhibition Rate	Compound	Inhibition Rate
1	96.1%	**9**	76.1%
3	11.7%	**10**	3.4%
8	100%	**11**	24.5%

Table 2. The IC_{50} values of compounds **1**, **8**, and **9** against HCT 116 cells (μM).

Compound	IC_{50}
1	0.02 ± 0.003
8	20 ± 0.006
9	16.3 ± 0.003

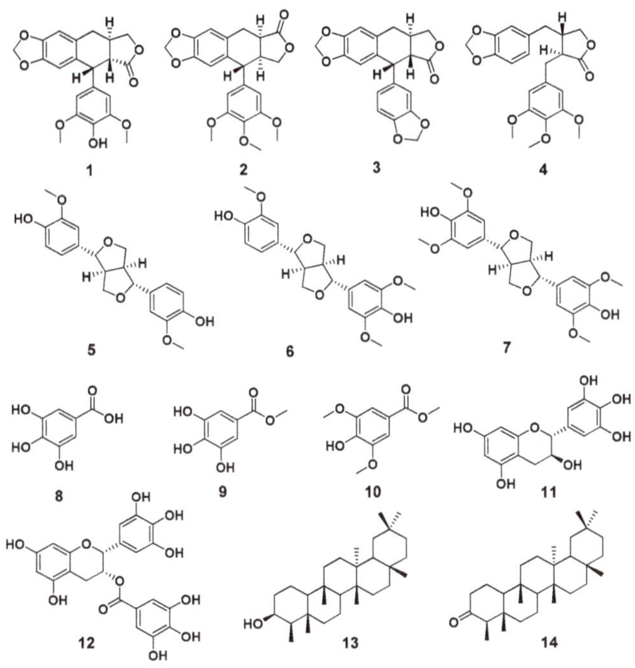

Figure 1. The chemical structures of compounds **1–14**.

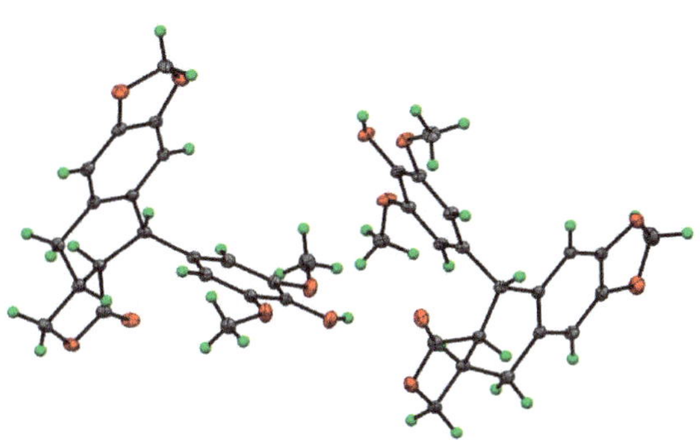

Figure 2. The ORTEP drawing of compound **1**.

2.3. *Compound* **1** *Induced Apoptosis in HCT116 Cancer Cells*

It has been previously reported that 4′-demethyl-4-deoxypodophyllotoxin (**1**) could induce cell cycle arrest with a significantly increased G2/M peak after treatment of HCT-116 cells [33]. We further investigated whether the antiproliferative activities of **1** against HCT116 cancer cells were associated with apoptosis, which is the most common mechanism of action of anticancer agents. As shown in Figure 3, the percentages of apoptotic cells were analyzed by flow cytometric analysis. While the percentages of apoptotic cells did not change significantly when **1** was applied at concentrations lower than its IC_{50} value, the number of apoptotic cells was remarkably increased when **1** was applied at 26 nM (slightly higher than its IC_{50} value). These findings suggest that the anti-proliferative effect of **1** in HCT116 cells could be derived from apoptosis.

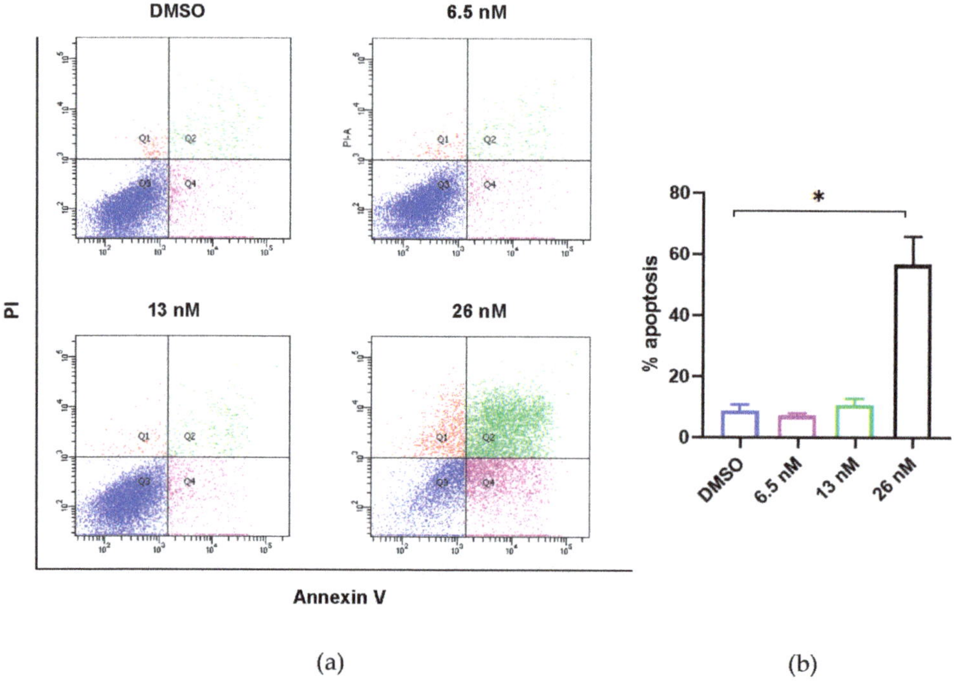

Figure 3. Effects of compound **1** on induction of apoptosis of HCT116 cancer cells. (**a**) Flow cytometric analysis of HCT116 cancer cells using Annexin V/PI staining. Cells were treated with **1** at the indicated concentrations for 48 h prior to labelling with annexin V and PI. Q1 shows the percentage of necrotic cells, and Q2 and Q4 represent late and early apoptotic cells, respectively. Q3 shows the population of normal cells. (**b**) Graphical representation of the percentage of apoptotic cells (sum of Q2 and Q4) from three independent replicates' data as the mean ± SEM; * $p < 0.05$.

2.4. *Compound* **1** *Reduced Cellular Mobilization of HCT116 Cancer Cells*

To fully elucidate the anticancer properties and examine whether compound **1** also affects cellular mobilization of HCT116 cancer cells, scratch wound assay was utilized after applying indicated concentrations of **1**. In Figure 4, we observed that the treatment with **1** at 3.25 and 6.5 nM significantly inhibited the mobilization of HCT116 cells in a dose-dependent manner, whose healing effects were obvious after 24 h. We thus conclude that the healing effects of HCT116 cells have been inhibited when treated with **1** from 0 to 6.5 nM in vitro.

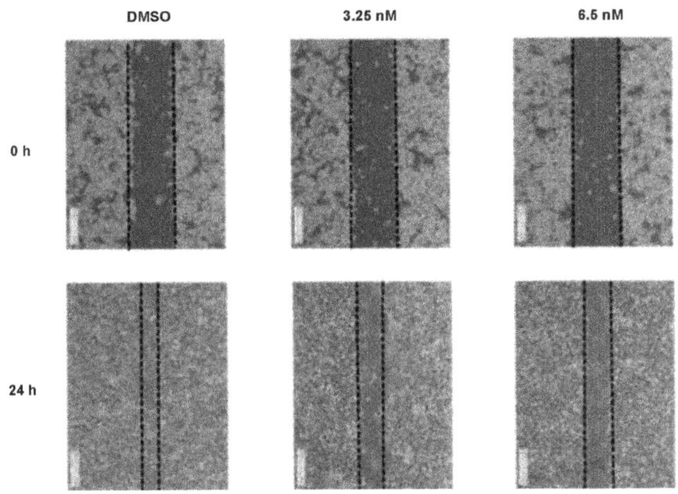

Figure 4. Compound **1** reduced cellular mobilization in HCT116 cells treated with various concentrations of **1**. Images were acquired at 0, 24 h after scratching. The dotted lines define the areas lacking cells. Data were collected from 3 independent experiments and only presentative pictures are shown.

2.5. *Compound* **1** *Disrupted Microtubules of HCT116 Cancer Cells*

As compound **1** shared similar chemical structures with podophyllotoxin, which is known as a microtubule-destabilizing agent, we further evaluated the effects of **1** on microtubule dynamics by immunofluorescent staining. As shown in Figure 5, HCT116 cells exhibited disruption of the microtubule network after the treatment of **1** for 48 h, especially at concentrations of 13 and 26 nM. Thus, the results confirmed our hypothesis that compound **1** has the ability to disrupt the microtubule assembly, indicating that microtubule dynamics may be one of the targets of the anti-proliferation and anti-migration effects of **1**.

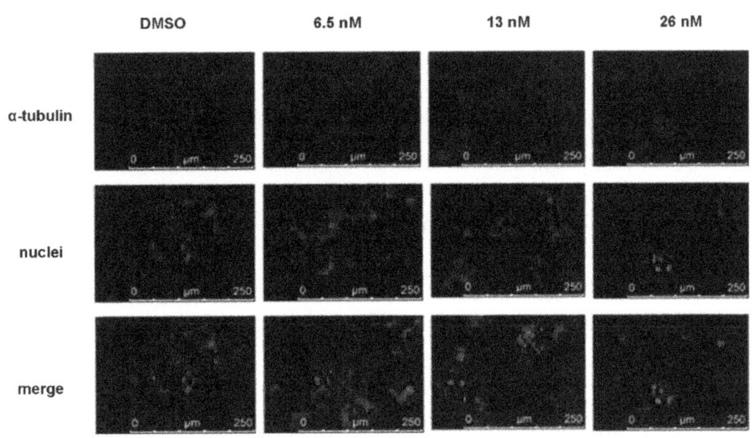

Figure 5. Immunofluorescence study of disruption of the microtubule network in HCT116 cells. Microtubules were visualized with an anti-α-tubulin antibody (red). The nuclei of cells were stained with 4′, 6-diamidino-2-phenylindole and thus showed a blue color.

2.6. Effects of Compound 1 on Apoptosis and Migration Regulatory Proteins of HCT116 Cancer Cells

To test the effects of compound **1** on apoptosis and migration regulatory proteins in HCT116 cancer cells, we performed western blotting analysis. Figure 6 showed that **1** could decrease the protein expression of pAKT and MMP2, while increasing the protein expression of p21. These regulations are all dose dependent, and the effects are most significant at the concentration of a 13 nM treatment. Collectively, our results suggest that the antitumor effect of **1** is majorly associated with the p21-mediated apoptosis and MMP2-mediated migration, which are both regulated by their upstream, phosphorylated AKT.

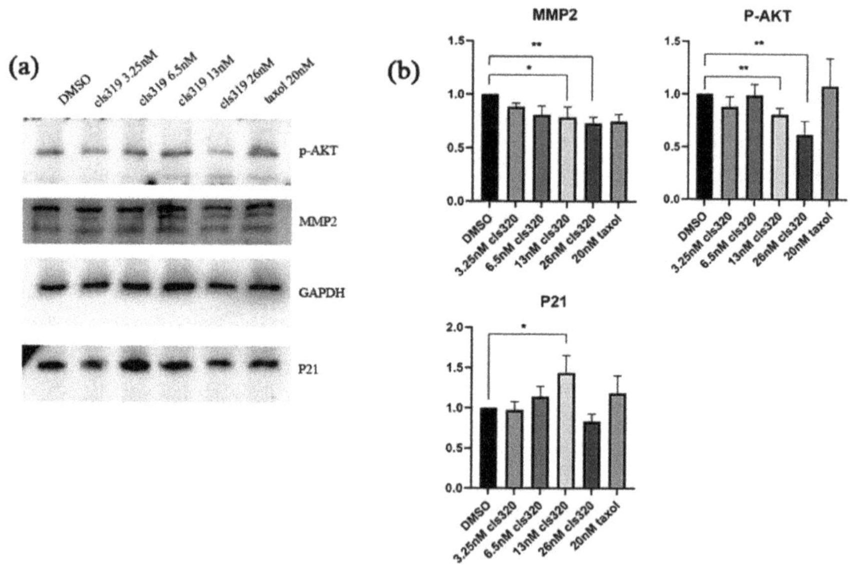

Figure 6. Western blot analysis shows effects of compound **1** on apoptosis and migration regulatory proteins in HCT116 cancer cells (**a**). The downregulation of p-AKT and MMP2 and upregulation of P21 are shown in a statistic graph (**b**), which indicating the inhibition of apoptosis and cell migrate. Data were collected from 3 independent experiments and only representative pictures are shown. (All data were analyzed by GraphPad Prism. Unpaired Student's *t* test was used when comparing treatment group with control group within the same cell type. $p < 0.05$ was considered statistically significant (* $p < 0.05$, ** $p < 0.01$).

3. Discussion

In previous studies, the investigations on *Bridelia* plant genus mainly focused on the validation of the folk or ethnic medical application using the crude extracts [10–13]. In the present study, we in-depth investigated the major phytochemical profile of the roots of *Bridelia balansae* by bioassay guided separation. The 14 compounds that were identified belong to aryltetralin [18–22] and 2-aryltetrahydrofuran [23–25] lignans, organic tannic acids [26–28], flavanones [29,30], and triterpenoids [31]. Although these compounds are known, these findings expanded the diversity of the plant sources of these bioactive compounds. More importantly, the absolute configuration of the lactone ring is difficult to confirm by the regular HRMS and NMR spectroscopies in terms of aryltetralin skeletons in some extreme cases. The confirmation of the absolute structure of compound **1** by X-ray crystallography method for the first time is beneficial to the structural determination of future unknown natural analogues. In addition, flavonoid–tannin conjugates are a rare subtype of small molecules in nature. As an important precursor of podophyllotoxin, aryltetralin lignans also exhibited a pronounced biological activity mainly as anti-tumor agents.

Considering the synthetic challenges due to multiple chiral centers, natural aryltetralin lignans isolated from the roots of *Bridelia balansae* can partially overcome the shortage of chemical sources, which will provide the substantial support of further pharmacological studies.

In the present study, we also evaluated the bioactivity of compound **1** in depth, including the elucidation of its anti-proliferative property, and its induction of apoptosis and migration using cellular platforms. Previously, it has been reported that 4′-demethyl-4-deoxypodophyllotoxin (**1**) could induce cell cycle arrest with a significantly increased G2/M peak after treatment of HCT-116 cells [33]. Cell cycle arrest could either provide cancer cells the opportunity for DNA damage repair, or lead to cell death [34]. Among all the cell death pathways, apoptosis is one of the most common mechanisms to eliminate damaged cells that will further restrict tumorigenesis [35]. The data of our flow cytometric analysis show that the anti-proliferative effect of **1** on human colorectal carcinoma cells is primarily derived from apoptosis. The apoptosis is mediated by the upregulation of p21 and might be caused by the effects of microtubule disassembly of **1**. This result is consistent with podophyllotoxin, an aryltetralin lignan congener of **1** and a well-known microtubule-destabilizing agent [36,37]. In the present study, compound **1** was found to induce a significant apoptosis event. Thus, further research is needed to determine if the aryltetralin lignan-induced apoptotic pathway is linked to the mechanism of action of the microtubule-destabilizing agents, such as the clinically used anticancer drugs etoposide and teniposide, which are derived from podophyllotoxin. Notably, it has been reported that targeted microtubule disruption can selectively inhibit metastasis, although these therapies are currently focusing exclusively on tumor growth [38]. Therefore, the anti-migration effects of **1** might also be contributed to by its microtubule disruption effects. These effects collectively regulated MMP2, and thus induced the reduction of cell migration. We also noticed a significant dose-dependent decrease of the expression of phosphorylated AKT, which is the upstream of both p21 and MMP2. It has been well acknowledged that AKT signaling promotes tumor cell proliferation, growth, and metastasis by activating its downstream effectors, and thus the inhibitors targeting this signaling pathway have been considered as the most effective treatment strategy for cancer [39]. The wound healing (or scratch) assay is a method to measure two-dimensional cell migration [40]. It involves creating an artificial gap in a confluent cell monolayer at 0 h, and tracking the movement of HCT116 cells through microscopy imaging at both 0 and 24 h. The narrowing of the gap indicated that compound **1** was capable of inhibiting the migration of HCT116 cells. Collectively, our study provides a basis for the anti-cancer effects of compound **1** against HCT116 cancer cells, which might lead to further structural modification of its functional groups or more detailed investigations of its mechanism, serving as a potential lead compound in colon cancer treatment.

4. Materials and Methods

4.1. General

Flash silica gel (300 mesh, Qingdao Haiyang Chemical Co., Ltd., Qingdao, China) was used as a stationary phase in column chromatography. Thin-layer chromatography (TLC) (Qingdao Haiyang Chemical Co., Ltd., Qingdao, China) was used to monitor the fractionation progress. Analytical and semi-preparative HPLC were performed on an Agilent Technologies Series 1100 HPLC with a diode array detector (DAD, Agilent Technologies, Inc., Santa Clara, CA, USA) equipped with a semi-preparative Thermo-C18 (150 × 4.6 mm) column or Nawei-UniSil 10–120 C18 Ultra (250 × 21.2 mm) column. A Bruker Ascend 400 MHz spectrometer with standard parameters supplied by the vendor (Bruker, Karlsruhe, Germany) was applied to measure one-dimensional (1D) and two-dimensional (2D) NMR spectra. High-resolution mass spectrometry was performed on an Agilent QTOF-6542 (Santa Clara, CA, USA). X-ray crystallographic data were obtained on a Bruker D8 Venture X-Ray Diffractometer (Karlsruhe, Germany). The optical rotation of a compound was measured on a polarimeter (JACSCO, P1010, Tokyo, Japan). The melting point was detected by a melting point tester (Electrothermal, London, UK). Mili Q water and acetonitrile (ACN, HPLC grade, Duksan, Republic of Korea) were used as a mobile phase of HPLC. Methanol (HPLC grade, Duksan, Republic of Korea), ethyl acetate (EA, HPLC grade, Duksan, Republic of Korea), petroleum ether (PE, HPLC grade, Duksan, Republic of Korea), and dichloromethane (DCM, HPLC grade, Duksan, Republic of Korea) were used for the fractionation and isolation of compounds.

4.2. Plant Material

The roots of *B. balansae* were collected from Jianfengling National Nature Reserve, Hainan Province in 1 August 2014. The collected plant materials were authenticated by Prof. CHEN Hubiao from School of Chinese Medicine, Hong Kong Baptist University. A voucher specimen (No. SHABB20140801) was deposited at School of Chinese Medicine, Hong Kong Baptist University, Hong Kong, China.

4.3. Compound Isolation and Structure Identification

The dry weight of the root part was 5.0 kg. The root part of plant material was then ground and extracted with 50 L methanol at room temperature to afford 1.0 kg crude extract, which was partitioned with ethyl acetate (EA) 3 times to produce an EA extract (300 g). The EA extract was further fractionated on a silica gel column, eluted with petroleum ether (PE), PE/DCM (1:1), DCM/methanol (MeOH) (1:0–0:1) gradient to afford 6 fractions (Fractions 1–6) (Figure S8). Compounds **2** (20 mg), **9** (3 mg), **13** (3 mg), and **14** (10 mg) were isolated from Fraction 3 by using HPLC with an isocratic elution (55% ACN: 45% H_2O as mobile phase, 3 mL/min) (Figure S9). Compounds **1** (30 mg), **3** (5 mg), **4** (6 mg), **5** (3 mg), **6** (2 mg), **7** (3 mg), **8** (2 mg), **10** (2 mg), and **11** (6 mg) were isolated from the major active Fraction 4 by using HPLC with an isocratic elution (50% ACN: 50% H_2O as mobile phase, 3 mL/min) (Figure S10), and compound **12** (7 mg) was isolated from Fraction 5 (Figure S11).

Compound **1**. Colourless crystal (acetone); mp 248.5–249.0 °C; $[\alpha]^{20}_D = -96°$ (*c* 0.3, $CHCl_3$); ^1H NMR (400 MHz, $CDCl_3$) *δ*: 6.65 (s, 1H), 6.51 (s, 1H), 6.34 (s, 2H), 5.93 (dd, *J* = 10.4, 1.2, 2H), 5.42 (brs, 1H), 4.58 (d, *J* = 2.8, 1H), 4.41–4.44 (m, 1H), 3.88–3.92 (m, 1H), 3.77 (s, 6H), 3.04–3.07 (m, 1H), 2.70–2.77 (m, 3H). ^{13}C NMR (100 MHz, $CDCl_3$): *δ*: 175.0, 146.9, 146.7, 146.4, 133.8, 131.8, 130.8, 128.3, 110.5, 108.5, 107.9, 101.2, 72.1, 56.4, 47.6, 43.6, 33.1, 32.7. HRESIMS *m*/*z*: 385.1287 $[M + H]^+$; calcd for $C_{21}H_{21}O_7$ 385.1282 [18].

Compound **2**. White powder; ^1H NMR (400 MHz, $CDCl_3$) *δ*: 6.66 (s, 1H), 6.51 (s, 1H), 6.34 (s, 2H), 5.92 (dd, *J* = 9.2, 1.2, 2H), 4.59 (d, *J* = 2.8, 1H), 4.43–4.47 (m, 1H), 3.89–3.94 (m, 1H), 3.79 (s, 3H), 3.74 (s, 6H), 3.03–3.10 (m, 1H), 2.69–2.80 (m, 3H). ^{13}C NMR (100 MHz, $CDCl_3$): *δ*: 174.9, 152.5, 147.0, 146.7, 136.9, 136.3, 130.6, 128.3, 110.5, 108.5, 108.2, 101.2, 72.1, 60.8, 56.2, 47.5, 43.7, 33.1, 32.7. HRESIMS *m*/*z*: 399.1444 $[M + H]^+$; calcd for $C_{22}H_{23}O_7$ 399.1438 [19].

Compound **3**. White powder; ^1H NMR (400 MHz, $CDCl_3$) *δ*: 6.66 (d, *J* = 7.2, 2H), 6.60 (d, *J* = 7.6, 2H), 6.47 (s, 1H), 5.92 (s, 2H), 5.89 (dd, *J* = 6.0, 1.2z, 2H), 4.56 (d, *J* = 4.4, 1H), 4.42–4.46 (m, 1H), 3.92 (t, *J* = 8.0, 1H), 3.06 (d, *J* = 10.4, 1H), 2.71–2.77 (m, 3H). ^{13}C NMR (100 MHz, $CDCl_3$): *δ*: 174.8, 147.2, 146.9, 146.8, 146.5, 134.5, 131.0, 128.2, 124.2, 111.1, 110.4, 108.5, 107.7, 101.2, 100.9, 72.1, 47.3, 43.2, 33.1, 32.6. HRESIMS *m*/*z*: 353.1021 $[M + H]^+$; calcd for $C_{20}H_{17}O_6$ 353.1020 [20].

Compound **4**. White powder; ^1H NMR (400 MHz, $CDCl_3$) *δ*: 6.69 (d, *J* = 7.2 Hz, 1H), 6.46 (d, *J* = 7.6 Hz, 1H), 6.35 (s, 2H), 6.19 (s, 1H), 5.93 (dd, *J* = 4.0, 1.2 Hz, 2H), 4.17 (dd, *J* = 9.2, 7.2, 1H), 3.85–3.89 (m, 1H), 3.74–3.83 (brs, 9H), 2.89–2.91 (m, 1H), 2.50–2.60 (m, 4H). ^{13}C NMR (100 MHz, $CDCl_3$): *δ*: 178.6, 153.3, 147.9, 146.4, 133.3, 131.5, 121.5, 108.8, 108.3, 106.2, 105.5, 101.1, 71.2, 60.9, 56.1, 46.5, 41.0, 38.4, 35.3. HRESIMS *m*/*z*: 401.1599 $[M + H]^+$; calcd for $C_{22}H_{25}O_7$ 401.1595 [21].

Compound **5**. White powder; ^1H NMR (400 MHz, $CDCl_3$) *δ*: 6.88–6.90 (m, 4H), 6.81 (dd, *J* = 2.0, 8.0 Hz, 2H), 5.62 (brs, 2H), 4.73 (d, *J* = 4.4 Hz, 2H), 4.22–4.26 (m, 2H), 3.89 (brs, 6H), 3.85 (dd, J = 3.6, 9.2 Hz, 2H), 3.08–3.11 (m, 2H). ^{13}C NMR (100 MHz, $CDCl_3$): 146.7 (2C),

145.2 (2C), 132.9 (2C), 118.9 (2C), 114.3 (2C), 108.6 (2C), 85.9 (2C), 71.7 (2C), 55.9 (2C), 54.1 (2C. H RESIMS *m*/*z*: 359.1493 [M + H]$^+$; calcd for $C_{20}H_{23}O_6$ 359.1489 [22].

Compound **6**. White powder; ^{1}H NMR (400 MHz, $CDCl_3$) δ: 6.81–6.83 (m, 2H), 6.76 (dd, *J* = 1.2, 8.0 Hz, 1H), 6.51 (s, 2H), 4.65 (dd, *J* = 3.6, 7.6 Hz, 2H), 4.16–4.21 (m, 2H), 3.80–3.85 (brs, 11H), 3.03 (brs, 2H). ^{13}C NMR (100 MHz, $CDCl_3$): 147.2 (2C), 146.7, 145.2, 134.4, 132.9, 132.1, 118.9, 114.3, 108.6, 102.7 (2C), 86.1, 85.8, 71.8, 71.6, 56.4 (2C), 55.9, 54.4, 54.1. HRESIMS *m*/*z*: 389.1593 [M + H]$^+$; calcd for $C_{21}H_{25}O_7$ 389.1595 [23].

Compound **7**. White powder; ^{1}H NMR (400 MHz, $CDCl_3$) δ: 6.59 (brs, 4H, Ar-H), 5.60 (brs, 2H), 4.73 (d, *J* = 3.6 Hz, 2H), 4.28 (t, *J* = 6.0, 2H), 3.82–3.92 (brs, 14H), 3.10 (brs, 2H). ^{13}C NMR (100 MHz, $CDCl_3$): 147.2 (4C), 134.3 (2C), 132.0 (2C), 102.7 (4C), 86.0 (2C), 71.8 (2C), 56.4 (4C), 54.3 (2C). HRESIMS *m/z*: 419.1707 [M + H]$^+$; calcd for $C_{22}H_{27}O_8$ 419.1700 [24].

Compound **8**. Brown powder; ^{1}H NMR (400 MHz, acetone-d_6) δ: 7.06 (s, 2H). ^{13}C NMR (100 MHz, $CDCl_3$): δ: 170.9, 146.4 (2C), 139.4, 122.7, 110.3 (2C). HRESIMS *m*/*z*: 171.0280 [M + H]$^+$; calcd for $C_7H_7O_5$ 171.0288 [25].

Compound **9**. Brown powder; ^{1}H NMR (400 MHz, acetone-d_6) δ: 7.07 (brs, 2H), 3.83 (s, 3H). ^{13}C NMR (100 MHz, acetone-*d6*): δ: 169.1, 146.5, 139.8, 121.5, 110.1, 52.4. HRESIMS *m*/*z*: 185.0447 [M + H]$^+$; calcd for $C_8H_9O_5$ 185.0444 [26].

Compound **10**. Brown powder; ^{1}H NMR (400 MHz, acetone-d_6) δ: 7.32 (s, 2H), 3.94 (s, 6H), 3.89 (s, 3H). ^{13}C NMR (100 MHz, $CDCl_3$): 166.8, 146.6, 139.2, 127.1, 106.6, 56.4, 52.1. HRESIMS *m/z*: 213.0781 [M + H]$^+$; calcd for $C_{10}H_{13}O_5$ 213.0757 [27].

Compound **11**. White powder; ^{1}H NMR (400 MHz, MeOD) δ: 6.41 (s, 2H), 5.93 (d, *J* = 2.0, 1H), 5.87 (d, *J* = 2.4, 1H), 4.54 (d, *J* = 7.2, 1H), 3.97 (dd, *J* = 12.8, 7.6 Hz, 1H), 3.30–3.31 (m, 1H), 2.80 (dd, *J* = 16.0, 5.6, 1H), 2.50 (dd, *J* = 16.0, 7.6, 1H). ^{13}C NMR (100 MHz, $CDCl_3$): δ: 157.8, 157.6, 156.8, 146.9 (2C), 134.0, 131.6, 107.2 (2C), 100.8, 96.3, 95.6, 82.9, 68.8, 28.1. HRESIMS *m*/*z*: 307.0809 [M + H]$^+$; calcd for $C_{15}H_{15}O_7$ 307.0812 [28].

Compound **12**. White powder; ^{1}H NMR (400 MHz, $CDCl_3$) δ: 6.95 (s, 2H), 6.50 (s, 2H), 5.96 (s, 2H), 5.52 (brs, 1H), 4.97 (s, 2H), 2.96 (dd, *J* = 17.2, 4.4, 1H), 2.82 (dd, *J* = 17.2, 2.4, 1H). ^{13}C NMR (100 MHz, $CDCl_3$): δ: 167.7, 157.9, 157.8, 157.3, 146.7 (2C), 146.3 (2C), 139.8, 133.8, 130.8, 121.5, 110.3 (2C), 106.9 (2C), 99.5, 96.6, 95.9, 78.6, 69.9, 26.9. HRESIMS *m*/*z*: 459.0920 [M + H]$^+$; calcd for $C_{22}H_{19}O_{11}$ 459.0922 [29].

Compound **13**. White solid; ^{1}H NMR (400 MHz, $CDCl_3$) δ: 3.73 (brs, 1H), 2.35–2.45 (m, 1H), 2.30–2.35 (m, 1H), 2.20–2.30 (m, 1H), 1.93–2.00 (m, 1H), 1.87–1.92 (m, 1H), 1.70–1.78 (m, 2H), 1.65–1.70 (m, 1H), 1.43–1.50 (m, 6H), 1.32–1.42 (m, 10H), 1.24–1.32 (m, 8H), 1.20–1.24 (m, 1H), 1.15 (d, *J* = 4.4 Hz, 6H), 1.05 (s, 3H), 0.98–1.03 (m, 6H), 0.92–0.97 (m, 6H), 0.84–0.89 (m, 6H), 0.72 (s, 3H). HRESIMS *m*/*z*: 439.4089 [M + H]$^+$; calcd for $C_{30}H_{53}O$ 429.4091 [30].

Compound **14**. White solid; ^{1}H NMR (400 MHz, $CDCl_3$) δ: 2.36–2.42 (m, 1H), 2.27–2.34 (m, 1H), 2.22–2.27 (m, 1H), 1.93–2.00 (m, 1H), 1.72–1.78 (m, 1H), 1.62–1.71 (m, 1H), 1.52–1.58 (m, 5H), 1.43–1.53 (m, 5H), 1.32–1.42 (m, 6H), 1.23–1.32 (m, 3H), 1.20–1.23 (m, 1H), 1.18 (s, 3H), 1.05 (s, 3H), 1.01 (s, 3H), 0.99 (s, 3H), 0.95 (s, 3H), 0.88 (d, *J* = 6.8 Hz, 3H), 0.87 (s, 3H), 0.72 (s, 3H). ^{13}C NMR (100 MHz, $CDCl_3$): 213.3, 59.5, 58.2, 53.1, 42.8, 42.2, 41.6, 41.3, 39.7, 39.3, 38.3, 37.5, 36.0, 35.6, 35.4, 35.0, 32.8, 32.4, 31.8, 30.5, 30.0, 28.2, 22.3. HRESIMS *m*/*z*: 427.3931 [M + H]$^+$; calcd for $C_{30}H_{51}O$ 427.3934 [30].

4.4. *X-ray Crystallographic Data of* **1**

Crystals of **1** were obtained from acetone solvent at room temperature. Single-crystal X-ray crystallographic analyses of **1** were obtained on a Bruker D8 Venture X-ray Diffractometer (Karlsruhe, Germany).

Crystallographic data of **1**. $C_{21}H_{20}O_7$, M = 384.37, a = 7.8438(7) Å, b = 9.4648(9) Å, c = 12.1581(12) Å; α = 93.206(3)°, β = 91.396(3)°, γ = 90.214(3), V = 900.93(15) Å^3, T = 297(2) K, space group P1, Z = 2, μ (Cu Kα) = 0.894 mm^{-1}; 18,230 reflections collected,

6148 independent reflections (R_{int} = 0.1127). The final R1 value was 0.0940 [I > 2σ(I)]. The final wR (F2) value was 0.2233 [I > 2σ (I)]. The final R1 value was 0.1209 (all data). The final wR (F2) value was 0.2710 (all data). The goodness of fit on F2 was 1.141. Flack parameter = 0.13(17). The crystal structure of **1** has been deposited in the CCDC database and the deposit number is CCDC 2287689. The DOI link can be accessed through DOI: 10.5517/ccdc.csd.cc2gsjf7 (accessed on 9 August 2023).

4.5. Cell Proliferation Assay

The human colon cancer cell line HCT116 was purchased from the American Type Culture Collection (Manassas, VA, USA) and maintained in Dulbecco's modified Eagle's medium (DMEM), supplemented with 10% fetal bovine serum (FBS) at 37 °C, 5% CO_2. The effects of isolated compounds on [17] colon cancer cell viability were determined by sulforhodamine B (SRB) assay [41]. The cells were seeded at a density of 5000 cells/well in 96-well microculture plates, in the absence or presence of different concentrations of a compound, incubated for 48 h. Thereafter, 50 μL of cold 50% trichloroacetic acid was added to each well to fix proteins and incubated at 4 °C for at least 1 h. The plate was washed with tap water four times and dried. For each well, 100 μL 0.4% SRB in 1% acetic acid was added to do the staining and incubated for 10 min at room temperature. After the SRB was discarded in the sink, the plate was washed by 1% acetic acid four times and dried. For the optical density (OD) value reading, 200 μL of 10 mM tris base (pH 10) was added to each well and shaken for 30 min or more. The optical absorbance at wavelength 515 nm was detected by a microplate spectrophotometer (BIO-RAD, Benchmark Plus). Percentage growth inhibition was calculated as [OD (cells + samples) − OD (Day 0 cells)]/[OD (cells + 10% DMSO) − OD (Day 0 cells)] = % survival, cytotoxicity = 1 − % survival. The IC_{50} values of the results were analyzed using GraphPad Prism version 8 (GraphPad Software, San Diego, CA, USA). Paclitaxel was used as the positive reference.

4.6. Flow Cytometry

The cells were seeded in a 6-well plate (250,000 cells) and incubated for 24 h at 37 °C with 5% CO_2. After settling, cells were further incubated in the absence or presence of **1** at the concentrations of 6.5, 13, and 26 nM to induce apoptosis for 48 h. The cells were harvested after the incubation period and washed in cold phosphate-buffered saline. The washed cell pellets were resuspended in 100 μL 1× annexin-binding buffer per assay, with 5 μL of FITC annexin V and 1 μL of the 100 μg/mL PI working solution added. The cells were incubated at room temperature for 15 min. After the incubation period, 400 μL of 1× annexin-binding buffer was added and mixed gently on ice. The stained cells were then analyzed by FACS Calibur™ (BD Biosciences, San Jose, CA, USA).

4.7. Wound Healing Assay

HCT116 cells were seeded in 6-well plates and cultured with full medium with or without compound **1** treatment until the confluent before the experiment was started. The seeding density was 300,000 cells with 2 mL medium per well. A 10 μL pipette tip was used to make a straight scratch, simulating a wound. After the scratch, the medium was changed into DMEM without FBS.

The wound edges were allowed to be imaged using a 5× objective and were focused on using the focus knob on the microscope. The positions desired to take were selected and imaged by a microscope (Leica DMIRB). After 24 h, the wound edges were imaged again with the same selected positions.

4.8. Immunofluoresent Staining

HCT116 cells were seeded in 6-well plates on slides and cultured with full medium with or without compound **1** at different concentrations for 48 h. The seeding density was 300,000 cells with 2 mL medium per well. After fixation in ice-cold acetone: MeOH (1:1, *v*/*v*) for 20 min, cells were incubated with anti-α-tubulin antibody followed by fluorescent-conjugated secondary antibody, visualized, and imaged by a microscope (Leica AF6000, Hong Kong SAR, China).

4.9. Western Blot Analysis

Proteins were extracted from HCT-116 cells using an ice-cold lysis buffer containing protease and phosphatase inhibitors (Thermo Fisher Scientific, Waltham, MA, USA). Cell lysates were loaded to 10% SDS-polyacrylamide gel for electrophoresis separation and transferred onto polyvinylidene fluoride membranes by wet electroblotting. The membranes were blocked with 5% non-fat dry milk in Tris-buffered saline containing 0.1% Tween 20 for 1 h at room temperature, followed by incubation with primary antibodies overnight at 4 °C, including p21 (Abcam, Cambridge, UK), MMP2 (CST), GAPDH (Bio-Rad, Herkleys, CA, USA). Subsequently, the membranes were incubated with secondary antibodies of interest for 1 h at room temperature and further visualized.

5. Conclusions

Our bioassay-guided phytochemical investigation of the roots of *B. balansae* led to the identification of 14 compounds. Among them, **3** has not been reported in nature previously, and **1** is the first aryltetralin lignan compound identified from this plant species. In addition, the absolute configuration of **1** was validated by X-ray crystallography for the first time, and its distinguished cytotoxic effect on HCT116 cells with an IC_{50} value at 20 nM was induced via an apoptosis induction mechanism. Compound **1** could also significantly decrease the migration rate of HCT116 cells, indicating its potential application against cancer metastasis. Collectively, the present study on the phytochemical and biological profile of *B. balansae* has determined the plant as a useful source to produce promising anticancer lead compounds.

Supplementary Materials: The following supporting information can be downloaded at: https://www.mdpi.com/article/10.3390/molecules28166165/s1, Figure S1: The proton NMR spectrum of compound 1; Figure S2: The ^{13}carbon NMR spectrum of compound **1**; Figure S3: The DEPT135 NMR spectrum of compound **1**; Figure S4: The HSQC NMR spectrum of compound **1**; Figure S5: The HMBC NMR spectrum of compound **1**; Figure S6: The ^{1}H-^{1}H COSY NMR spectrum of compound **1**; Figure S7: The NOESY NMR spectrum of compound **1**; Figure S8: The extraction and fraction flow chart of air-dried root of *B. Balansae*; Figure S9: The isolation flow chart of sub fraction F3; Figure S10: The isolation flow chart of sub fraction F4; Figure S11: The isolation flow chart of sub fraction F5.

Author Contributions: Conceptualization: M.-Z.W. and H.-J.Z.; compound extraction, isolation, and structure determination: W.-J.X. and M.-Z.W.; in vitro experiments: L.Z., Y.-X.D., Y.-X.X. and C.F.K.; data analysis: L.Z., W.-J.X. and K.-L.L.; funding acquisition: H.-J.Z.; supervision: H.-J.Z.; writing—original draft: L.Z. and W.-J.X.; writing—review and editing: Y.-X.D., Z.O. and H.-J.Z. All authors have read and agreed to the published version of the manuscript.

Funding: This work is supported by the HKBU Initiation Grant for Faculty Niche Research Areas (IG-FNRA) 2021/22 (RC-IGFNRA/21–22/SCM/01), Research Grant Council of the Hong Kong Special Administrative Region, China (Projects No. HKBU12103021) and the grants from the Shenzhen Technology University (Natural Science Foundation of Top Talent of SZTU).

Institutional Review Board Statement: Not applicable.

Informed Consent Statement: Not applicable.

Data Availability Statement: Data are contained within the article or Supplementary Materials.

Conflicts of Interest: The authors declare no conflict of interest.

Sample Availability: Not applicable.

References

1. Dias, D.A.; Urban, S.; Roessner, U. A historical overview of natural products in drug discovery. *Metabolites* **2012**, *2*, 303–336. [CrossRef] [PubMed]
2. Atanasov, A.G.; Zotchev, S.B.; Dirsch, V.M.; Supuran, C.T. Natural products in drug discovery: Advances and opportunities. *Nat. Rev. Drug Discov.* **2021**, *20*, 200–216. [CrossRef] [PubMed]
3. Patil, S.D.; Chaudhari, M.A.; Sapkale, P.V.; Chaudhari, R.B. A recent review on anticancer herbal drugs. *J. Drug Discov. Ther.* **2013**, *1*, 77–84.
4. Waltenberger, B.; Mocan, A.; Šmejkal, K.; Heiss, E.H.; Atanasov, A.G. Natural Products to Counteract the Epidemic of Cardiovascular and Metabolic Disorders. *Molecules* **2016**, *21*, 807. [CrossRef]
5. Tintore, M.; Vidal-Jordana, A.; Sastre-Garriga, J. Treatment of multiple sclerosis—Success from bench to bedside. *Nat. Rev. Neurol.* **2019**, *15*, 53–58. [CrossRef]
6. Harvey, A.L.; Edrada-Ebel, R.; Quinn, R.J. The re-emergence of natural products for drug discovery in the genomics era. *Nat. Rev. Drug Discov.* **2015**, *14*, 111–129. [CrossRef]
7. Newman, D.J.; Cragg, G.M. Natural Products as Sources of New Drugs over the Nearly Four Decades from 01/1981 to 09/2019. *J. Nat. Prod.* **2020**, *83*, 770–803. [CrossRef]
8. Chabner, B.A. NCI-60 cell line screening: A radical departure in its time. *J. Natl. Cancer Inst.* **2016**, *108*, djv388. [CrossRef]
9. Shoemaker, R.H. The NCI60 human tumour cell line anticancer drug screen. *Nat. Rev. Cancer* **2006**, *6*, 813–823. [CrossRef]
10. Ngueyem, T.A.; Brusotti, G.; Caccialanza, G.; Finzi, P.V. The genus Bridelia: A phytochemical and ethnopharmacological review. *J. Ethnopharmacol.* **2009**, *124*, 339–349. [CrossRef]
11. Karou, S.D.; Tchacondo, T.; Tchibozo, M.A.D.; Abdoul-Rahaman, S.; Anani, K.; Koudouvo, K.; Batawila, K.; Agbonon, A.; Simpore, J.; de Souza, C. Ethnobotanical study of medicinal plants used in the management of diabetes mellitus and hypertension in the Central Region of Togo. *Pharm. Biol.* **2011**, *49*, 1286–1297. [CrossRef] [PubMed]
12. Yeboah, G.N.; Owusu, F.W.A.; Archer, M.A.; Kyene, M.O.; Kumadoh, D.; Ayertey, F.; Mintah, S.O.; Atta-Adjei Junior, P.; Appiah, A.A. Bridelia ferruginea Benth; An ethnomedicinal, phytochemical, pharmacological and toxicological review. *Heliyon* **2022**, *8*, e10366. [CrossRef]
13. Maroyi, A. Utilization of Bridelia mollis as herbal medicine, nutraceutical and functional food in southern Africa: A review. *Trop. J. Pharm. Res.* **2019**, *18*, 203–209. [CrossRef]
14. Murthy, H.N.; Dalawai, D.; Mamatha, U.; Angadi, N.B.; Dewir, Y.H.; Al-Suhaibani, N.A.; El-Hendawy, S.; Al-Ali, A.M. Bioactive constituents and nutritional composition of *Bridelia stipularis* L. Blume fruits. *Int. J. Food Prop.* **2021**, *24*, 796–805. [CrossRef]
15. Pettit, G.R.; Searcy, J.D.; Tan, R.; Cragg, G.M.; Melody, N.; Knight, J.C.; Chapuis, J.C. Antineoplastic Agents. 585. Isolation of Bridelia ferruginea Anticancer Podophyllotoxins and Synthesis of 4-Aza-podophyllotoxin Structural Modifications. *J. Nat. Prod.* **2016**, *79*, 507–518.
16. Ochwang'i, D.O.; Kimwele, C.N.; Oduma, J.A.; Gathumbi, P.K.; Kiama, S.G.; Efferth, T. Cytotoxic activity of medicinal plants of the Kakamega County (Kenya) against drug-sensitive and multidrug-resistant cancer cells. *J. Ethnopharmacol.* **2018**, *215*, 233–240. [CrossRef]
17. Huong, D.T.; Kamperdick, C.; Van Sung, T. Homogentisic acid derivatives from Miliusa balansae. *J. Nat. Prod.* **2004**, *67*, 445–447. [CrossRef]
18. Jackson, D.E.; Dewick, P.M. Aryltetralin lignans from Podophyllum hexandrum and Podophyllum peltatum. *Phytochemistry* **1984**, *23*, 1147–1152. [CrossRef]
19. Xu, H.; Wang, J.; Sun, H.; Lv, M.; Tian, X.; Yao, X.; Zhang, X. Semisynthesis and quantitative structure- activity relationship (QSAR) study of novel aromatic esters of 4′-demethyl-4-deoxypodophyllotoxin as insecticidal agents. *J. Agric. Food Chem.* **2009**, *57*, 7919–7923. [CrossRef]
20. Moujir, L.S.; Seca, A.M.L.; Silva, A.M.S.; Barreto, M.C. Cytotoxic Activity of Diterpenes and Extracts of Juniperus brevifolia. *Planta Med.* **2008**, *74*, 751–753. [CrossRef]
21. Wada, K.; Munakata, K. (−) Parabenzlactone, a new piperolignanolide isolated from Parabenzoin trilobum nakai. *Tetrahedron Lett.* **1970**, *11*, 2017–2019. [CrossRef]
22. Enders, D.; Lausberg, V.; Del Signore, G.; Berner, O.M. A general approach to the asymmetric synthesis of lignans:(−)-methyl piperitol,(−)-sesamin,(−)-aschantin,(+)-yatein,(+)-dihydroclusin,(+)-burseran, and (−)-isostegane. *Synthesis* **2002**, *2002*, 515–522. [CrossRef]
23. Deyama, T.I.T.; Kitagawa, S.; Nishibe, S. The Constituents of Eucommia ulmoicles OLIV. V. Isolation of Dihydroxydehydrodiconiferyl Alcohol Isomers and Phenolic Compounds (Organic, Chemical). *Chem. Pharm. Bull.* **1987**, *35*, 1785–1789.
24. Haajanen, K.; Botting, N.P. Synthesis of multiply ^{13}C-labeled furofuran lignans using ^{13}C-labeled cinnamyl alcohols as building blocks. *Steroids* **2006**, *71*, 231. [CrossRef] [PubMed]
25. Das, B.V.B.; Kashinatham, A. Studies on phytochemicals.Part XXV. (+)-Syringaresinol from Parthenium hysterophorus. *Fitoterapia* **1999**, *70*, 2.
26. Gottlieb, H.E.; Kumar, S.; Sahai, M.; Ray, A.B. Ethyl brevifolin carboxylate from Flueggea microcarpa. *Phytochemistry* **1991**, *30*, 2435–2438. [CrossRef]
27. Mostafa, M.; Nahar, N.; Mosihuzzaman, M.; Sokeng, S.D.; Fatima, N.; Rahman, A.U.; Choudhary, M.I. Phosphodiesterase-I inhibitor quinovic acid glycosides from Bridelia ndellensis. *Nat. Prod. Res.* **2006**, *20*, 686–692. [CrossRef]

28. Russell, K.M.; Molan, P.C.; Wilkins, A.L.; Holland, P.T. Identification of some antibacterial constituents of New Zealand manuka honey. *J. Agric. Food Chem.* **1990**, *38*, 4. [CrossRef]
29. Foo, L.; Newman, R.; Waghorn, G.; McNabb, W.; Ulyatt, M. Proanthocyanidins from Lotus corniculatus. *Phytochemistry* **1996**, *41*, 617–624. [CrossRef]
30. Yang, L.-L.; Chang, C.C.; Chen, L.G.; Wang, C.C. Antitumor principle constituents of *Myrica rubra* var. acuminata. *J. Agric. Food Chem.* **2003**, *51*, 2974–2979. [CrossRef]
31. Li, L.; Huang, X.; Sattler, I.; Fu, H.; Grabley, S.; Lin, W. Structure elucidation of a new friedelane triterpene from the mangrove plant *Hibiscus tiliaceus*. *Magn. Reson. Chem.* **2006**, *44*, 5. [CrossRef]
32. Charlton, J.L.; Ploude, G.L.; Koh, K.; Secco, A.S. Asemmetric synthesis of podophyllotoxin analogs. *Can. J. Chem.* **1990**, *68*, 2022–2027. [CrossRef]
33. Zhang, X.; Zhang, J.; Su, M.; Zhou, Y.; Chen, Y.; Li, J.; Lu, W. Design, synthesis and biological evaluation of 4′-demethyl-4-deoxypodophyllotoxin derivatives as novel tubulin and histone deacetylase dual inhibitors. *RSC Adv.* **2014**, *4*, 40444–40448. [CrossRef]
34. Schwartz, G.K.; Shah, M.A. Targeting the cell cycle: A new approach to cancer therapy. *J. Clin. Oncol.* **2005**, *23*, 9408–9421. [CrossRef]
35. Strasser, A.; Cory, S.; Adams, J.M. Deciphering the rules of programmed cell death to improve therapy of cancer and other diseases. *EMBO J.* **2011**, *30*, 3667–3683. [CrossRef]
36. Schilstra, M.J.; Martin, S.R.; Bayley, P.M. The effect of Podophyllotoxin on Microtubule Dynamics. *J. Biol. Chem.* **1989**, *264*, 8827–8834. [CrossRef]
37. Mukhtar, E.; Adhami, V.M.; Mukhtar, H. Targeting microtubules by natural agents for cancer therapy. *Mol. Cancer Ther.* **2014**, *13*, 275–284. [CrossRef] [PubMed]
38. Thompson, K.N.; Ju, J.A.; Ory, E.C.; Pratt, S.J.P.; Lee, R.M.; Mathias, T.J.; Chang, K.T.; Lee, C.J.; Goloubeva, O.G.; Bailey, P.C.; et al. Microtubule disruption reduces metastasis more effectively than primary tumor growth. *Breast Cancer Res.* **2022**, *24*, 13. [CrossRef]
39. He, Y.; Sun, M.M.; Zhang, G.G.; Yang, J.; Chen, K.S.; Xu, W.W.; Li, B. Targeting PI3K/Akt signal transduction for cancer therapy. *Signal Transduct. Target Ther.* **2021**, *6*, 425. [CrossRef] [PubMed]
40. Lin, H.-H.; Robertson, K.L.; Don, S.S.L.; Taylor, S.R.; Farkas, M.E. Chemical modulation of circadian rhythms and assessment of cellular behavior via indirubin and derivatives. In *Methods in Enzymology*; Academic Press: Cambridge, MA, USA, 2020; Volume 639, pp. 115–140.
41. Vichai, V.; Kirtikara, K. Sulforhodamine B colorimetric assay for cytotoxicity screening. *Nat. Protoc.* **2006**, *1*, 1112–1116. [CrossRef] [PubMed]

Disclaimer/Publisher's Note: The statements, opinions and data contained in all publications are solely those of the individual author(s) and contributor(s) and not of MDPI and/or the editor(s). MDPI and/or the editor(s) disclaim responsibility for any injury to people or property resulting from any ideas, methods, instructions or products referred to in the content.

molecules

MDPI

Article

Cepharanthine Exerts Antioxidant and Anti-Inflammatory Effects in Lipopolysaccharide (LPS)-Induced Macrophages and DSS-Induced Colitis Mice

Guangxin Chen [1,2,3,*,†], Da Wen [3,†], Lin Shen [3], Yazhi Feng [3], Qiuhong Xiong [3], Ping Li [3] and Zhonghua Zhao [3,*]

1 State Key Laboratory of Animal Nutrition, College of Animal Science and Technology, China Agricultural University, Beijing 100193, China
2 State Key Laboratory of Biological Feed, Ministry of Agriculture and Rural Afairs, Boen Biotechnology Co., Ltd., Ganzhou 341000, China
3 Institute of Biomedical Sciences, Shanxi University, Taiyuan 030006, China; wdbio@outlook.com (D.W.); 19581550438@163.com (L.S.); fyz_99@163.com (Y.F.); qxiong@sxu.edu.cn (Q.X.); pingli@sxu.edu.cn (P.L.)
* Correspondence: chengx@sxu.edu.cn (G.C.); zhzhao@sxu.edu.cn (Z.Z.)
† These authors contributed equally to this work.

Abstract: Cepharanthine (CEP), a biscoclaurine alkaloid extracted from *Stephania cepharantha* Hayata, has been widely used for the treatment of various acute and chronic diseases, including leukopenia, and snake bites. Here, our objective was to investigate the anti-oxidative stress and anti-inflammatory response effects of CEP in lipopolysaccharide (LPS)-induced macrophages as well as dextran sulfate sodium (DSS)-induced colitis mice. Our findings demonstrated that supplementation with CEP effectively mitigates body weight loss and elevation of disease activity index (DAI), reduces the malondialdehyde (MDA) content to 2.45 nM/mL while increasing the reduced glutathione (GSH) content to 35.53 μg/mL, inhibits inflammatory response, and maintains proper intestinal epithelium tight junctions in DSS-induced wild type (WT) mice. However, it failed to provide protective effects in DSS-induced transcription factor nuclear factor erythroid 2-related factor 2 (NRF2) knockout ($NRF2^{-/-}$) mice. GSH content decreased to 10.85 μg/10^6 cells following LPS treatment, whereas supplementation with CEP increased the GSH content to 12.26 μg/10^6 cells. Moreover, CEP effectively attenuated ROS production in LPS-induced macrophages. Additionally, CEP exhibited inhibitory effects on pro-inflammatory cytokines and mediators in LPS-induced macrophages. Furthermore, we observed that supplementation with CEP promoted the expression of NRF2/heme oxygenase 1 (HO-1)/NADPH quinone oxidoreductase-1 (NQO-1) as well as the phosphorylation of the adenosine monophosphate-activated protein kinase alpha 1 (AMPK-α1)/protein kinase B (AKT)/glycogen synthase kinase-3 beta (GSK-3β) signaling pathway in macrophages while inhibiting the phosphorylation of the extracellular signal-regulated kinase (ERK)/c-Jun N-terminal kinase (JNK), and nuclear factor-kappa B p65 (NF-κB p65) signaling pathway in LPS-induced macrophages. Although CEP did not demonstrate inhibitory effects on oxidative stress or promote the expression of HO-1/NQO-1, it effectively activated the phosphorylation of the AMPK-α1/AKT/GSK-3β signaling pathway which is an upstream regulator of NRF2 in LPS-induced primary peritoneal macrophages from $NRF2^{-/-}$ mice. In summary, our findings suggest that CEP exerts protective effects against oxidative stress and inflammatory response by activating the AMPK-α1/AKT/GSK-3β/NRF2 signaling pathway while concurrently inhibiting the activation of mitogen activated protein kinases (MAPKs) and the NF-κB p65 signaling pathway. These results not only elucidate the mechanisms underlying CEP's protective effects on colon oxidative stress and inflammation but also provide evidence supporting NRF2 as a potential therapeutic target for IBD treatment.

Keywords: inflammatory bowel disease; CEP; NRF2; inflammatory response; oxidative stress

check for updates

Citation: Chen, G.; Wen, D.; Shen, L.; Feng, Y.; Xiong, Q.; Li, P.; Zhao, Z. Cepharanthine Exerts Antioxidant and Anti-Inflammatory Effects in Lipopolysaccharide (LPS)-Induced Macrophages and DSS-Induced Colitis Mice. *Molecules* **2023**, *28*, 6070. https://doi.org/10.3390/molecules28166070

Academic Editors: Xun Song, Chenyang Li and Yifu Guan

Received: 2 August 2023
Revised: 10 August 2023
Accepted: 10 August 2023
Published: 15 August 2023

Copyright: © 2023 by the authors. Licensee MDPI, Basel, Switzerland. This article is an open access article distributed under the terms and conditions of the Creative Commons Attribution (CC BY) license (https://creativecommons.org/licenses/by/4.0/).

1. Introduction

Inflammatory bowel diseases (IBDs) are chronic, progressive, and relapsing inflammatory disorders that affect the gastrointestinal tract. The two primary types of IBDs are ulcerative colitis (UC) and Crohn's disease [1,2]. These conditions are characterized by a dysregulated immune response leading to intestinal inflammation and tissue damage [3,4]. The etiology of IBDs is multifactorial, host genetics, microbial factors, environmental factors, overuse of antibiotics, hormone regulation, and chronic immune activation all contribute to their development [3,5]. Although traditional general anti-inflammatory and anti-immunosuppressive drugs, such as 5-aminosalicylates, corticosteroids, and thiopurines have demonstrated efficacy in many IBD patients, the introduction of biologic drugs targeting specific pathways such as the anti-TNF-α, anti-IL-12/23, and Janus kinase (JAK) signaling pathways has provided relief to a subset of patients [6,7]. However, there remains a small proportion of IBD patients who exhibit resistance to all currently available medications. Therefore, it is imperative to explore novel treatment targets and develop new therapeutic agents.

Reactive oxygen species (ROS), generated and released by immune cells, play a crucial role as signaling molecules to enhance the immunological functions of immune cells [8]. An appropriate concentration of ROS contributes to mitogenic response and defense against infections [8], however, excessive release of ROS and related products in the local microenvironment can lead to extensive cellular and molecular damage, initiating an intestinal inflammatory response and tissue destruction [9], ultimately impairing intestinal absorption [10]. The gastrointestinal tract has the most opportunities for interaction with exogenous factors, including food antigens and microbes, compared to any other tissues [11]. Although the gastrointestinal tract possesses a robust immune system to maintain host health, consistent exposure to extraneous stimulation could exacerbate the immune system's response, resulting in overproduction of free radicals or weakening of the endogenous antioxidant system [12]. The imbalance between free radicals and antioxidant systems is commonly referred to as oxidative stress. It represents an imbalance between pro-oxidation and anti-oxidation processes that are associated with inflammatory responses. To date, increasing clinical and experimental evidence has demonstrated that oxidative stress is implicated in the pathogenesis of IBDs [13,14]. Therefore, targeting intestinal tissue oxidative stress may offer potential therapeutic benefits for treating IBD.

Nuclear factor erythroid 2-related factor 2 (NRF2) belongs to the cap 'n' collar transcription factor family [15]. It is activated by various stimuli, including oxidative, electrophilic, proteotoxic stress, and small natural molecules. NRF2 plays a pivotal role in safeguarding and restoring cellular homeostasis [16]. Previous studies have demonstrated that NRF2 activation maintains cellular redox homeostasis and regulates inflammatory response in neurodegenerative disease [17], diabetes [16], and kidney disease [18]. In fact, NRF2 activation has exhibited promising outcomes against lung, liver, eye, gastrointestinal, metabolic, neurodegenerative as well as autoimmune diseases where oxidative stress and inflammation contribute to pathogenesis [19]. Pharmacological agents targeting NRF2 such as dimethyl fumarate have been employed for treating multiple sclerosis and psoriasis, and bardoxolone methyl has entered clinical phase II/III trials for treating CTD-PAH, pulmonary hypertension-ILD, Alport syndrome, and autosomal dominant polycystic kidney disease [19]. The protective effects of pharmacological NRF2 activators have been observed in numerous human disease models with benefits seen in human intervention trials. Therefore, NRF2 is considered a suitable drug target for screening against oxidative stress and inflammation during various disease pathogeneses [19].

Previous studies have demonstrated the anti-oxidative property of various natural products, including curcumin analogues and extracts from *isodon suzhouensis*, which provide in vivo and in vitro antioxidant effects [20–22]. In this study, we aim to investigate the impact of CEP on oxidative stress and inflammatory response in macrophages and mice. CEP is a biscoclaurine alkaloid derived from *Stephania cepharantha* Hayata that exhibits diverse biological properties. It effectively ameliorates leukopenia and immune thrombocytopenic purpura by maintaining proper leukocyte levels [23,24]. Moreover, it alleviates inflammation and pain caused by snake bites [25]. Additionally, CEP demonstrates various pharmacological activities including anti-inflammatory response through a reduction in NLRP3 inflammasome activation [26] and the inhibition of MAPKs and NF-κB p65 signaling pathway [27,28], antioxidant effects to combat oxidative stress [26], prevention of oxidative stress-induced DNA damage [29], inhibition of autophagy by blocking autophagosome-lysosome fusion and lysosomal cathepsin B/D maturation [30], as well as antiviral activity when combined with nelfinavir to reduce viral load accumulation and facilitate virus elimination [31]. Furthermore, CEP enhances the sensitivity of host or cells to anticancer agents [30,32], although it does not directly affect cancer itself. Intestinal inflammation, oxidative stress, and destruction of intestinal epithelial barrier are well-known symptoms associated with IBD. Notably, CEP possesses inhibitory effects on the inflammatory response and oxidative stress. However, the impact of CEP on IBD remains understudied. Therefore, our objective is to explore the effects of CEP on oxidative stress and the inflammatory response in LPS-induced macrophages as well as DSS-induced colitis mice.

2. Results

2.1. CEP Ameliorates IBD Characteristics in DSS-Induced Colitis Mice

The essential features of IBD encompass a diminished survival rate, increased body weight loss, shortened colon length, and an elevated clinical score. To investigate the potential role of CEP in colitis, we assessed these crucial characteristic indices in DSS-induced WT and NRF2$^{-/-}$ mice. Pre-treatment with CEP ameliorated weight loss in DSS-induced WT mice, however, this effect was not observed in NRF2$^{-/-}$ mice (Figure 1A). While DSS dramatically reduced colon length in both WT and NRF2$^{-/-}$ mice, supplementation with CEP failed to mitigate this parameter (Figure 1B). Furthermore, DSS augmented the disease activity index (DAI) in both WT and NRF2$^{-/-}$ mice. Nevertheless, supplementation with CEP significantly decreased the DAI only in WT mice but not in NRF2$^{-/-}$ mice (Figure 1C). Histopathological examination evaluated the severity of inflammation and ulceration in colitis-afflicted rodents. Crypt loss, mucosal layer destruction, and edema were exacerbated both in DSS-induced WT and NRF2$^{-/-}$ mice. Conversely, pre-treatment with CEP effectively alleviated intestinal tissue damage only in DSS-induced WT mice, no effects on NRF2$^{-/-}$ mice were observed (Figure 1D).

2.2. CEP Inhibits Oxidative Stress and Inflammatory Response in DSS-Induced Colitis Mice

Currently, mounting clinical and experimental evidence strongly supports the involvement of oxidative stress in the pathogenesis of colitis. Therefore, we initially assessed serum oxidative stress levels in DSS-induced mice. Our results demonstrated that DSS treatment significantly increased MDA levels to 5.51 nM/mL in serum from WT mice, while pre-treatment with CEP dramatically reduced MDA level to 2.45 nM/mL in WT mice, however it had no impact on NRF2$^{-/-}$ mice (Figure 2A). Next, we evaluated change in GSH content in DSS-induced WT and NRF2$^{-/-}$ mice. DSS markedly decreased GSH content to 35.53 μg/mL in the serum of WT mice, whereas CEP significantly increased the GSH level to 55.02 μg/mL in WT mice; however, it also had no effect on NRF2$^{-/-}$ mice (Figure 2B). Inflammatory response is a prominent characteristic of colitis, therefore, pro-inflammatory cytokines were measured in DSS-induced WT and NRF2$^{-/-}$ mice. Our findings showed that DSS increased the expression of pro-inflammatory cytokines TNF-α, IL-1β, IL-6, IL-9, IL-13, and IL-23 in both WT and NRF2$^{-/-}$ mice. Pre-treatment with CEP inhibited the expression of pro-inflammatory cytokines in DSS-induced mice but failed to inhibit their rise in DSS-induced NRF2$^{-/-}$ mice (Figure 2C–H).

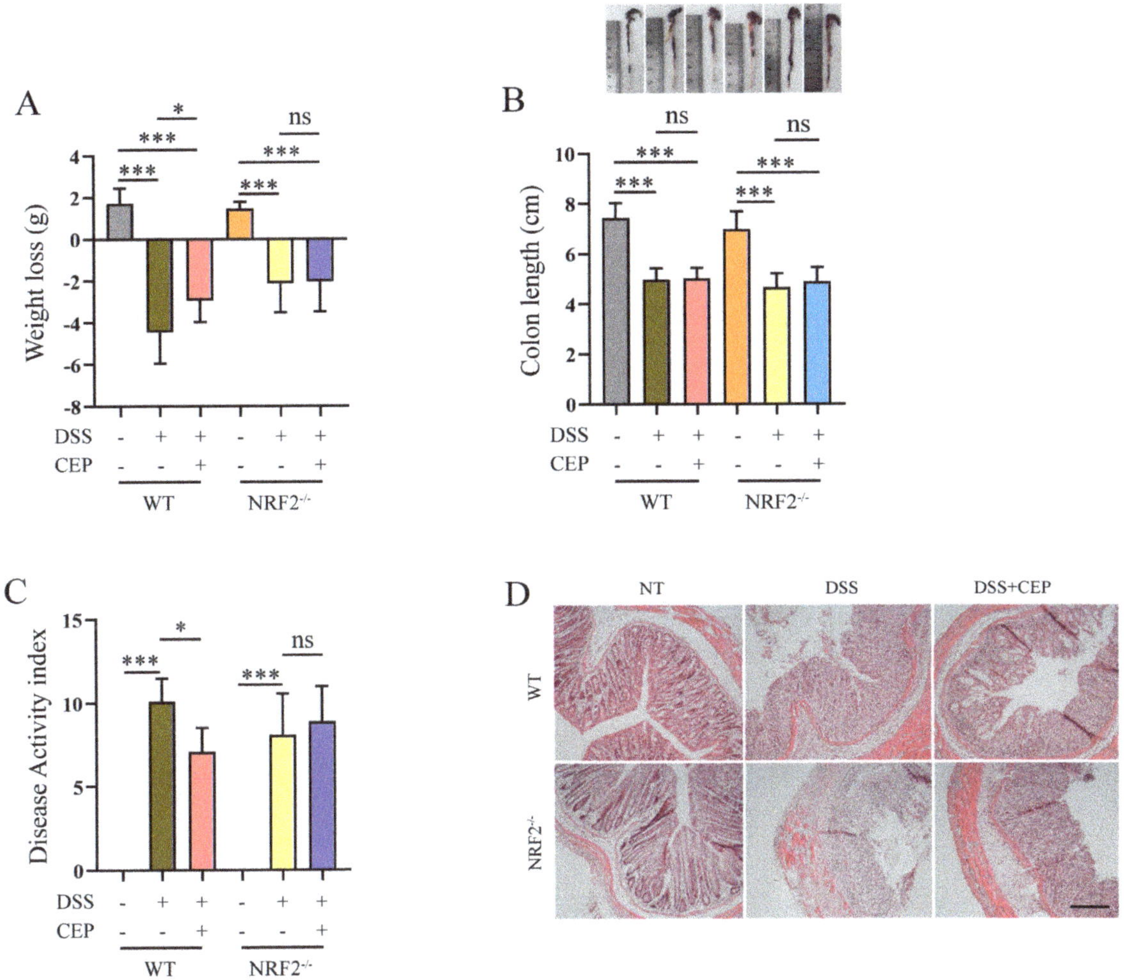

Figure 1. The ameliorative effects of CEP on DSS-induced colitis mice. WT and NRF2$^{-/-}$ mice were pre-treated with CEP for one week followed by treatment with DSS for an additional week. (**A**) The weight loss (initial body weight minus trial termination body weight) of DSS- and CEP-treated WT and NRF2$^{-/-}$ mice, "ns" representative no significance (n = 8–12). (**B**) The colon length of DSS- and CEP-treated WT and NRF2$^{-/-}$ mice, "ns" representative no significance (n = 8–12). (**C**) The disease activity index (DAI) of DSS- and CEP-treated WT and NRF2$^{-/-}$ mice, "ns" representative no significance (n = 8–12). (**D**) H&E staining was performed on colon tissue from DSS- and CEP-treated WT and NRF2$^{-/-}$ mice, magnification shown is 10×, and the scale bar represents 200 μm (n = 3–5). The results shown are means ± SEM, * $p < 0.05$, and *** $p < 0.001$.

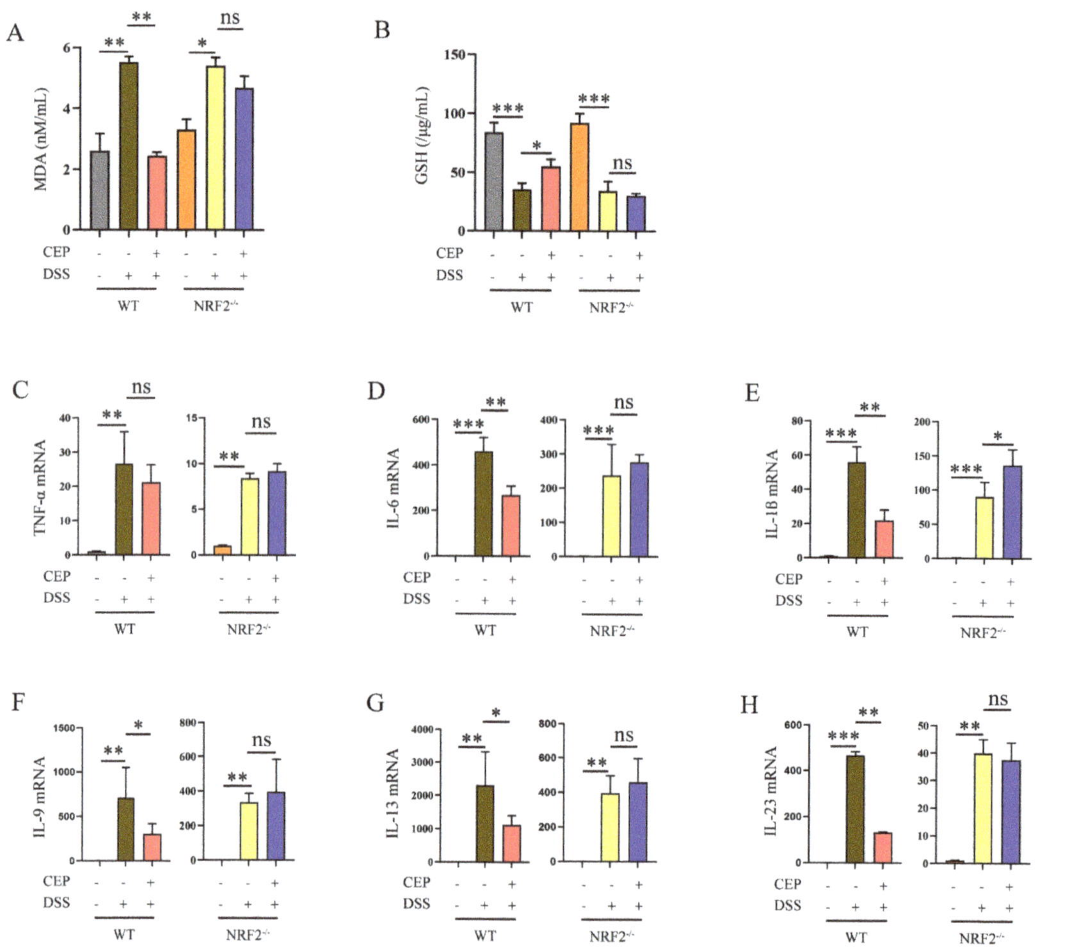

Figure 2. CEP inhibits oxidative stress and inflammatory response in DSS-induced colitis mice. (**A**) The MDA level in serum from CEP and DSS treatment WT and NRF2$^{-/-}$ mice, "ns" representative no significance (n = 3). (**B**) The GSH level in serum from CEP and DSS treatment WT and NRF2$^{-/-}$ mice, "ns" representative no significance (n = 3). (**C–H**). qRT-PCR analyzed the expression of the pro-inflammatory cytokines TNF-α, IL-6, IL-1β, IL-9, IL-13, and IL-23 in colon tissue homogenate, "ns" representative no significance (n = 3). The results shown are means ± SEM, * $p < 0.05$, ** $p < 0.01$, and *** $p < 0.001$.

2.3. *CEP Maintains Proper Intestinal Epithelium Barrier in DSS-Induced Colitis Mice*

Immunohistochemistry analysis revealed a decrease in mucin 2 (MUC2) protein expression in both DSS-induced WT and NRF2$^{-/-}$ mice. In WT mice, CEP promoted the expression of MUC2 in colon tissue, while no effects were observed in NRF2$^{-/-}$ mice (Figure 3A). Tight junction proteins play a crucial role in maintaining the integrity of the intestinal epithelial barrier. Therefore, we evaluated the levels of ZO-1, Claudin-1, and Occludin in colon tissue from WT and NRF2$^{-/-}$ mice. Immunohistochemistry results demonstrated that DSS significantly disrupted the structure of colon tissue and reduced the expression of ZO-1, Claudin-1, and Occludin in both DSS-induced WT and NRF2$^{-/-}$ mice. However, pre-treatment with CEP effectively mitigated the colon tissue damage by restoring its structural integrity and enhancing the expression of ZO-1, Claudin-1, and Occludin specifically in WT mice but had no effect on NRF2$^{-/-}$ mice.

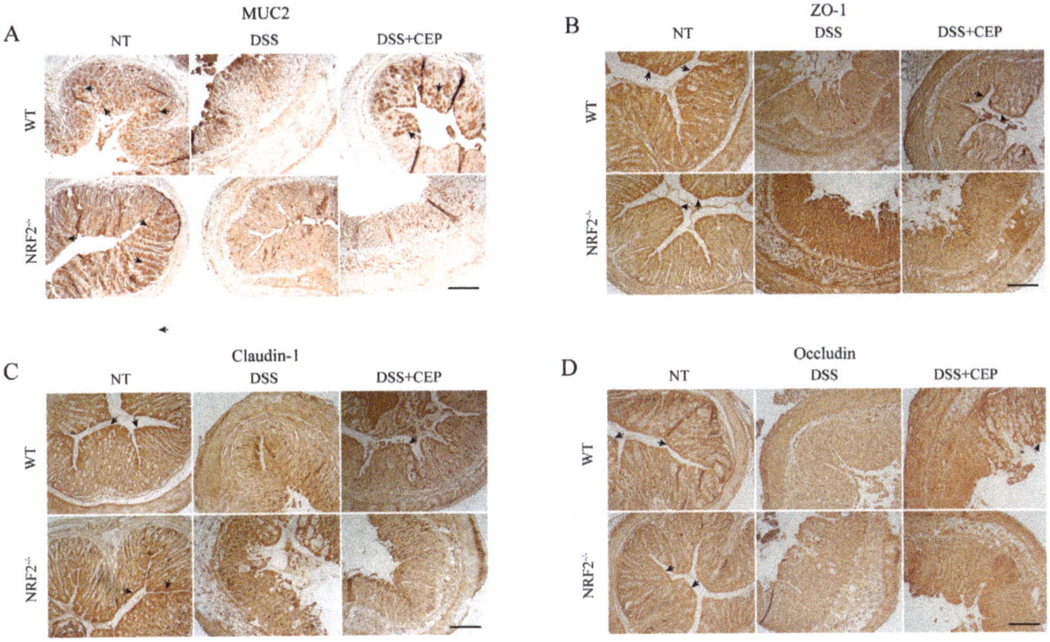

Figure 3. CEP maintains proper intestinal epithelium barrier in DSS-induced colitis mice. (**A**) Immunohistochemistry analysis was performed to evaluate the expression of MUC2 in the colon tissue from DSS- and CEP-treated WT and NRF2$^{-/-}$ mice treated with DSS and CEP, magnification shown is 10×, the scale bar represents 200 μm, and the arrow marked the location and abundance of the protein (n = 3–5). (**B–D**) Immunohistochemistry analysis was conducted to assess the expression of ZO-1 (**B**), Claudin-1 (**C**), and Occludin (**D**) in the colon tissue from DSS- and CEP-treated WT and NRF2$^{-/-}$ mice, magnification shown is 10×, the scale bar represents 200 μm, and the arrow marked the location and abundance of the protein (n = 3–5).

2.4. *CEP Inhibits Inflammatory Response and Oxidative Stress in LPS-Induced Macrophages*

Next, we investigated the impact of CEP on inflammatory response and oxidative stress in LPS-induced macrophages. The concentration of CEP (20 μM) for macrophage treatment was determined based on the results obtained from the CCK8 assay (Figure 4A). qRT-PCR analysis revealed a significant inhibition of IL-6 and IL-1β expression in LPS-induced macrophages following pre-treatment with CEP (Figure 4B,C). Western blot analysis demonstrated that COX-2 and iNOS expression were suppressed by pre-treatment with CEP in LPS-induced macrophages (Figure 4D–F). These findings indicate that supplementation with CEP effectively attenuates the inflammatory response in LPS-induced macrophages. Furthermore, we evaluated the effects of CEP on oxidative stress. Our results showed that GSH levels were significantly reduced to 10.85 μg/10^6 cells upon exposure to LPS, while supplementation with CEP increased it to 12.26 μg/10^6 cells in macrophages (Figure 4G). Flow cytometry and cell fluorescence data indicated a substantial increase in ROS levels upon exposure to LPS, which were dramatically decreased by treatment with CEP in macrophages (Figure 4H,I).

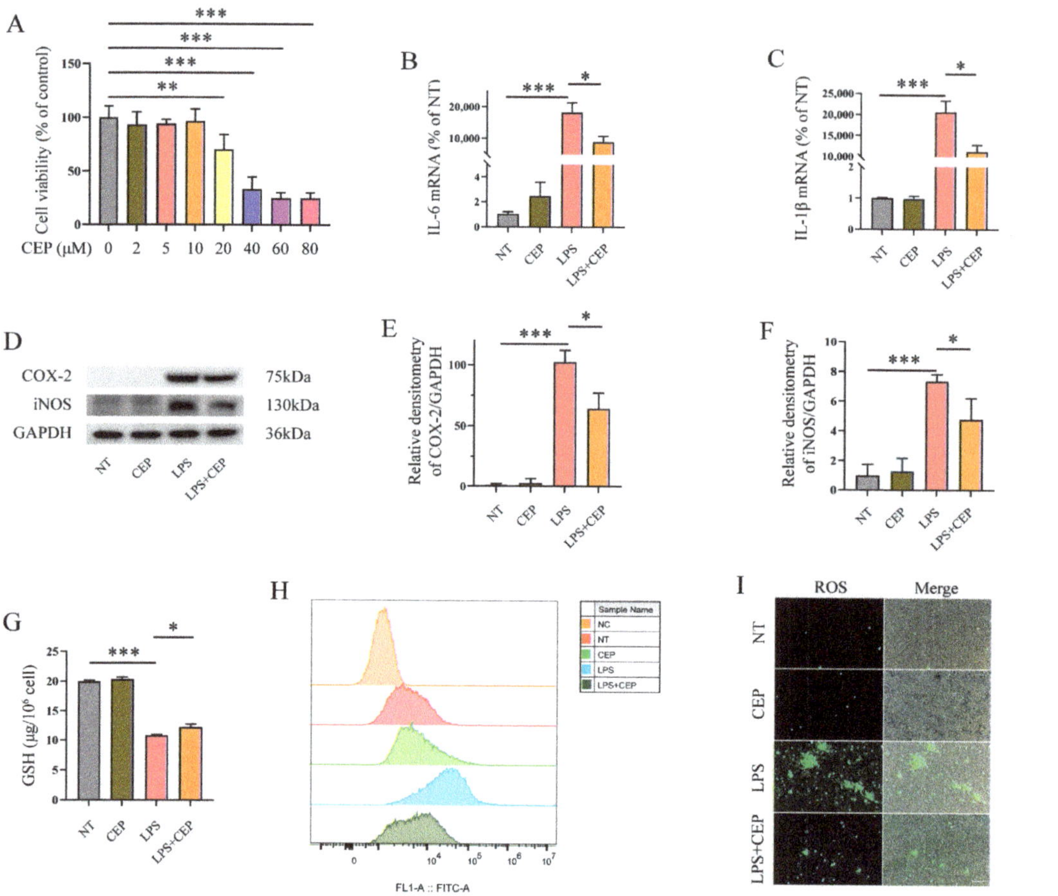

Figure 4. CEP inhibits inflammatory response and oxidative stress in LPS-induced macrophages. (**A**) The CCK8 assay analyzed the impact of CEP on macrophage viability ($n = 5$). (**B**,**C**) qRT-CPR analyzed the expression of the pro-inflammatory cytokines IL-6 (**B**) and IL-1β (**C**) in LPS and CEP treated macrophages ($n = 3$). (**D**–**F**) Western blot analysis of the impact of CEP on the expression of COX-2 (**E**) and iNOS (**F**) in LPS-induced macrophages ($n = 3$). (**G**) The GSH level in the medium of CEP and LPS induced-macrophages ($n = 3$). (**H**,**I**) Flow cytometry and cell fluorescence analyzed the production of ROS in LPS-induced macrophages ($n = 3$). The results shown are means ± SEM, * $p < 0.05$, ** $p < 0.01$, and *** $p < 0.001$.

2.5. The Effects of CEP on NRF2/HO-1/NQO-1, and AMPK-α1/AKT/GSK-3β Signaling Pathways

NRF2, is the most crucial transcription factor against oxidative stress and plays a pivotal role in anti-oxidants and inflammation relief [33]. Therefore, we investigated the effects of CEP on NRF2 and its related signaling pathway. The results indicated that treatment with CEP promoted the expression of NRF2/HO-1/NQO-1 in a time-dependent manner (Figure 5A–D), suggesting that CEP may possess anti-oxidative properties. Previous studies have demonstrated that GSK-3β is a novel regulator of NRF2, therefore, the anti-oxidative property of NRF2 may be associated with the phosphorylation of the AMPK-α1/AKT/GSK-3β signaling pathway [34]. Consequently, we explored the impact of CEP on the AMPK-α1/AKT/GSK-3β signaling pathway in macrophages. Our findings revealed that CEP enhanced the phosphorylation of the AMPK-α1/AKT/GSK-3β signaling pathway (Figure 5E–H).

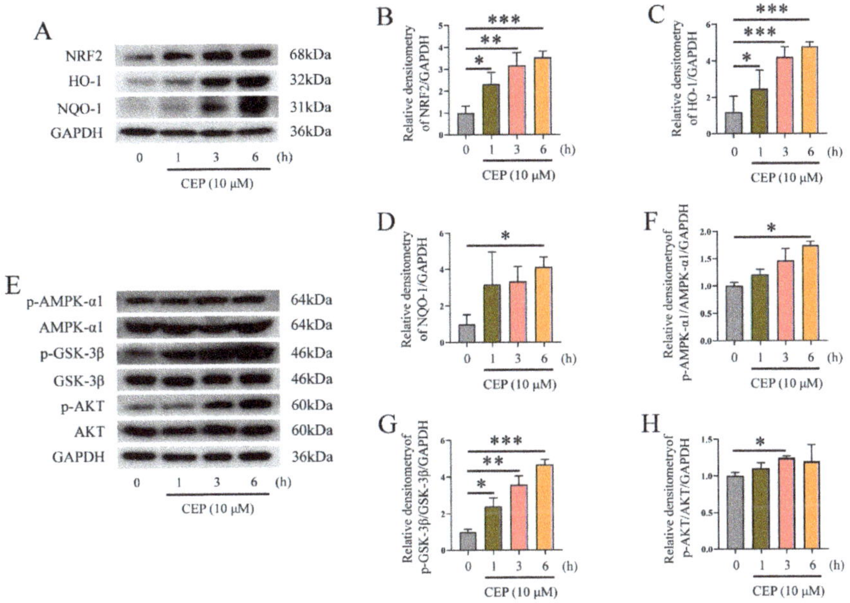

Figure 5. The effects of CEP on the NRF2/HO-1/NQO-1, and AMPK-α1/AKT/GSK-3β signaling pathways. (**A–D**) Western blot analysis of the impact of CEP on the expression of NRF2 (**B**), HO-1 (**C**), and NQO-1 (**D**) signaling pathway in LPS-induced macrophages ($n = 3$). (**E–H**) Western blot analysis of the impact of CEP on the phosphorylation of the AMPK-α1 (**F**), AKT (**G**), and GSK-3β (**H**) signaling pathways in LPS-induced macrophages ($n = 3$). The results shown are means ± SEM, * $p < 0.05$, ** $p < 0.01$, and *** $p < 0.001$.

2.6. CEP Inhibits the Phosphorylation of MAPKs and the NF-κB Signaling Pathway

The Western blot results revealed a significant upregulation of the ERK and JNK signaling pathway phosphorylation upon LPS stimulation, whereas pretreatment with CEP effectively attenuated the phosphorylation effects (Figure 6A–C). Additionally, CEP exhibited inhibitory effects on the NF-κB p65 signaling pathway in LPS-induced macrophages (Figure 6D,E).

2.7. The Effects of CEP on Oxidative Stress in LPS-Induced NRF2$^{-/-}$ Mice Primary Peritoneal Macrophages

Our above findings have demonstrated that CEP exerts a protective role against oxidative stress and inflammatory response by activating NRF2 and its related signaling pathway. To further elucidate the underlying mechanism of CEP, we measured its effects on LPS-induced macrophages derived from NRF2$^{-/-}$ mice. Western blot analysis revealed that while CEP failed to enhance the expression of the HO-1/NQO-1 signaling pathway in these cells (Figure 7A–C), it did promote the phosphorylation of the AMPK-α1/AKT/GSK-3β signaling pathway (Figure 7D–G). These results suggest that CEP activates the phosphorylation of the AMPK-α1/AKT/GSK-3β signaling pathway, subsequently facilitating NRF2 activation, and thereby inhibiting oxidative stress in macrophages. To further validate these observations, we evaluated the impact of CEP on LPS-induced primary peritoneal macrophages obtained from NRF2$^{-/-}$ mice. Our data indicated that CEP failed to suppress MDA production and had no effect on the reduced GSH levels in LPS-induced primary peritoneal macrophages obtained from NRF2$^{-/-}$ mice (Figure 7H,I).

Collectively, our experiments revealed that although CEP activates the phosphorylation of the AMPK-α1/AKT/GSK-3β signaling pathway, it is evident that NRF2 plays a pivotal role in protecting against oxidative stress in macrophages.

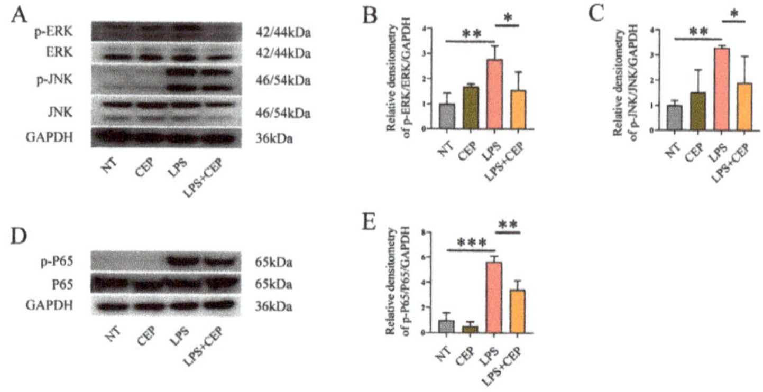

Figure 6. CEP inhibits the phosphorylation of MAPKs and the NF-κB signaling pathway. (**A–C**) Western blot analysis of the impact of CEP on the phosphorylation of ERK (**B**), and JNK (**C**) signaling pathways in LPS-induced macrophages ($n = 3$). (**D,E**) Western blot analysis of the impact of CEP on the phosphorylation of the NF-κB p65 signaling pathway in LPS-induced macrophages ($n = 3$). The results shown are means ± SEM, * $p < 0.05$, ** $p < 0.01$, and *** $p < 0.001$.

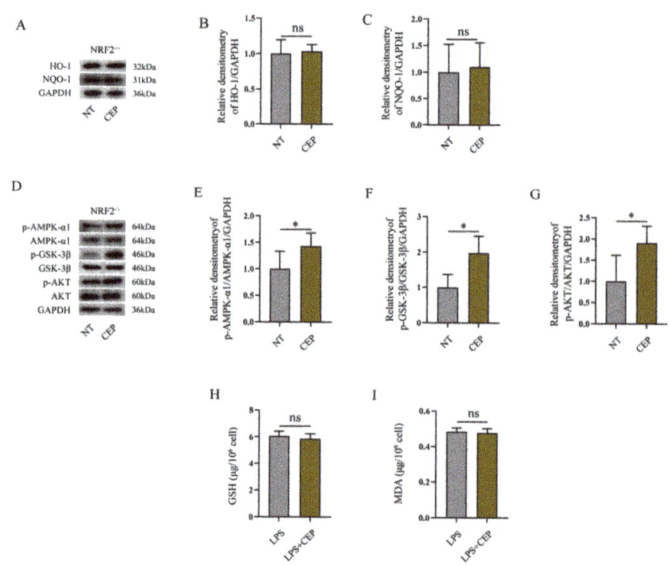

Figure 7. The effects of CEP on oxidative stress in LPS-induced NRF2$^{-/-}$ mice primary peritoneal macrophages. (**A–C**) Western blot analysis of the impact of CEP on the expression of HO-1 (**B**), and NQO-1 (**C**) signaling pathway in primary peritoneal macrophages from NRF2$^{-/-}$ mice, "ns" representative no significance ($n = 3$). (**D–G**) Western blot analysis of the impact of CEP on the phosphorylation of AMPK-α1, GSK-3β, and AKT signaling pathways in primary peritoneal macrophages from NRF2$^{-/-}$ mice ($n = 3$). (**H**) The GSH level in medium of CEP and LPS induced-primary peritoneal macrophages from NRF2$^{-/-}$ mice, "ns" representative no significance ($n = 3$). (**I**) The MDA level in medium of CEP and LPS induced-primary peritoneal macrophages from NRF2$^{-/-}$ mice, "ns" representative no significance, ($n = 3$). The results shown are means ± SEM, * $p < 0.05$.

3. Discussion

IBD is a highly prevalent chronic gastrointestinal disease globally, and is characterized by a complex pathogenesis involving various factors such as host genetics, intestinal microbiota, environmental triggers, lifestyle factors, and diet [35]. The interaction among these multiple components disrupts the balance within the intestinal mucosa while also altering immune responses [36], leading to an increase in permeability of the intestinal epithelium. The invasion of enterocoel through the epithelial layer triggers immune cell activation and induces the production of pro-inflammatory cytokines. Chronic local inflammation further enhances the generation and release of ROS by immune cells, which serve as crucial signaling molecules in immunological functions [8]. However, excessive production of ROS and related products in the actively inflamed mucosal microenvironment leads to collateral damage, including intestinal inflammation and tissue destruction [9]. Accumulating evidence suggests that oxidative stress plays a pivotal role in the pathogenesis of IBD. Therefore, reducing ROS levels and inhibiting oxidant stress may have significant therapeutic implications for IBD.

CEP is a biscoclaurine alkaloid derived from *Stephania cepharantha* Hayata, which has been extensively utilized in Japan for over 40 years for the treatment of acute and chronic diseases [37]. CEP exhibits diverse pharmacological activities, including anti-inflammatory effects by reducing NLRP3 inflammasome activation [26] and inhibiting MAPKs and NF-κB p65 signaling pathways [27,28], as well as anti-oxidative stress, anti-autophagy, and anti-viral effects [26,30,31]. Furthermore, CEP enhances the sensitivity of host cells to anti-cancer agents [30,32], without exerting any direct impact on cancer itself. The objective of this study is to investigate the effects of CEP on oxidative stress and inflammatory response both in vivo and in vitro.

The essential features of IBD include a depressed survival rate, increased body weight loss, shortened colon length, and elevated clinical score [38]. Our findings demonstrated that supplementation with CEP significantly improves weight loss, DAI, and colon tissue impairment in DSS-induced WT mice but not in $NRF2^{-/-}$ mice. This is consistent with previous studies showing that CEP inhibits the increased weight loss, rise of DAI and histological score in DSS-induced WT colitis mice [39,40]. Transient middle cerebral artery occlusion (tMCAO) increases levels of ROS and MDA, while decreasing SOD levels in mice. However, supplementation with CEP significantly ameliorates oxidative stress in the tMCAO-induced mouse model [26]. Furthermore, CEP inhibits inflammatory response in LPS-induced acute lung injury and the DSS-induced colitis mouse model [28,39]. In this study, we found that supplementation with CEP inhibits oxidative stress and inflammatory response in DSS-induced WT mice, but not in $NRF2^{-/-}$ mice. Additionally, we found that supplementation with CEP maintains proper intestinal epithelium barrier in DSS-induced WT mice. Our results demonstrated that CEP provides a protective role against oxidative stress and inflammatory response in DSS-induced mice. Moreover, CEP had no significant effects on the characteristics observed in DSS-induced $NRF2^{-/-}$ colitis mice, suggesting that the protective functions of CEP on oxidative stress and inflammatory response are in a NRF2-dependent manner.

To further investigate the effects of CEP on oxidative stress and inflammatory response, we assessed its impact on LPS-induced inflammatory response and oxidative stress in macrophages. Our findings demonstrated that 10 μM CEP effectively attenuates the expression of pro-inflammatory cytokines IL-6 and IL-1β, as well as pro-inflammatory mediators iNOS and COX-2 in LPS-stimulated macrophages. Moreover, supplementation with CEP was observed to enhance the reduced GSH level while reducing ROS production in LPS-treated macrophages. These results are consistent with previous studies highlighting the potent anti-inflammatory and anti-oxidant properties of CEP [28,41]. Collectively, our findings along with prior research, provide compelling evidence for the inhibitory effects of CEP on both inflammatory responses and oxidative stress.

Next, we explored the potential mechanism of CEP on oxidative stress and inflammatory response. NRF2 is a pivotal transcription factor that plays a crucial role in combating oxidative stress and inflammatory response [33]. Therefore, we examined the impact of CEP on NRF2 and its associated signaling pathway. Our results demonstrated that supplementation with CEP significantly upregulated the expression of NRF2/HO-1/NQO-1 in a time-dependent manner, indicating the essential role of NRF2 in mediating the anti-oxidative and anti-inflammatory effects of CEP. AMPK activation regulates both catabolism and anabolism, while also maintaining redox balance and modulating inflammatory response [42,43]. Moreover, AMPK promotes phosphorylation events involving PI3K/AKT [44] and GSK-3β, which inhibits mitochondrial oxidative stress [45]. Previous studies have suggested GSK-3β as a novel regulator of NRF2, therefore, it is plausible that the anti-oxidative properties of NRF2 are associated with the involvement of phosphorylation events mediated by the AMPK-α1/AKT/GSK-3β signaling pathway [34]. Our findings revealed that CEP enhanced phosphorylation levels within the AMPK-α1/AKT/GSK-3β signaling pathway. Overall, our study suggests that CEP exerts potent anti-oxidative effects in macrophages through activation of the AMPK-α1/AKT/GSK-3β/NRF2 signaling pathway.

Although activation of NRF2 is generally believed to counter inflammation in various inflammatory diseases [46], natural products may also have impacts on other signaling pathways. Our results demonstrated that supplementation with CEP significantly inhibits the phosphorylation of the ERK, JNK, and NF-κB p65 signaling pathways in LPS-treated macrophages. This is consistent with previous studies showing that CEP inhibits the activation of the MAPKs and NR-κB p65 signaling pathways [28,47,48]. These findings suggest that the effects of CEP on inflammatory response are not limited to NRF2 but also depend on its inhibition of the ERK, JNK and NF-κB P65 signaling pathways.

Our findings in mice demonstrated that CEP fails to inhibit the oxidative stress and inflammatory response in NRF2-deficient mice, suggesting the crucial role of NRF2 in mediating its effects on anti-inflammatory response and oxidative stress. To validate these results, we investigated the impact of CEP on primary peritoneal macrophages from NRF2$^{-/-}$ mice. The outcomes revealed that CEP failed to enhance the expression of the HO-1/NQO-1 signaling pathway; however, it activated the phosphorylation of the AMPK-α1/AKT/GSK-3β signaling pathway in primary peritoneal macrophages from NRF2$^{-/-}$ mice. Moreover, CEP had no influence on MDA levels and reduced GSH levels in LPS-induced primary peritoneal macrophages from NRF2$^{-/-}$ mice. These findings suggest that while NRF2 plays a significant role in the antioxidant activity of CEP, it also activates the phosphorylation of the AMPK-α1/AKT/GSK-3β pathway which is an upstream regulator of NRF2.

4. Materials and Methods

4.1. Animal Model

WT and NRF2$^{-/-}$ C57BL/6 mice aged 6–8 weeks-old were utilized in this study. All mice were housed under controlled conditions at a temperature of 22–23 °C on a 12 h light/dark cycle, with ad libitum access to food and water. The mice were randomly divided into six groups, each containing 8–12 individuals: NT group; DSS group, where the animals received a solution of 3% (*w*/*v*) DSS via drinking water for one week; DSS + CEP group, where the mice were administered CEP through intragastric gavage at a dosage of 10 mg/kg, based on previous research [48], followed by DSS treatment for another week. All experimental procedures involving animal subjects have been approved by the Committee of Scientific Research at Shanxi University.

4.2. Disease Activity Index

Throughout the experiment of DSS-induced colitis, mice were closely monitored for changes in body weight, stool consistency, and fecal occult blood. Disease activity index (DAI) scores were determined using a previously established scoring system [49].

4.3. Hematoxylin and Eosin (H&E) Staining

The colon tissue samples obtained from mice were washed with phosphate-buffered saline (PBS), fixed in a solution containing 4% paraformaldehyde, embedded in paraffin, sectioned into 5 μM slices, and subsequently stained with hematoxylin and eosin.

4.4. Cell Culture

The murine macrophage cell line RAW264.7 was obtained from BeNa Culture Collection and cultured in Dulbecco's modified Eagle's medium (Gibco, Grand Island, NY, USA) supplemented with 10% fetal bovine serum (Sorfa, Beijng, China). The cells were incubated at a temperature of 37 °C with 5% CO_2 in a humidified chamber.

4.5. Cell Viability Assay

RAW264.7 macrophages were seeded in 96-well plates and pre-treated with various concentrations of CEP (2, 5, 10, 20, 40, 60, and 80 μM) for a duration of 24 h. Following the incubation, cck8 solution (1 mg/mL) of the Cell Counting Kit (CCK8, 40203ES60, Yeasen, Shanghai, China) was added to each well at a volume of 10 μL/well and allowed to react for a period of 30 min. Subsequently, dimethyl sulfoxide (DMSO) was introduced into each well followed by reading of absorbance using a microplate reader set at an wavelength of 450 nm.

4.6. Western Blot

Total protein was extracted from RAW264.7 macrophages and primary peritoneal macrophages using radio immunoprecipitation lysis buffer (RIPA, Solarbio, Beijing, China) supplemented with phenylmethylsulfonyl fluoride (PMSF) and phosphatase inhibitors (Solarbio, Beijing, China). The protein was collected, and its concentration was measured using a Pierce BCA protein assay kit (Thermo, Rockford, IL, USA). Equal amounts of proteins were separated by electrophoresis on SDS-PAGE gels and subsequently transferred onto PVDF membranes. Blocking was with TBST containing non-fat milk powder, the membranes were incubated overnight at 4 °C with primary antibodies. The primary antibodies against NRF2, p-AKT, AKT, iNOS (Cell Signaling Technology, Danvers, MA, USA), and HO-1, NQO-1, COX-2, p-GSK-3β, GSK-3β, p-AMPK-α1, AMPK-α1, p-ERK, EKR, p-JNK, JNK, p-P38, P38, p-NF-κB p65, NF-κB p65, and GAPDH (ABclonal, Wuhan, China) were diluted to a ratio of 1:2000. Before adding secondary antibodies (goat anti-rabbit or goat anti-mouse) diluted to a concentration of 1:10,000 (ABclonal, Wuhan, China), the membranes were washed four times with TBST for 15 min each time. Membranes were visualized using an Amersham Imager 600 (a gel imaging system from GE Co., Fairfield, CT, USA) after applying enhanced chemiluminescence. For detailed information of WB procedures, refer to our previous study [50].

4.7. Real-Time Quantitative PCR

Total RNA was extracted from cells/colon tissue using Trizol reagent (TransGen Biotech) following the manufacturer's protocol. The isolated RNA was treated with DNase I (Sigma, St. Louis, MO, USA), quantified, and reverse transcribed into cDNA using TransScript first-strand cDNA synthesis SuperMix (TransGen Biotech, Beijing, China). Real-time quantitative PCR was performed in the ABI PRISM®7500 real-time PCR system (Applied Biosystems, Foster City, CA, USA) utilizing SYBR® Premix Ex Taq™ II (Tli RNase H Plus) (TaKaRa, Dalian, China). The primer sequences are shown below. TNF-α, F: 5′-GCAACTGCTGCACGAAATC-3′, R: 5′-CTGCTTGTCCTCTGCCCAC-3′; IL-1β, F: 5′-GTTCCCATTAGACAACTGCACTACAG-3′,R: 5′-GTCGTTGCTTGGTTCTCCTTGTA-3′; IL-6, F: 5′-CCAGAAACCGCTATGAAGTTCC-3′, R: 5′-GTTGGGAGTGGTATCCTCTGTGA-3′; IL-9, F: 5′-ATGTTGGTGACATACATCCTTGC-3′, R: 5′-TGACGGTGGATCATCCTTCAG-3; IL-13, F: 5′-TGAGCAACATCACACAAGACC-3′, R: 5′-GGCCTTGCGGTTACAGAGG-3′; IL-23, F: 5′-CAGCAGCTCTCTCGGAATCTC-3′, R: 5′-TGGATACGGGGCACATTATTTTT-3′; β-actin, F: 5′-GTCAGGTCATCACTATCGGCAAT-3′, R: 5′-AGAGGTCTTTACGGATGTC

AACGT-3′. Each of the samples were analyzed in triplicate, in a similar manner to our previous study [51].

4.8. Reactive Oxygen Species (ROS) Detection

RAW264.7 macrophages and primary peritoneal macrophages were cultured in 6-well plates and pre-treated with a concentration of 10 μM CEP for 1 h. Subsequently, the cells were exposed to LPS for a period of 12 h. Intracellular levels of ROS were quantified using the Reactive Oxygen Species Assay Kit (Solarbio, CA1410, Beijing, China) using a DCFH-DA ROS probe following the manufacturer's instructions. The fluorescence intensity was then measured using flow cytometry and a fluorescence microscope. Detailed experimental procedures are outlined in the kit instructions.

4.9. MDA and Reduced GSH Detection

The MDA content in the serum of mice and primary peritoneal macrophages was quantified using a Malondialdehyde (MDA) Content Assay Kit (Solarbio, BC0025, Beijing, China). The GSH content in the serum of mice, RAW264.7 macrophages and primary peritoneal macrophages was determined using the Reduced Glutathione (GSH) Content Assay Kit (Solarbio, BC1175, Beijing, China). Detailed experimental procedures are outlined in the kit instructions.

4.10. Statistical Analysis

In this study, the data were presented as mean ± SEM. Statistical analysis was performed using GraphPad Prism software version 7.00. Group comparisons were conducted using one-way ANOVA followed by the least significant difference test (* $p < 0.05$, ** $p < 0.01$, *** $p < 0.001$). Two-group comparisons were analyzed using Student's t-test (* $p < 0.05$). Each group in this study was replicated three times.

5. Conclusions

In conclusion, our study based on in vitro and in vivo data demonstrates that CEP exerts protective effects against oxidative stress and inflammatory responses by modulating the activation of the AMPK-α1/AKT/GSK-3β/NRF2 signaling pathway. Specifically, NRF2 plays a crucial role in safeguarding against oxidative stress and inflammatory response in LPS-induced macrophages and DSS-induced mice. Furthermore, CEP also triggers the phosphorylation of the JNK/ERK and NF-κB p65 signaling pathways, which are crucial for mediating inflammatory response. Our findings highlight the potential of CEP in ameliorating intestinal oxidative stress and inflammatory response. Moreover, the anti-oxidative and anti-inflammatory properties of CEP may have implications for other diseases characterized by inflammatory response and oxidative stress; however, further extensive experiments are warranted to validate this hypothesis. Nevertheless, this study substantiates the antioxidant and anti-inflammatory effects of CEP while elucidating its potential mechanisms of action.

Author Contributions: Conceptualization, G.C. and Z.Z.; methodology, P.L. and Q.X.; software, Y.F. and L.S.; validation, P.L. and Q.X.; formal analysis, Z.Z.; investigation, G.C. and D.W.; data curation, D.W.; writing—original draft preparation, G.C.; writing—review and editing, Z.Z.; supervision, Z.Z.; funding acquisition, G.C. All authors have read and agreed to the published version of the manuscript.

Funding: This research was funded by "National Nature Science Foundation of China, grant number 32202698", "China Postdoctoral Science Foundation, grant number 2022M710476", and "Central Guidance on Local Science and Technology Development Fund of Shanxi Province, grant number YDZJSX20231A005".

Institutional Review Board Statement: All methods were carried out in accordance with relevant guidelines and regulations and were approved by the "Institutes of Biomedical Sciences, Shanxi University". All experiments were approved by the Committee of Scientific Research in Shanxi University (SXULL2019004).

Informed Consent Statement: Not applicable.

Data Availability Statement: All data are contained within the article.

Conflicts of Interest: The authors declare no conflict of interest.

Sample Availability: Not applicable.

References

1. Ungaro, R.; Mehandru, S.; Allen, P.B.; Peyrin-Biroulet, L.; Colombel, J.F. Ulcerative colitis. *Lancet* **2017**, *389*, 1756–1770. [CrossRef] [PubMed]
2. Baumgart, D.C.; Sandborn, W.J. Crohn's disease. *Lancet* **2012**, *380*, 1590–1605. [CrossRef] [PubMed]
3. Khor, B.; Gardet, A.; Xavier, R.J. Genetics and pathogenesis of inflammatory bowel disease. *Nature* **2011**, *474*, 307–317. [CrossRef] [PubMed]
4. Sartor, R.B. Microbial influences in inflammatory bowel diseases. *Gastroenterology* **2008**, *134*, 577–594. [CrossRef] [PubMed]
5. Kaser, A.; Zeissig, S.; Blumberg, R.S. Inflammatory bowel disease. *Annu. Rev. Immunol.* **2010**, *28*, 573–621. [CrossRef] [PubMed]
6. Ho, G.T.; Cartwright, J.A.; Thompson, E.J.; Bain, C.C.; Rossi, A.G. Resolution of Inflammation and Gut Repair in IBD: Translational Steps Towards Complete Mucosal Healing. *Inflamm. Bowel Dis.* **2020**, *26*, 1131–1143. [CrossRef] [PubMed]
7. Soendergaard, C.; Bergenheim, F.H.; Bjerrum, J.T.; Nielsen, O.H. Targeting JAK-STAT signal transduction in IBD. *Pharmacol. Ther.* **2018**, *192*, 100–111. [CrossRef] [PubMed]
8. Pereira, C.; Gracio, D.; Teixeira, J.P.; Magro, F. Oxidative Stress and DNA Damage: Implications in Inflammatory Bowel Disease. *Inflamm. Bowel Dis.* **2015**, *21*, 2403–2417. [CrossRef]
9. Campbell, E.L.; Colgan, S.P. Control and dysregulation of redox signalling in the gastrointestinal tract. *Nat. Rev. Gastroenterol. Hepatol.* **2019**, *16*, 106–120. [CrossRef]
10. Zhu, H.; Li, Y.R. Oxidative stress and redox signaling mechanisms of inflammatory bowel disease: Updated experimental and clinical evidence. *Exp. Biol. Med.* **2012**, *237*, 474–480. [CrossRef]
11. Bhattacharyya, A.; Chattopadhyay, R.; Mitra, S.; Crowe, S.E. Oxidative stress: An essential factor in the pathogenesis of gastrointestinal mucosal diseases. *Physiol. Rev.* **2014**, *94*, 329–354. [CrossRef] [PubMed]
12. Mousavi, T.; Hadizadeh, N.; Nikfar, S.; Abdollahi, M. Drug discovery strategies for modulating oxidative stress in gastrointestinal disorders. *Expert Opin. Drug Discov.* **2020**, *15*, 1309–1341. [CrossRef] [PubMed]
13. Tian, T.; Wang, Z.; Zhang, J. Pathomechanisms of Oxidative Stress in Inflammatory Bowel Disease and Potential Antioxidant Therapies. *Oxidative Med. Cell. Longev.* **2017**, *2017*, 4535194. [CrossRef] [PubMed]
14. Dudzinska, E.; Gryzinska, M.; Ognik, K.; Gil-Kulik, P.; Kocki, J. Oxidative Stress and Effect of Treatment on the Oxidation Product Decomposition Processes in IBD. *Oxidative Med. Cell. Longev.* **2018**, *2018*, 7918261. [CrossRef] [PubMed]
15. Toki, T.; Itoh, J.; Kitazawa, J.; Arai, K.; Hatakeyama, K.; Akasaka, J.; Igarashi, K.; Nomura, N.; Yokoyama, M.; Yamamoto, M.; et al. Human small Maf proteins form heterodimers with CNC family transcription factors and recognize the NF-E2 motif. *Oncogene* **1997**, *14*, 1901–1910. [CrossRef] [PubMed]
16. Matzinger, M.; Fischhuber, K.; Heiss, E.H. Activation of Nrf2 signaling by natural products-can it alleviate diabetes? *Biotechnol. Adv.* **2018**, *36*, 1738–1767. [CrossRef] [PubMed]
17. Fao, L.; Mota, S.I.; Rego, A.C. Shaping the Nrf2-ARE-related pathways in Alzheimer's and Parkinson's diseases. *Ageing Res. Rev.* **2019**, *54*, 100942. [CrossRef] [PubMed]
18. Yamawaki, K.; Kanda, H.; Shimazaki, R. Nrf2 activator for the treatment of kidney diseases. *Toxicol. Appl. Pharmacol.* **2018**, *360*, 30–37. [CrossRef]
19. Cuadrado, A.; Rojo, A.I.; Wells, G.; Hayes, J.D.; Cousin, S.P.; Rumsey, W.L.; Attucks, O.C.; Franklin, S.; Levonen, A.L.; Kensler, T.W.; et al. Therapeutic targeting of the NRF2 and KEAP1 partnership in chronic diseases. *Nat. Rev. Drug Discov.* **2019**, *18*, 295–317. [CrossRef]
20. Hussain, H.; Ahmad, S.; Shah, S.W.A.; Ullah, A.; Rahman, S.U.; Ahmad, M.; Almehmadi, M.; Abdulaziz, O.; Allahyani, M.; Alsaiari, A.A.; et al. Synthetic Mono-Carbonyl Curcumin Analogues Attenuate Oxidative Stress in Mouse Models. *Biomedicines* **2022**, *10*, 2597. [CrossRef]
21. Hussain, H.; Ahmad, S.; Shah, S.W.A.; Ullah, A.; Almehmadi, M.; Abdulaziz, O.; Allahyani, M.; Alsaiari, A.A.; Halawi, M.; Alamer, E. Investigation of Antistress and Antidepressant Activities of Synthetic Curcumin Analogues: Behavioral and Biomarker Approach. *Biomedicines* **2022**, *10*, 2385. [CrossRef] [PubMed]
22. Wang, W.; Li, H.; Lv, J.; Khan, G.J.; Duan, H.; Zhu, J.; Bao, N.; Zhai, K.; Xue, Z. Determination of the Anti-Oxidative Stress Mechanism of Isodon suzhouensis Leaves by Employing Bioinformatic and Novel Research Technology. *ACS Omega* **2023**, *8*, 3520–3529. [CrossRef] [PubMed]

23. Kanamori, S.; Hiraoka, M.; Fukuhara, N.; Oizumi, Y.; Danjo, A.; Nakata, K.; Owaki, K.; Nishimura, Y. Clinical Efficacy of Cepharanthin(R)for Radiotherapy-Induced Leukopenia—A Nationwide, Multicenter, and Observational Study. *Gan Kagaku Ryoho. Cancer Chemother.* **2016**, *43*, 1075–1079.
24. Tabata, R.; Tabata, C.; Tazoh, A.; Nagai, T. Low dose cepharanthine ameliorates immune thrombocytopenic purpura associated with multiple myeloma. *Int. Immunopharmacol.* **2012**, *13*, 242–244. [CrossRef] [PubMed]
25. Hifumi, T.; Yamamoto, A.; Morokuma, K.; Ogasawara, T.; Kiriu, N.; Hasegawa, E.; Inoue, J.; Kato, H.; Koido, Y.; Takahashi, M. Surveillance of the clinical use of mamushi (*Gloydius blomhoffii*) antivenom in tertiary care centers in Japan. *Jpn. J. Infect. Dis.* **2011**, *64*, 373–376. [CrossRef] [PubMed]
26. Zhao, J.; Piao, X.; Wu, Y.; Liang, S.; Han, F.; Liang, Q.; Shao, S.; Zhao, D. Cepharanthine attenuates cerebral ischemia/reperfusion injury by reducing NLRP3 inflammasome-induced inflammation and oxidative stress via inhibiting 12/15-LOX signaling. *Biomed. Pharmacother. = Biomed. Pharmacother.* **2020**, *127*, 110151. [CrossRef] [PubMed]
27. Paudel, K.R.; Karki, R.; Kim, D.W. Cepharanthine inhibits in vitro VSMC proliferation and migration and vascular inflammatory responses mediated by RAW264.7. *Toxicol. Vitr. Int. J. Publ. Assoc. BIBRA* **2016**, *34*, 16–25. [CrossRef]
28. Huang, H.; Hu, G.; Wang, C.; Xu, H.; Chen, X.; Qian, A. Cepharanthine, an alkaloid from Stephania cepharantha Hayata, inhibits the inflammatory response in the RAW264.7 cell and mouse models. *Inflammation* **2014**, *37*, 235–246. [CrossRef]
29. Halicka, D.; Ita, M.; Tanaka, T.; Kurose, A.; Darzynkiewicz, Z. Biscoclaurine alkaloid cepharanthine protects DNA in TK6 lymphoblastoid cells from constitutive oxidative damage. *Pharmacol. Rep. PR* **2008**, *60*, 93–100.
30. Tang, Z.H.; Cao, W.X.; Guo, X.; Dai, X.Y.; Lu, J.H.; Chen, X.; Zhu, H.; Lu, J.J. Identification of a novel autophagic inhibitor cepharanthine to enhance the anti-cancer property of dacomitinib in non-small cell lung cancer. *Cancer Lett.* **2018**, *412*, 1–9. [CrossRef]
31. Ohashi, H.; Watashi, K.; Saso, W.; Shionoya, K.; Iwanami, S.; Hirokawa, T.; Shirai, T.; Kanaya, S.; Ito, Y.; Kim, K.S.; et al. Potential anti-COVID-19 agents, cepharanthine and nelfinavir, and their usage for combination treatment. *iScience* **2021**, *24*, 102367. [CrossRef] [PubMed]
32. Ikeda, R.; Che, X.F.; Yamaguchi, T.; Ushiyama, M.; Zheng, C.L.; Okumura, H.; Takeda, Y.; Shibayama, Y.; Nakamura, K.; Jeung, H.C.; et al. Cepharanthine potently enhances the sensitivity of anticancer agents in K562 cells. *Cancer Sci.* **2005**, *96*, 372–376. [CrossRef] [PubMed]
33. Kobayashi, E.; Suzuki, T.; Yamamoto, M. Roles nrf2 plays in myeloid cells and related disorders. *Oxidative Med. Cell. Longev.* **2013**, *2013*, 529219. [CrossRef] [PubMed]
34. Xing, H.Y.; Cai, Y.Q.; Wang, X.F.; Wang, L.L.; Li, P.; Wang, G.Y.; Chen, J.H. The Cytoprotective Effect of Hyperoside against Oxidative Stress Is Mediated by the Nrf2-ARE Signaling Pathway through GSK-3beta Inactivation. *PLoS ONE* **2015**, *10*, e0145183. [CrossRef] [PubMed]
35. Fiocchi, C. Inflammatory bowel disease: Etiology and pathogenesis. *Gastroenterology* **1998**, *115*, 182–205. [CrossRef] [PubMed]
36. Bourgonje, A.R.; Feelisch, M.; Faber, K.N.; Pasch, A.; Dijkstra, G.; van Goor, H. Oxidative Stress and Redox-Modulating Therapeutics in Inflammatory Bowel Disease. *Trends Mol. Med.* **2020**, *26*, 1034–1046. [CrossRef] [PubMed]
37. Furusawa, S.; Wu, J. The effects of biscoclaurine alkaloid cepharanthine on mammalian cells: Implications for cancer, shock, and inflammatory diseases. *Life Sci.* **2007**, *80*, 1073–1079. [CrossRef]
38. Hung, S.P.; Sheu, M.J.; Ma, M.C.; Hu, J.T.; Sun, Y.Y.; Lee, C.C.; Chung, Y.C.; Tsai, Y.J.; Wang, J.Y.; Chen, C.L. Runx1-deficient afferents impair visceral nociception, exacerbating dextran sodium sulfate-induced colitis. *Brain Behav. Immun.* **2014**, *35*, 96–106. [CrossRef]
39. Zhang, M.N.; Xie, R.; Wang, H.G.; Wen, X.; Wang, J.Y.; He, L.; Zhang, M.H.; Yang, X.Z. Cepharanthine Alleviates DSS-Induced Ulcerative Colitis via Regulating Aconitate Decarboxylase 1 Expression and Macrophage Infiltration. *Molecules* **2023**, *28*, 1060. [CrossRef]
40. Wang, H.G.; Zhang, M.N.; Wen, X.; He, L.; Zhang, M.H.; Zhang, J.L.; Yang, X.Z. Cepharanthine ameliorates dextran sulphate sodium-induced colitis through modulating gut microbiota. *Microb. Biotechnol.* **2022**, *15*, 2208–2222. [CrossRef]
41. Bailly, C. Cepharanthine: An update of its mode of action, pharmacological properties and medical applications. *Phytomed. Int. J. Phytother. Phytopharm.* **2019**, *62*, 152956. [CrossRef]
42. O'Neill, L.A.; Hardie, D.G. Metabolism of inflammation limited by AMPK and pseudo-starvation. *Nature* **2013**, *493*, 346–355. [CrossRef] [PubMed]
43. Levine, Y.C.; Li, G.K.; Michel, T. Agonist-modulated regulation of AMP-activated protein kinase (AMPK) in endothelial cells. Evidence for an AMPK -> Rac1 -> Akt -> endothelial nitric-oxide synthase pathway. *J. Biol. Chem.* **2007**, *282*, 20351–20364. [CrossRef] [PubMed]
44. Zheng, T.; Yang, X.; Wu, D.; Xing, S.; Bian, F.; Li, W.; Chi, J.; Bai, X.; Wu, G.; Chen, X.; et al. Salidroside ameliorates insulin resistance through activation of a mitochondria-associated AMPK/PI3K/Akt/GSK3beta pathway. *Br. J. Pharmacol.* **2015**, *172*, 3284–3301. [CrossRef] [PubMed]
45. Horike, N.; Sakoda, H.; Kushiyama, A.; Ono, H.; Fujishiro, M.; Kamata, H.; Nishiyama, K.; Uchijima, Y.; Kurihara, Y.; Kurihara, H.; et al. AMP-activated protein kinase activation increases phosphorylation of glycogen synthase kinase 3β and thereby reduces cAMP-responsive element transcriptional activity and phosphoenolpyruvate carboxykinase C gene expression in the liver. *J. Biol. Chem.* **2008**, *283*, 33902–33910. [CrossRef] [PubMed]

46. Ahmed, S.M.; Luo, L.; Namani, A.; Wang, X.J.; Tang, X. Nrf2 signaling pathway: Pivotal roles in inflammation. *Biochim. Et Biophys. Acta. Mol. Basis Dis.* **2017**, *1863*, 585–597. [CrossRef] [PubMed]
47. Kudo, K.; Hagiwara, S.; Hasegawa, A.; Kusaka, J.; Koga, H.; Noguchi, T. Cepharanthine exerts anti-inflammatory effects via NF-kappaB inhibition in a LPS-induced rat model of systemic inflammation. *J. Surg. Res.* **2011**, *171*, 199–204. [CrossRef]
48. Ershun, Z.; Yunhe, F.; Zhengkai, W.; Yongguo, C.; Naisheng, Z.; Zhengtao, Y. Cepharanthine attenuates lipopolysaccharide-induced mice mastitis by suppressing the NF-kappaB signaling pathway. *Inflammation* **2014**, *37*, 331–337. [CrossRef]
49. Murano, M.; Maemura, K.; Hirata, I.; Toshina, K.; Nishikawa, T.; Hamamoto, N.; Sasaki, S.; Saitoh, O.; Katsu, K. Therapeutic effect of intracolonically administered nuclear factor kappa B (p65) antisense oligonucleotide on mouse dextran sulphate sodium (DSS)-induced colitis. *Clin. Exp. Immunol.* **2000**, *120*, 51–58.
50. Chen, G.; Ran, X.; Li, B.; Li, Y.; He, D.; Huang, B.; Fu, S.; Liu, J.; Wang, W. Sodium Butyrate Inhibits Inflammation and Maintains Epithelium Barrier Integrity in a TNBS-induced Inflammatory Bowel Disease Mice Model. *eBioMedicine* **2018**, *30*, 317–325. [CrossRef]
51. Chen, G.; Fu, S.; Feng, W.; Huang, B.; Xu, S.; Wang, W.; Liu, J. AMP010014A09 in Sus Scrofa Encodes an Analog of G Protein-Coupled Receptor 109A, Which Mediates the Anti-Inflammatory Effects of Beta-Hydroxybutyric Acid. *Cell. Physiol. Biochem.* **2017**, *42*, 1420–1430. [CrossRef]

Disclaimer/Publisher's Note: The statements, opinions and data contained in all publications are solely those of the individual author(s) and contributor(s) and not of MDPI and/or the editor(s). MDPI and/or the editor(s) disclaim responsibility for any injury to people or property resulting from any ideas, methods, instructions or products referred to in the content.

molecules

MDPI

Communication

Coptisine Inhibits Influenza Virus Replication by Upregulating p21

Ming-Feng He [1,†], Jian-Hui Liang [2,†], Yan-Ni Shen [2,3,†], Chao-Wei Zhang [4], Kuang-Yang Yang [1], Li-Chu Liu [1], Qian Xie [2], Chun Hu [3], Xun Song [5,*] and Yan Wang [2,*]

1 Foshan Hospital of Traditional Chinese Medicine, Foshan 528000, China; hemf@fshtcm.com.cn (M.-F.H.); yangky@fshtcm.com.cn (K.-Y.Y.); liulc@fshtcm.com.cn (L.-C.L.)
2 Center for Translation Medicine Research and Development, Shenzhen Institutes of Advanced Technology, Chinese Academy of Sciences, Shenzhen 518055, China; jh.liang@siat.ac.cn (J.-H.L.); yn.shen@siat.ac.cn (Y.-N.S.); qian.xie@szu.edu.cn (Q.X.)
3 Key Laboratory of Structure-Based Drug Design & Discovery, Ministry of Education, School of Pharmaceutical Engineering, Shenyang Pharmaceutical University, Shenyang 110016, China; chunhu@syphu.edu.cn
4 School of Pharmaceutical Science, Shenzhen University, Shenzhen 518000, China; 2110245018@email.szu.edu.cn
5 College of Pharmacy, Shenzhen Technology University, Shenzhen 518118, China
* Correspondence: songxun@sztu.edu.cn (X.S.); yan.wang@siat.ac.cn (Y.W.); Tel.: +86-0755-26417985 (Y.W.)
† These authors contributed equally to this work.

Abstract: The activation of innate antiviral immunity is a promising approach for combatting viral infections. In this study, we screened Chinese herbs that activated human immunity and identified coptisine as a potent inhibitor of the influenza virus with an EC_{50} of 10.7 μM in MDCK cells. The time of an addition assay revealed that pre-treatment with coptisine was more effective at reducing viral replication than co-treatment or post-treatment. Our bulk RNA-sequencing data showed that coptisine upregulated the p21 signaling pathway in MDCK cells, which was responsible for its antiviral effects. Specifically, coptisine increased the expression of p21 and FOXO1 in a dose-dependent manner while leaving the MELK expression unchanged. Docking analysis revealed that coptisine likely inhibited MELK activity directly by forming hydrogen bonds with ASP-150 and GLU-87 in the catalytic pocket. These findings suggest that coptisine may be a promising antiviral agent that regulates the p21 signaling pathway to inhibit viral replication.

Keywords: coptisine; influenza virus; antiviral; p21

check for updates

Citation: He, M.-F.; Liang, J.-H.; Shen, Y.-N.; Zhang, C.-W.; Yang, K.-Y.; Liu, L.-C.; Xie, Q.; Hu, C.; Song, X.; Wang, Y. Coptisine Inhibits Influenza Virus Replication by Upregulating p21. *Molecules* **2023**, *28*, 5398. https://doi.org/10.3390/molecules28145398

Academic Editors: Laura De Martino and Keykavous Parang

Received: 19 May 2023
Revised: 9 June 2023
Accepted: 27 June 2023
Published: 14 July 2023

Copyright: © 2023 by the authors. Licensee MDPI, Basel, Switzerland. This article is an open access article distributed under the terms and conditions of the Creative Commons Attribution (CC BY) license (https://creativecommons.org/licenses/by/4.0/).

1. Introduction

The influenza virus, as a member of the Orthomyxoviridae family [1], cause acute respiratory infections in humans and are responsible for 300,000–500,000 deaths each year worldwide, with a mortality rate of approximately 0.3–0.5% [2]. These viruses can be divided into four types (A, B, C, and D) based on their core proteins [3,4]. Seasonal influenza viruses that circulate in the population are typically Type A (H1N1 and H3N2 subtypes) and Type B (Yamagata and Victoria strains). Influenza A virus (IAV) has a large number of hosts in nature and is more prone to mutations or reassortments, resulting in a rapid spread in the population [5]. While vaccination and antiviral drugs are the primary treatments for influenza, the emergence of viral resistance [6,7] and adverse side effects [8,9] have made the discovery of new antiviral drugs a pressing concern.

The principle of action for synthetic antiviral drugs is mainly to target enzymes or proteins in relation to replication and release in the virus [10–12], thereby preventing the virus from continuing to infect other normal cells. Unlike synthetic drugs, traditional Chinese medicine often indirectly inhibits viruses by activating innate antiviral immunity in the host cells [13–15]. Therefore, we screened the effective ingredients of some Chinese herbs that could activate human immunity [16,17], hoping to obtain some phytochemicals with anti-influenza virus activity from them.

In this study, we first identified the anti-IAV activity of coptisine in vitro and elucidated its potential mechanism of action for inhibiting IAV infection. Coptisine (Figure 1A) is a bioactive isoquinoline alkaloid derived from *Coptis chinensis Franch*. The antiviral properties of *Coptis chinensis* have been reported [18], but the bioactive components and pharmacological mechanism need further exploration. We first confirmed the anti-influenza virus activity of coptisine in vitro through a CPE reduction experiment and plaque experiment. Next, we speculated that coptisine could resist influenza infection by activating host immunity and then designed a time addition experiment to confirm our hypothesis. Finally, we elucidated the signaling pathways activated by coptisine at the molecular level through RNA sequencing and a series of biochemical experiments. Our results indicate that coptisine could be used as a MELK inhibitor to manage influenza virus infection.

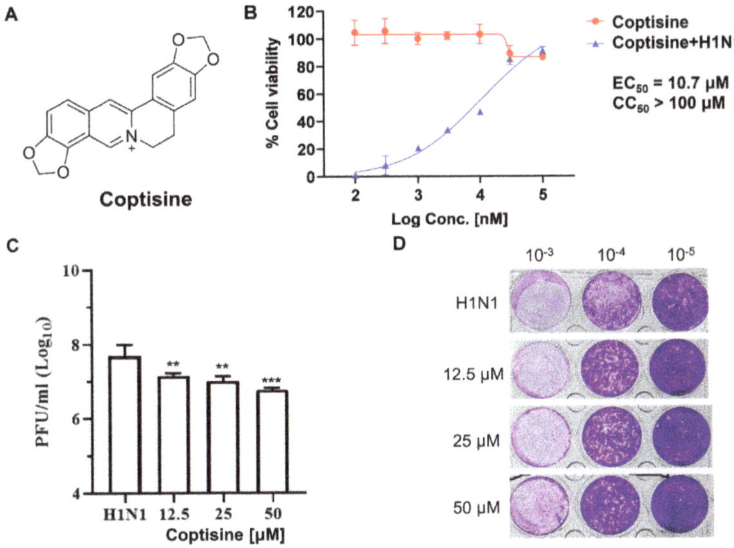

Figure 1. In vitro inhibitory activity of coptisine against H1N1 influenza virus infection. (**A**) Chemical structure of coptisine. (**B**) For the cytopathic effect assay, the H1N1 virus was incubated with coptisine at 4 °C for 30 min before infecting MDCK cells at 37 °C for 2 h. Subsequently, the viral inoculum was removed and replaced by adding coptisine. CPE was observed at 24 h post−incubation by microscopy. For the cytotoxicity test, coptisine was applied to MDCK cells and incubated for 24 h at the indicated concentrations. Then, cell viability was measured by the CCK−8 reagent. The EC_{50} and CC_{50} values were calculated using the inhibitor dose−response function in Prism 5. (**C**) After treating the cells as described in B, the viral titer was detected by a plaque assay and shown as a log value. Bars represent the mean and SD of three biological replicates, with dots representing the values of the replicates. Statistical significance was analyzed with Dunnett's test (** $p < 0.01$; *** $p < 0.005$, compared with H1N1−treated) (**D**) Representative images of viral plaques in each histogram are shown in (**C**).

2. Results

2.1. Antiviral Activity of Coptisine against H1N1 In Vitro

The influenza virus infection could cause typical cell cytopathic effects (CPE) [1]. To determine the antiviral activity of Coptisine, a combination of Coptisine and the H1N1 virus was inoculated into MDCK cells. After the virus was removed, Coptisine was supplemented for 24 h. The results show that Coptisine significantly reduced CPE formation on MDCK cells induced by an H1N1 virus in a dose-dependent manner, with an EC_{50} of 12.04 μM (Figure 1B).

To further evaluate the anti-influenza virus efficacy of Coptisine, we assessed the effect of Coptisine on the virus titer in the cell supernatant using a plaque reduction assay 48 h after administration. The results showed that the plaque of each group treated with different concentrations of Coptisine was significantly reduced compared to the control group, indicating that Coptisine could reduce H1N1 replication in the MDCK cell (Figure 1C,D). These in vitro experiments demonstrate that Coptisine has significant antiviral efficacy against the H1N1 influenza virus.

2.2. *Administering Coptisine before Infection Is More Effective in Exerting Its Antiviral Effects*

Drawing on the traditional Chinese medical science theory and previous research on TCM's antiviral mechanisms [19], we postulated that coptisine's antiviral action involved directly activating the innate antiviral immunity of host cells. We conducted a time-of-addition assay (Figure 2A) using a concentration of 5 × EC_{50} coptisine and designed various medication modes, including pre-treatment, co-treatment, and post-treatment. Our findings demonstrate that coptisine treatment improved the cell viability reduction caused by viral infections at 24-, 48-, and 72 h post-infection (hpi). A microscopic observation (Figure 2B) reveals that the morphology and cell density of coptisine-treated cells were superior to those of the H1N1 infection group at different time points post-infection. The histogram of cell viability (Figure 2C) indicated no significant differences among the three groups at 24 and 48 hpi. However, at 72 h post-infection, the pre-treatment group's cell viability was significantly higher than the other groups, suggesting that administering coptisine before infection was more effective when exerting its antiviral effects. Taken together, our results suggest that coptisine, when added 24 h before infection, could alleviate the influenza virus infection's damage and suppress IAV infection, indicating that coptisine might activate the innate antiviral immunity of host cells.

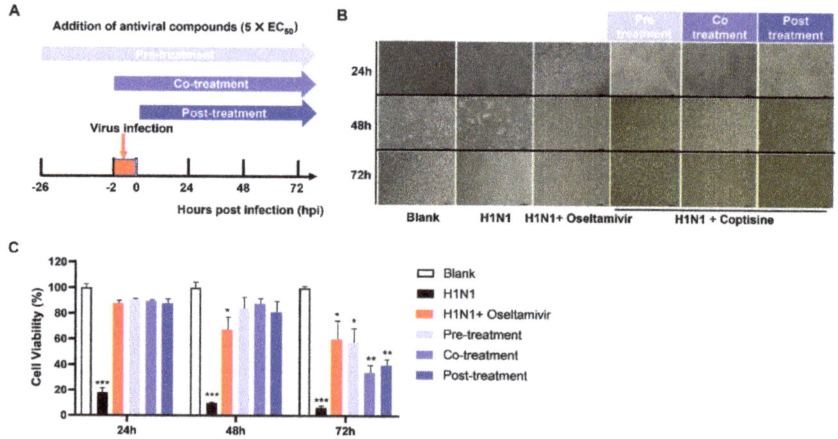

Figure 2. Time−of−addition assay used to identify the steps of the H1N1 virus life cycle, which could be inhibited by coptisine. (**A**) Schematic outline of the time−of−addition assay. MDCK cells were inoculated with H1N1 and treated with ~10 times the EC_{50} of coptisine or oseltamivir. Cell viability was measured at 24, 48, and 72 hpi. (**B**) MDCK cells were infected with H1N1, and oseltamivir or coptisine was added at different time points, as shown in (**A**). Microscopic images of cells were observed at 24, 48, and 72 hpi. (**C**) Measuring cell viability at different time points after treating MDCK cells with different protocols, and the results are presented as histograms. Each bar represents the mean ± SD of three independent experiments. Statistical significance was analyzed with Dunnett's test (* $p < 0.05$; ** $p < 0.01$; *** $p < 0.005$, compared with blank).

2.3. Coptisine Upregulates the Expression of CDKNI1A and FOXO1

After confirming that coptisine could resist influenza viruses by activating the host's innate immune response, we aimed to further explore the relevant genes and signaling pathways involved in the innate immune response against viruses activated by coptisine. Therefore, we performed bulk RNA sequencing to identify the critical genes regulated by coptisine. The volcano plot in Figure 3A displays the distribution of differential gene expressions between the two samples. Among the genes significantly upregulated by coptisine, CDKN1A could be associated with host immunity. The protein p21 encoded by CDKN1A plays a crucial role in maintaining the balance of innate immune activity [20,21]. The study of Ma C et al. also showed that p21 is influenza limiting factor [22]. Figure 3B summarizes the enriched pathways, highlighting pathways related to CDKN1A in red, including the cell cycle, FoxO signaling pathway, Epstein–Barr virus infection, and some cancers. It is worth noting that the FoxO signaling pathway could regulate cellular antiviral response [23]. The heatmap displays that the expression levels of upstream and downstream genes of CDKN1A could be upregulated by coptisine in MDCK cells (Figure 3C). The RNA-Seq data were deposited in GEO (GSE232189). To further demonstrate the above conclusions, we validated the expression of CDKN1A and its upstream regulator, FOXO1, at the mRNA and protein levels. The RT-qPCR results (Figure 3D) show that coptisine significantly upregulated the mRNA expression of CDKN1A at lower concentrations (2.5 μM), but only when the concentration reached 10 μM increased the expression of FOXO1 significantly. The following Western blot (Figure 3E) showed consistent results with RT-qPCR.

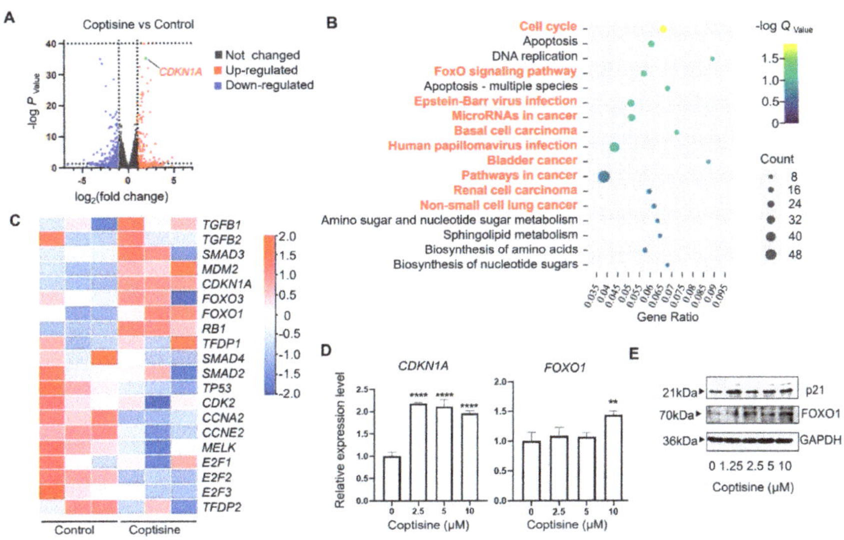

Figure 3. The key genes and signaling pathways modulated by coptisine were explored by bulk RNA sequencing. (**A**) Volcano plot displaying differentially expressed genes between coptisine and control groups (n = 3 biological replicates). Significant genes were called via Cuffdiff. The red dots represent the up−regulated expressed transcripts between coptisine and control; the blue dots represent the transcripts whose expression was down−regulated. (**B**) Significantly enriched modulated KEGG pathways. The depth of the color represents the adjusted p-value, while the area of the circle represents gene counts. The pathway marked in red font is related to *CDKN1A*. (**C**) Heat map of differential expression of genes associated with *CDKN1A* between the control and coptisine groups (n = 3 biologically independent cell samples; FDR < 0.001). Statistical tests were embedded in

Cuffdiff. Each row represents a gene; red means an increased gene expression, blue means decreased gene expression, and the darker the color, the more obvious the trend is. Heatmap analysis shows that coptisine could significantly upregulate or downregulate the expression of certain genes compared to the controls. The RNA-Seq data were deposited in GEO (GSE232189) (**D**,**E**) MDCK was cultured for 24 h in the presence of different concentrations of coptisine. The total RNA was isolated to detect the expression level of *CDKN1A* and *FOXO1* by quantitative PCR (**D**). The total protein was collected for Western blot to quantify p21 and FOXO1 (**E**). Each error bar represents the mean ± SD of triplicate wells assayed for each sample. Statistical significance was analyzed with Dunnett's test (** $p < 0.01$; **** $p < 0.001$, compared with 0 μM).

2.4. Coptisine Might Target against MELK to Inhibit Influenza Virus

MELK (maternal embryonic leucine zipper kinase), a cell cycle-dependent protein kinase belonging to the KIN1/PAR-1/MARK family, is a crucial target of coptisine to inhibit the influenza virus [24]. Previous research has shown that MELK could induce p21 protein expression through the novel substrate FOXO1 after its inhibition [25]. We conducted experiments to determine whether coptisine inhibited MELK expression and found that it did not decrease MELK at the gene and protein levels in MDCK cells treated with coptisine (50 μM) for 24 h (Supplementary Figure S1A,B). However, molecular docking revealed that coptisine could form hydrogen bonds with specific amino acid residues, including ASP-150 and GLU-87, in the catalytic pocket of MELK, indicating coptisine's interaction with MELK (Supplementary Figure S1C). Therefore, we proposed that coptisine potently inhibited MELK activity, inducing FOXO1 and its downstream p21 expression, which could regulate the host's immune system to resist the influenza virus infection (Supplementary Figure S1D).

3. Discussion

This study aimed to identify active compounds from traditional Chinese medicines that could activate the autoimmune system for antiviral drug development. Our results showed that coptisine exhibited strong anti-influenza virus (H1N1) activities in vitro, as confirmed by CPE reduction and plaque assays. The time of the addition assay revealed that coptisine exerted antiviral activity by activating the host's innate antiviral immunity. RNA sequencing analysis indicated that coptisine upregulated the expression of *CDKN1A*: a gene related to the host's immune system. RT-qPCR and Western blot assays further confirmed that coptisine upregulated the expression of CDKN1A and its upstream regulator, FOXO1. Molecular docking results showed that coptisine inhibited MELK activity, an upstream regulator of FOXO1, leading to the upregulation of FOXO1 and p21, and thereby, exerting antiviral activity. In conclusion, our study demonstrated that coptisine, an isoquinoline alkaloid derived from the traditional Chinese herbal medicine *Coptis chinensis Franch*, could effectively improve the host's antiviral ability by regulating the p21 signaling pathway. According to The Plant List (http://www.theplantlist.org/, accessed on 26 June 2023), there are a total of 13 species in the genus *Coptis*, including *Coptis asplenifolia* Salisb., *Coptis chinensis* Franch., *Coptis deltoidea* C.Y.Cheng and P.K.Hsiao, *Coptis japonica* (Thunb.) Makino, *Coptis laciniata* A.Gray, *Coptis minamitaniana* Kadota, *Coptis occidentalis* (Nutt.) Torr. and A.Gray, *Coptis omeiensis* (C.Chen) C.Y.Cheng, *Coptis quinquefolia* Miq., *Coptis quinquesecta* W.T.Wang, *Coptis teeta* Wall., *Coptis trifolia* (L.) Salisb., and *Coptis trifoliolata* (Makino) Makino. If coptisine can be found in these species, these plants have a good chance of exhibiting antiviral activity through the upregulation of p21 in the host cell.

4. Materials and Methods

4.1. Cells and Viruses

Madin-Darby canine kidney (MDCK) cells obtained from ATCC were maintained in Dulbecco's Modified Eagle Medium (DMEM) (Gibco: C11995500BT) and supplemented with 10% inactivated fetal bovine serum (FBS) (Gibco: 10270106) and 1% penicillin/streptomycin

(Gibco: 15140-122) at 37 °C and 5% CO_2. Coptisine was obtained from Alfa Biotechnology Co., Ltd. (Chengdu, China; Lot. AF21091551, purity, >99.5%). The influenza A virus strain A/PR/8/34 was propagated in MDCK cells, and its titer was determined as previously described [16]. The viruses were stored at −80 °C until further use.

4.2. Cytotoxicity Assay

MDCK cells were seeded in 96-well culture plates at a density of 1.0×10^4 cells/well and were cultured at 37 °C with 5% CO_2 overnight. The cells were then treated with different concentrations of coptisine obtained from the National Institutes for Food and Drug Control (Beijing, China) for 24 h. After that, 10 µL of the CCK-8 reagent (Dojindo, Rockville, MD, USA, CK04) was added to each well, followed by 2 h of incubation. The absorbance was measured at 450 nm using a microplate reader (Rayto, Shenzhen, China).

4.3. Cytopathic Effect Assay

MDCK cells were seeded in 12-well plates at a density of 2.0×10^5 cells/well and incubated at 37 °C with 5% CO_2 overnight. The next day, different concentrations of coptisine and A/PR/8/34 (H1N1) (MOI = 0.5) were mixed at 4 °C for 30 min before being added to the MDCK cells at 37 °C for 2 h. After that, the supernatant was removed, and the cells were washed twice with PBS (Solarbio, P1010, Beijing, China). Different concentrations of chrysin were added to the cells, which were then cultured for 24 h at 37 °C. The cytopathic effect was observed under a microscope.

4.4. Plaque Reduction Assay

Confluent MDCK cells in 12-well plates were washed with PBS and inoculated with a gradient dilution of the virus solution (MOI = 0.5) [Opti-MEM (Thermo Fisher Scientific, Catalog number: 31985070, Waltham, MA, USA) with 0.3% BSA and 2.5 µg/mL TPCK-treated trypsin (Sigma–Aldrich, Catalog number: T1426, St. Louis, MO, USA)] for 2 h. After removing the virus inoculum, cell monolayers were overlaid with 1.5 mL of the semisolid medium (1.5% CMCNa:DMEM = 1:1, with 0.3% BSA and 2.5 µg/mL TPCK-treated trypsin) and incubated at 37 °C for 30 h. The cell monolayers were then fixed with 4% paraformaldehyde and stained with 0.2% (*w*/*v*) crystal violet (Sigma, C0775). The size of the plaques was counted and measured using Image J software 1.50 version.

4.5. RNA Sequencing

To perform RNA sequencing, MDCK cells were initially seeded in a 6-well plate and allowed to incubate for 24 h prior to drug treatment. Subsequently, the cells were collected and sent to the BGI Genomics institution for RNA extraction and bulk mRNA sequencing. Libraries were constructed according to the manufacturer's instructions (Illumina, San Diego, CA, USA), and paired-end sequencing was conducted using the Illumina Hiseq2000 sequencer (Illumina, USA). Both library construction and RNA-seq were conducted at BGI. The raw sequencing reads (fastq) quality was checked using SOAPnuke (v1.5.2), and alignment was performed by Bowtie2 (v2.2.5). The expression level of genes was calculated using RSEM (v1.2.12), while DESeq2 (v1.4.5) was used for differential expression analysis with a p-value < 0.05.

4.6. Western Blot Analysis

MDCK cells were treated with drugs according to the protocol and then washed with ice-cold PBS before being lysed with a RIPA buffer (Thermo Scientific, Catalog number: 89900, Waltham, MA, USA) for protein extraction. The extracted protein was quantified using the BCA assay kit (Beyotime, Catalog number: P0012, Shanghai, China) and then mixed with a 5 × protein loading buffer (Meilunbio, Catalog number: MA0003-D, Dalian, China). The samples were boiled at 95 °C for 5 min, and an equivalent quantity of protein samples was loaded onto a 10% sodium dodecyl sulfate-polyacrylamide gel electrophoresis gel (Beyotime, Catalog number: P0012A, Shanghai, China). The protein was subsequently transferred to a polyvinylidene fluoride (PVDF) membrane (Merck Millipore, Catalog number: IPVH15150, Billerica, MA, USA). This nonspecific binding was blocked with a 3% BSA solution, and the membrane was incubated with a primary antibody overnight at 4 °C. The primary antibodies are listed here: Rabbit Anti-MELK antibody (Bioss, Catalog number: bs-12201R, Woburn, MA, USA), Rabbit Anti-p21 antibody (Bioss, Catalog number: bsm-60698R, Woburn, MA, USA), Rabbit Anti-FOXO1 antibody (Bioss, Catalog number: bs-23175R, Woburn, MA, USA), Anti-GAPDH Rabbit polyclonal antibody(Sangon Biotech, Catalog number: D110016, Shanghai, China), and the secondary antibody was the Goat Anti-Rabbit IgG antibody (HRP) (GeneTex, Catalog number: GTX213110, Irvine, CA, USA). After washing with a Tris-buffered solution containing 0.05% Tween 20, the membrane was incubated with secondary antibodies for two hours at room temperature. After three additional washes, the protein bands were visualized using an ECL reagent kit (Solarbio, Catalog number: PE0010, Beijing, China) according to the manufacturer's instructions.

4.7. RNA Extraction and RT-qPCR

To extract the total RNA, cells were harvested, and RNA was isolated using the RNAiso Plus reagent (Takara Bio, Shiga, Japan) following the manufacturer's protocol. The concentration and purity of RNA were measured using NanoDrop 2000 (Thermo Scientific, Wilmington, DE, USA). A total of 2 μg of RNA was reverse transcribed using a PrimeScript RT Master Mix (Takara Bio, Japan). A real-time quantitative polymerase chain reaction (RT-qPCR) was carried out to determine the mRNA levels on a LightCycler96 real-time fluorescence qPCR instrument (Roche, Basle, Switzerland). The reaction mix (10 μL) consisted of forward and reverse primers (0.5 μL each), 3 μL of sterile deionized distilled water, 5 μL of SYBR Green Premix (Accubate Biology, Changsha, China), and cDNA templates (1 μL). GAPDH was used as an internal control. The primer sequences for the target genes (CDKN1A, FOXO1, and MELK) and GAPDH were as follows: GAPDH: forward: 5′-GTCATCATCTCTGCTCCTTCTG-3′, reverse: 5′-GCTGACAATCTTGAGGGAGTT-3′; CDKN1A: forward: 5′-ATCCCTCATGGCAGCAAG-3′, reverse: 5′-CTCGGTGACGAAGTC AAAGT-3′; FOXO1: forward: 5′-ATTCACCCAGCCCAAACTAC-3′, reverse: 5′-GAGTCCT GGTGCACAGTTATAC-3′; MELK: forward: 5′- GTGCTAGAGACAGCCAACAA-3′, reverse: 5′-AGGCGATCCTGGGAAATTATG-3′. All reactions were performed in triplicate. The relative expression was calculated using the $2^{-\Delta\Delta Ct}$ method.

4.8. Molecular Docking

The 3D structure of the MELK protein was downloaded from the PDB database (http://www.rcsb.org/, accessed on 1 March 2023) with the PDB code 4UMQ. AutoDock Tools (version 1.5.6) was used for protein isolation and modification. The docking analysis of the prepared ligands and target proteins was performed using Autodock Vina (The Scripps Research Institute, La Jolla, CA, USA). The molecular library ligands were placed into previously identified binding sites of the target protein using a grid box, and the docking binding energy was predicted. The docking results were visualized using Pymol 1.3 (DeLano Scientific, San Carlos, CA, USA).

4.9. Statistical Analysis

Graph Pad Software 9.0 (San Diego, CA, USA) was used to perform Dunnett's test to determine the statistical significance. The results were presented as the means ± SEM, and the *p*-value less than 0.05 were considered significant.

Supplementary Materials: The following supporting information can be downloaded at: https://www.mdpi.com/article/10.3390/molecules28145398/s1, Figure S1. Coptisine inhibits H1N1 virus infection by targeting MELK.

Author Contributions: M.-F.H. is an expert in antiviral research. He was responsible for the experimental design of Sections 2.1 and 2.2, as well as the funding acquisition. J.-H.L. is an expert in medicinal chemistry. He was responsible for the experimental design and implementation of

Sections 2.1, 2.3 and 2.4, formal analysis, methodology establishment and validation. Y.-N.S. is an expert in pharmacology. She was responsible for data acquisition, formal analysis, methodology establishment, and the first draft of writing. C.-W.Z. is an expert in viruses. He was responsible for the experimental data acquisition and methodology establishment of Sections 2.1 and 2.2. K.-Y.Y. is an expert in physiology. He was responsible for the establishment of the methodology of Section 2.3. L.-C.L. is an expert in pharmacy. He was responsible for the methodology establishment and visualization of Section 2.4. Q.X. is an expert in pharmacy, she was responsible for project management. C.H. is an expert in pharmacy, he was responsible for providing Section 2.4 computing resources. X.S. is an expert in pharmacy, he was in charge of project supervision. Y.W. is an expert in pharmacy, and he was responsible for the conception and supervision of the project. All authors have read and agreed to the published version of the manuscript.

Funding: This work was supported by the Guangdong Basic and Applied Basic Research Fund (2020A1515011342 to Y.W.), the Shenzhen Science and Technology Innovation Fund (JCYJ20190807160601672 to Y.W.), and the Project for High Efficiency and Developed Hospital at Foshan City (202200087 to M.-F.H.; 202000189, 202200089 to K.-Y.Y.).

Institutional Review Board Statement: Not applicable.

Informed Consent Statement: Not applicable.

Data Availability Statement: The RNA-Seq data were deposited in GEO (GSE232189).

Conflicts of Interest: The authors declare no conflict of interest.

References

1. Krammer, F.; Smith, G.J.D.; Fouchier, R.A.M.; Peiris, M.; Kedzierska, K.; Doherty, P.C.; Palese, P.; Shaw, M.L.; Treanor, J.; Webster, R.G.; et al. Influenza. *Nat. Rev. Dis. Prim.* **2018**, *4*, 3. [CrossRef] [PubMed]
2. Petrova, V.N.; Russell, C.A. The evolution of seasonal influenza viruses. *Nat. Rev. Microbiol.* **2018**, *16*, 47–60. [CrossRef]
3. Su, S.; Fu, X.; Li, G.; Kerlin, F.; Veit, M. Novel Influenza D virus: Epidemiology, pathology, evolution and biological characteristics. *Virulence* **2017**, *8*, 1580–1591. [CrossRef] [PubMed]
4. Cheung, T.K.; Poon, L.L. Biology of influenza a virus. *Ann. N. Y. Acad. Sci.* **2007**, *1102*, 1–25. [CrossRef] [PubMed]
5. Shao, W.; Li, X.; Goraya, M.U.; Wang, S.; Chen, J.L. Evolution of Influenza A Virus by Mutation and Re-Assortment. *Int. J. Mol. Sci.* **2017**, *18*, 1650. [CrossRef] [PubMed]
6. Hayden, F.G. Antiviral resistance in influenza viruses—Implications for management and pandemic response. *N. Engl. J. Med.* **2006**, *354*, 785–788. [CrossRef]
7. Ison, M.G. Antivirals and resistance: Influenza virus. *Curr. Opin. Virol.* **2011**, *1*, 563–573. [CrossRef]
8. Izzedine, H.; Launay-Vacher, V.; Deray, G. Antiviral drug-induced nephrotoxicity. *Am. J. Kidney Dis. Off. J. Natl. Kidney Found.* **2005**, *45*, 804–817. [CrossRef]
9. McHutchison, J.G.; Manns, M.P.; Longo, D.L. Definition and management of anemia in patients infected with hepatitis C virus. *Liver Int. Off. J. Int. Assoc. Study Liver* **2006**, *26*, 389–398. [CrossRef]
10. Tao, J.; Wang, H.; Wang, W.; Mi, N.; Zhang, W.; Wen, Q.; Ouyang, J.; Liang, X.; Chen, M.; Guo, W.; et al. Binding mechanism of oseltamivir and influenza neuraminidase suggests perspectives for the design of new anti-influenza drugs. *PLoS Comput. Biol.* **2022**, *18*, e1010343. [CrossRef] [PubMed]
11. Kausar, S.; Said Khan, F.; Ishaq Mujeeb Ur Rehman, M.; Akram, M.; Riaz, M.; Rasool, G.; Hamid Khan, A.; Saleem, I.; Shamim, S.; Malik, A. A review: Mechanism of action of antiviral drugs. *Int. J. Immunopathol. Pharmacol.* **2021**, *35*, 20587384211002621. [CrossRef] [PubMed]
12. Świerczyńska, M.; Mirowska-Guzel, D.M.; Pindelska, E. Antiviral Drugs in Influenza. *Int. J. Environ. Res. Public Health* **2022**, *19*, 3018. [CrossRef] [PubMed]
13. Runfeng, L.; Yunlong, H.; Jicheng, H.; Weiqi, P.; Qinhai, M.; Yongxia, S.; Chufang, L.; Jin, Z.; Zhenhua, J.; Haiming, J.; et al. Lianhuaqingwen exerts anti-viral and anti-inflammatory activity against novel coronavirus (SARS-CoV-2). *Pharmacol. Res.* **2020**, *156*, 104761. [CrossRef]
14. Lee, D.Y.W.; Li, Q.Y.; Liu, J.; Efferth, T. Traditional Chinese herbal medicine at the forefront battle against COVID-19: Clinical experience and scientific basis. *Phytomedicine* **2021**, *80*, 153337. [CrossRef]
15. Mohammed, F.S.; Uysal, İ.; Sevindik, M. A Review on Antiviral Plants Effective against Different Virus Types. *Prospect. Pharm. Sci.* **2023**, *21*, 1–21. [CrossRef]
16. Liu, Y.; Song, X.; Li, C.; Hu, H.; Li, W.; Wang, L.; Hu, J.; Liao, C.; Liang, H.; He, Z.; et al. Chrysin Ameliorates Influenza Virus Infection in the Upper Airways by Repressing Virus-Induced Cell Cycle Arrest and Mitochondria-Dependent Apoptosis. *Front. Immunol.* **2022**, *13*, 872958. [CrossRef]
17. Zhu, X.; Hu, Z.; Yu, T.; Hu, H.; Zhao, Y.; Li, C.; Zhu, Q.; Wang, M.; Zhai, P.; He, L.; et al. The Antiviral Effects of Jasminin via Endogenous TNF-alpha and the Underlying TNF-alpha-Inducing Action. *Molecules* **2022**, *27*, 1598. [CrossRef] [PubMed]

18. Lee, B.H.; Chathuranga, K.; Uddin, M.B.; Weeratunga, P.; Kim, M.S.; Cho, W.K.; Kim, H.I.; Ma, J.Y.; Lee, J.S. Coptidis Rhizoma extract inhibits replication of respiratory syncytial virus in vitro and in vivo by inducing antiviral state. *J. Microbiol.* **2017**, *55*, 488–498. [CrossRef]
19. Zhao, J.; Tian, S.; Lu, D.; Yang, J.; Zeng, H.; Zhang, F.; Tu, D.; Ge, G.; Zheng, Y.; Shi, T.; et al. Systems pharmacological study illustrates the immune regulation, anti-infection, anti-inflammation, and multi-organ protection mechanism of Qing-Fei-Pai-Du decoction in the treatment of COVID-19. *Phytomedicine* **2021**, *85*, 153315. [CrossRef]
20. Balomenos, D.; Martín-Caballero, J.; García, M.I.; Prieto, I.; Flores, J.M.; Serrano, M.; Martínez, A.C. The cell cycle inhibitor p21 controls T-cell proliferation and sex-linked lupus development. *Nat. Med.* **2000**, *6*, 171–176. [CrossRef]
21. Chen, H.; Li, C.; Huang, J.; Cung, T.; Seiss, K.; Beamon, J.; Carrington, M.F.; Porter, L.C.; Burke, P.S.; Yang, Y.; et al. CD4+ T cells from elite controllers resist HIV-1 infection by selective upregulation of p21. *J. Clin. Investig.* **2011**, *121*, 1549–1560. [CrossRef] [PubMed]
22. Ma, C.; Li, Y.; Zong, Y.; Velkov, T.; Wang, C.; Yang, X.; Zhang, M.; Jiang, Z.; Sun, H.; Tong, Q.; et al. p21 restricts influenza A virus by perturbing the viral polymerase complex and upregulating type I interferon signaling. *PLoS Pathog.* **2022**, *18*, e1010295. [CrossRef] [PubMed]
23. Lei, C.Q.; Zhang, Y.; Xia, T.; Jiang, L.Q.; Zhong, B.; Shu, H.B. FoxO1 negatively regulates cellular antiviral response by promoting degradation of IRF3. *J. Biol. Chem.* **2013**, *288*, 12596–12604. [CrossRef] [PubMed]
24. Jiang, P.; Zhang, D. Maternal embryonic leucine zipper kinase (MELK): A novel regulator in cell cycle control, embryonic development, and cancer. *Int. J. Mol. Sci.* **2013**, *14*, 21551–21560. [CrossRef] [PubMed]
25. Matsuda, T.; Kato, T.; Kiyotani, K.; Tarhan, Y.E.; Saloura, V.; Chung, S.; Ueda, K.; Nakamura, Y.; Park, J.H. p53-independent p21 induction by MELK inhibition. *Oncotarget* **2017**, *8*, 57938–57947. [CrossRef]

Disclaimer/Publisher's Note: The statements, opinions and data contained in all publications are solely those of the individual author(s) and contributor(s) and not of MDPI and/or the editor(s). MDPI and/or the editor(s) disclaim responsibility for any injury to people or property resulting from any ideas, methods, instructions or products referred to in the content.

molecules

MDPI

Article

Creating a Vaccine-like Supplement against Respiratory Infection Using Recombinant *Bacillus subtilis* Spores Expressing SARS-CoV-2 Spike Protein with Natural Products

Ben Chung-Lap Chan [1], Peiting Li [1], Miranda Sin-Man Tsang [1,2], Johnny Chun-Chau Sung [3], Keith Wai-Yeung Kwong [3], Tao Zheng [1], Sharon Sze-Man Hon [1,4], Ching-Po Lau [1], Wen Cheng [1], Fang Chen [1], Clara Bik-San Lau [1], Ping-Chung Leung [1] and Chun-Kwok Wong [1,4,5,*]

1 Institute of Chinese Medicine and State Key Laboratory of Research on Bioactivities and Clinical Applications of Medicinal Plants, The Chinese University of Hong Kong, Hong Kong, China; benchan99@cuhk.edu.hk (B.C.-L.C.); peiting@link.cuhk.edu.hk (P.L.); miranda.tsang@rmit.edu.au (M.S.-M.T.); zhengtao@cuhk.edu.hk (T.Z.); sharonhon@link.cuhk.edu.hk (S.S.-M.H.); chingpolau@cuhk.edu.hk (C.-P.L.); wencheng@cuhk.edu.hk (W.C.); fangchen@cuhk.edu.hk (F.C.); claralau@cuhk.edu.hk (C.B.-S.L.); pingcleung@cuhk.edu.hk (P.-C.L.)

2 China-Australia International Research Centre for Chinese Medicine, School of Health and Biomedical Sciences, STEM College, RMIT University, Bundoora, VIC 3083, Australia

3 Research Department, DreamTec Cytokines Limited, Hong Kong, China; johnnysung@dreamtec.hk (J.C.-C.S.); keithkwong@dreamtec.hk (K.W.-Y.K.)

4 Department of Chemical Pathology, The Chinese University of Hong Kong, Prince of Wales Hospital, Shatin, NT, Hong Kong, China

5 Li Dak Sum Yip Yio Chin R & D Centre for Chinese Medicine, The Chinese University of Hong Kong, Hong Kong, China

* Correspondence: ck-wong@cuhk.edu.hk

check for updates

Citation: Chan, B.C.-L.; Li, P.; Tsang, M.S.-M.; Sung, J.C.-C.; Kwong, K.W.-Y.; Zheng, T.; Hon, S.S.-M.; Lau, C.-P.; Cheng, W.; Chen, F.; et al. Creating a Vaccine-like Supplement against Respiratory Infection Using Recombinant *Bacillus subtilis* Spores Expressing SARS-CoV-2 Spike Protein with Natural Products. *Molecules* **2023**, *28*, 4996. https://doi.org/10.3390/molecules28134996

Academic Editors: Xun Song, Chenyang Li and Yifu Guan

Received: 2 June 2023
Revised: 19 June 2023
Accepted: 24 June 2023
Published: 26 June 2023

Copyright: © 2023 by the authors. Licensee MDPI, Basel, Switzerland. This article is an open access article distributed under the terms and conditions of the Creative Commons Attribution (CC BY) license (https://creativecommons.org/licenses/by/4.0/).

Abstract: Vaccination is the most effective method of combating COVID-19 infection, but people with a psychological fear of needles and side effects are hesitant to receive the current vaccination, and alternative delivery methods may help. *Bacillus subtilis*, a harmless intestinal commensal, has recently earned a strong reputation as a vaccine production host and delivery vector, with advantages such as low cost, safety for human consumption, and straightforward oral administration. In this study, we have succeeded generating "S spores" by engineering *B. subtilis* with spore coat proteins resembling the spike (S) protein of the ancestral SARS-CoV-2 coronavirus. With the addition of two immunostimulating natural products as adjuvants, namely *Astragalus membranaceus* (Fisch.) Bge (AM) and *Coriolus versicolor* (CV), oral administration of S spores could elicit mild immune responses against COVID-19 infection without toxicity. Mucosal IgA against the S protein was enhanced by co-feeding with AM and CV in an S spores-inoculated mouse model. Faster and stronger IgG responses against the S protein were observed when the mice were fed with S spores prior to vaccination with the commercial COVID-19 vaccine CoronaVac. In vitro studies demonstrated that AM, CV, and *B. subtilis* spores could dose-dependently activate both macrophages and dendritic cells by secreting innate immunity-related IL-1β, IL-6, and TNF-α, and some other proinflammatory chemokines and cytokines. In conclusion, the combination of S spores with AM and CV may be helpful in developing a vaccine-like supplement against respiratory infection.

Keywords: *Bacillus subtilis*; cytokines; dendritic cells; macrophages; SARS-CoV-2; *Astragalus membranaceus* (Fisch.) Bge (AM); *Coriolus versicolor* (CV)

1. Introduction

The Coronavirus Disease 2019 (COVID-19) pandemic has been ravaging the world for more than three years. Despite some vaccines having been developed and applied to combat and prevent the disease, COVID-19 continues to rage and remains unpredictable due to the emergence of the Omicron variant and its descendant subvariants with increasing infectivity [1–4]. The BQ and XBB subvariants of severe acute respiratory syndrome coronavirus 2 (SARS-CoV-2) Omicron are now spreading rapidly, possibly due to spike mutations that alter the antibody required by the host, resulting in immune escape [5]. Interventions are urgently needed to prevent the new emerging variant and/or to treat the disease. Genetic engineering of the spores of *Bacillus subtilis* (*B. subtilis*) has raised interest in vaccine development. The *B. subtilis* spores are desirable as a carrier for vaccines because the FDA generally recognizes them as safe, and the spores can act as an adjuvant to boost immune response, especially innate immunity [6,7]. *B. subtilis* has been studied for decades and developed as a bioactive supplement for immunomodulation [8]. Specific proteins with known targets for the induction of adaptive immune responses can be used as antigens for vaccine development. The ease of administration is another major advantage for the development of oral vaccination against infectious diseases. We have successfully generated *B. subtilis* spores expressing the SARS-CoV-2 receptor binding domain (RBD) of the spike (S) protein genetically fused to the surface-exposed spore coat proteins CotC of the *B. subtilis* spores [9], and a similar *B. subtilis* construct has recently been produced by another group [10].

To further enhance the immunoreactivity of *B. subtilis* in vaccine development, an additional agent may be essential, and beta (β)-glucan enriched natural products may be a good choice. In our earlier study using a murine subcutaneous immunization model with Globo H and GD3 carbohydrates conjugated to keyhole limpet hemocyanin (KLH) as an immunogen mixed with a panel of β-glucan-enriched natural products as immunological adjuvants, we found that *Astragalus membranaceus* (Fisch.) Bge (AM) and *Coriolus versicolor* (CV) are the most active immune stimulants with adjuvant activity [11–13]. We found that the bioactive polysaccharides isolated from AM and CV exhibited significant immunomodulatory effects by stimulating the proliferation of human peripheral blood mononuclear cells (PBMC) and enhancing innate immunity-related IL-1β and TNF-α production [14–17]. The immunostimulatory properties of AM and CV polysaccharides may promote trained immunity [18] with a long-term enhancement of innate immune responses and induce heterologous protection against infection. The objectives of this study were to investigate the immunopotentiating effects of *B. subtilis* spores, AM and CV, by examining the (i) in vitro innate immunity-boosting effects using human macrophages and intestinal HT-29 cells co-culture and monocyte-derived dendritic cells (DC), (ii) the in vivo oral vaccination-like activities using a mouse immunization model [19,20], and (iii) the use of *B. subtilis* spores, S spores, AM, and CV as an oral supplement for CoronaVac vaccination.

2. Results

2.1. MTT Cytotoxicity Test

No cytotoxicity was observed in human PBMC treated with AM, CV (0.1 to 10 mg/mL), native, and S protein *B. subtilis* spores (5 to 500 μg/mL) for 48 h (Figure 1).

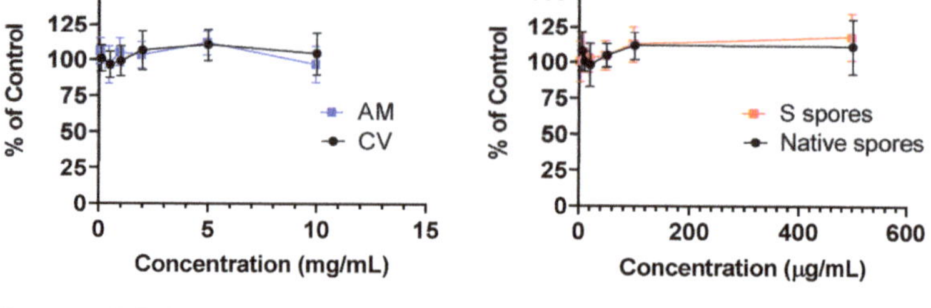

Figure 1. Cellular toxicity (MTT assay) of *Astragalus membranaceus* (Fisch.) Bge (AM), *Coriolus versicolor* (CV), native *B. subtilis* spores (Native spores), or genetically engineered *B. subtilis* (S spores) on human peripheral blood mononuclear cells (PBMC) (n = 4).

2.2. Co-Culture of Human Monocyte-Derived Macrophages with Intestine HT-29 Cells and Dendritic Cells (DC)

Human intestinal epithelial HT-29 cells were cultured to confluence, then human monocyte-derived macrophage suspensions (5×10^5/mL) were applied to HT-29 cells in the culture plate. The epithelial cells–monocyte co-cultures were incubated with or without native spores, S spores, AM, and CV, for a further 18 h. The concentrations of proinflammatory cytokines (IL-1β, IL-6, IL-8, IL-12, IFN-γ, and TNF-α) of innate immunity in the culture supernatant were quantitated (Figure 2). The production of IL-1β, IL-6, IL-8, IL-10, IFN-γ, and TNF-α, was dose-dependently increased when cells were incubated with AM, CV, or spores (Figure 2). The stimulatory effects of *B. subtilis* spores were more potent than AM and CV. CV (100–1000 μg/mL) produced significantly higher concentrations of IL-1β, IL-10, and TNF-α when compared to AM. Compared with macrophages and HT-29 culture separately without treatment, the cytokine production profiles were similar to the co-culture of macrophages with HT-29 cells (Figure A5). Results can, therefore, exclude the possibility of the effect of HT29 cells on macrophages in co-culture.

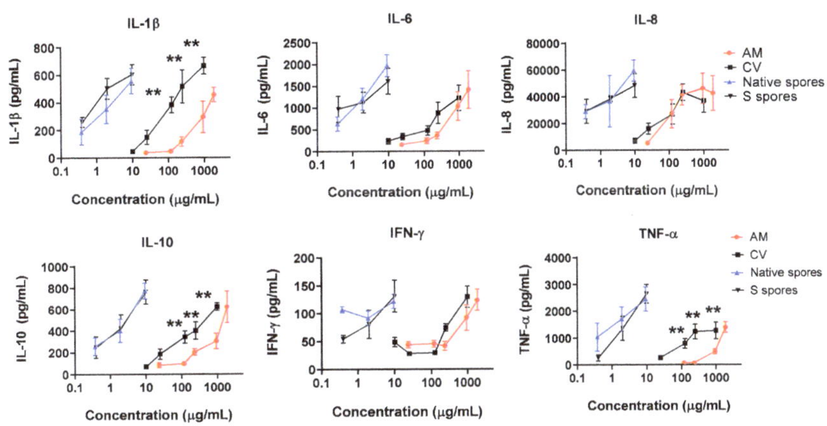

Figure 2. The immunostimulating effects of *Astragalus membranaceus* (Fisch.) Bge (AM), *Coriolus versicolor* (CV), native *B. subtilis* spores (native spores) or genetically engineered *B. subtilis* (S spores) on the production of proinflammatory cytokines (IL-1β, IL-6, IL-8, IL-10, IFN-γ, and TNF-α) from human macrophages and HT-29 cells co-culture (n = 4). ** $p < 0.01$ comparing to AM group.

IL-12 is one of the major cytokines secreted by activated DC. A significant and dose-dependent increase in IL-12, IL-1β, IL-10, and TNF-α was observed when cells were incubated with AM, CV, or spores. A small increase in anti-viral IFN-γ production was observed when a high dose of AM was used (2 mg/mL) (Figure 3). Similar to the co-culture of macrophage and HT-29 cells, the stimulatory effects of *B. subtilis* spores were more potent than AM and CV. CV (100–1000 μg/mL) produced significantly higher concentrations of IL-1β, IL-10, IL-12, and TNF-α when compared to AM. Taken together, *B. subtilis* or AM/CV could promote the immune response and has an immune-adjuvant effect on macrophages and DC, which are the principal antigen-presenting cells to B cells for antibody production in humoral immunity.

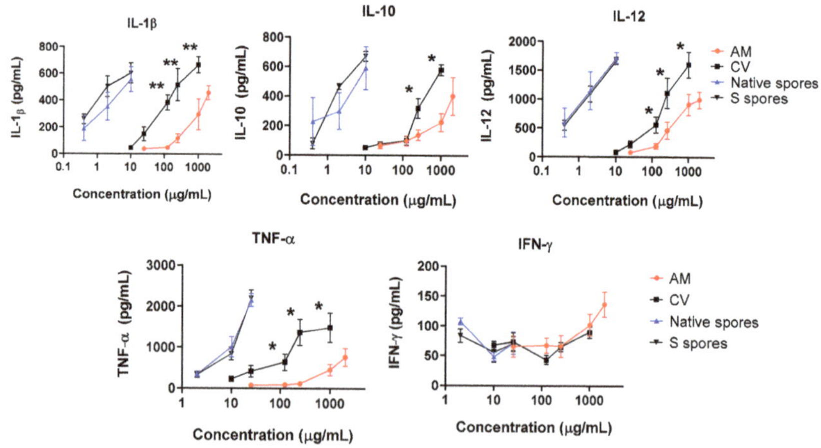

Figure 3. The immunostimulating effects of *Astragalus membranaceus* (Fisch.) Bge (AM), *Coriolus versicolor* (CV), native *B. subtilis* spores (native spores), or genetically engineered *B. subtilis* (S spores) on proinflammatory cytokines (IL-1β, IL-10, IL-12, IFN-γ, and TNF-α) productions from human monocyte-derived dendritic cells (DC) (n = 4). * $p < 0.05$ and ** $p < 0.01$ comparing with AM group.

2.3. Oral Vaccination of *B. subtilis* Spores, AM, and CV

Mice were well-tolerated to the *B. subtilis* or AM/CV after 3 feeding sessions. Clinical changes (body weight, hair loss, body temperature) and eating habits (food and water intake, diarrhea) were not significantly affected by spores and AM/CV treatment. After the first oral treatment with spores and AM/CV, a relatively higher serum antibody IgM titer against the SARS-CoV-2 S protein was detected in mice that received native spores and AM/CV when compared with the mice fed with S spores with or without co-treatment with AM/CV (Figure 4a). After oral vaccination 2 times, a small increase in serum IgG titer against the SARS-CoV-2 S protein was observed in mice receiving S protein-engineered *B. subtilis* together with AM/CV. Very small increases in IgA production (OD 0.05) against the SARS-CoV-2 S protein were detected in the sera of mice from all groups. Adjuvant activities on the serum antibody production were not observed in the groups receiving S spores and co-treated with AM/CV. From the feces collected from different time points (Figure 4b), significantly higher levels of IgA against S protein were observed from mice fed with S protein-engineered *B. subtilis* alone, co-treatment with AM/CV compared with PBS group and mice fed only with native spores or AM/CV after first oral vaccination (all $p < 0.05$). For intestine lavage of the sacrificed mice (Figure 4c), a stronger IgA secretion against spike protein was observed in the mice from the S spores treatment group when compared with the PBS and AM/CV treatment-only groups, and one mouse produced a strong IgA response against SARS-CoV-2 S protein (OD = 1.4). For the S spores and AM/CV co-treatment group, an enhanced effect on IgA production was observed in all 5 mice (OD ranged from 0.4–1.5). A small but detectable IgM response against the S protein (not statistically significant) was observed in the mice after the first and second oral vaccination with the native or S spores. This may be due to the cross-reactivity between the surface antigens of the spores and the ancestral SARS-CoV-2 S protein.

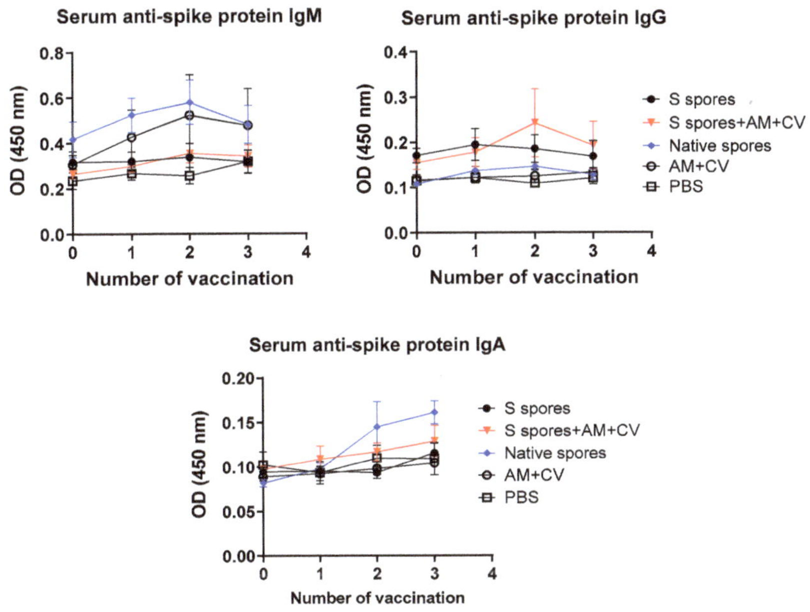

(**a**) serum IgA, IgM, and IgG against COVID-19 spike proteins

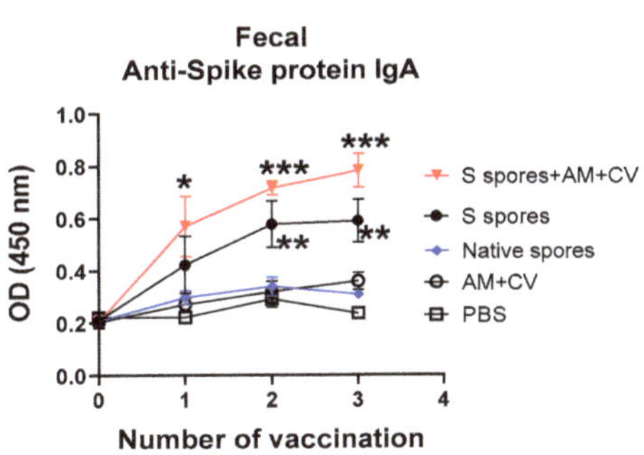

(**b**) Fecal

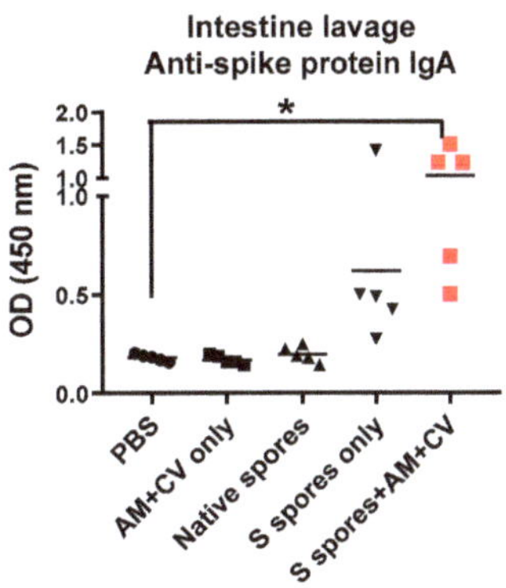

(**c**) intestine lavage IgA against COVID-19 spike proteins

Figure 4. Summary of the pilot animal study. (**a**) Serum IgM, IgG, and IgA against SARS-CoV-2 spike protein from mice (n = 5) orally administrated with *Astragalus membranaceus* (Fisch.) Bge (AM), *Coriolus versicolor* (CV), native *B. subtilis* spores (native spores), or genetically engineered *B. subtilis* (S spores) or PBS (control) for three courses (days 1–4, days 14–17, and days 28–31). The presence of specific IgA against SARS-CoV-2 Spike protein in (**b**) feces and (**c**) intestinal lavage. * $p < 0.05$; ** $p < 0.01$ and *** $p < 0.001$ compared to control groups.

The serum cytokine profiles of the mice (IL-1β, IL6, IL-10, GM-CSF, IFN-γ, and TNF-α) for the different treatment groups are shown in Figure 5. In the S spores group, the serum concentration of IL-1β in one mouse was 490 pg/mL, which was 2-fold higher than the PBS group. Overall, a significant increase in IL-1β production was observed in the group of mice treated with S spores only compared to the control group. The increase in cytokine secretion was not observed in the mice treated with S spores and AM/CV.

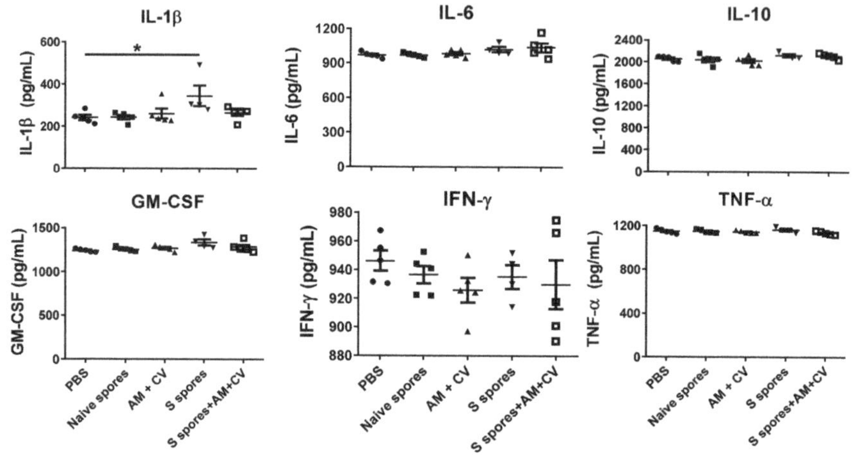

Figure 5. Serum cytokine profile of mice (n = 5) orally treated with *Astragalus membranaceus* (Fisch.) Bge (AM), *Coriolus versicolor* (CV), native *B. subtilis* spores (native spores), or genetically engineered *B. subtilis* (S spores) or PBS (control) for three courses (days 1–4, days 14–17, and days 28–31). * $p < 0.05$.

2.4. Toxicity Studies

Biochemical markers (serum concentrations of the liver enzyme aspartate aminotransferase (AST)) in mice from different groups were measured to determine whether there was any liver function impairment caused by *B. subtilis* spores, AM, and CV (Figure 6) compared to the PBS group. There was no significant difference in the serum concentrations of liver AST in all treatment groups. It was concluded that the liver functions of the mice were not affected by oral administration of *B. subtilis* spores, AM, or CV.

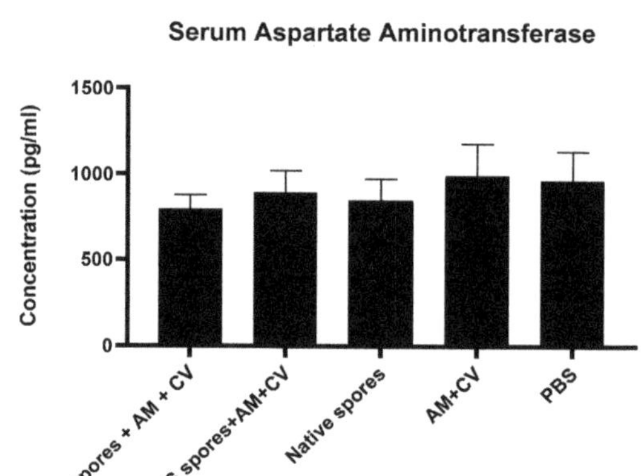

Figure 6. The concentrations of serum aspartate aminotransferase (AST) from mice treated with orally administrated *Astragalus membranaceus* (Fisch.) Bge (AM), *Coriolus versicolor* (CV), native *B. subtilis* spores (native spores), or genetically engineered *B. subtilis* (S spores) or PBS (control) for three courses (days 1–4, days 14–17, and days 28–31). Results were interpolated from the AST standard curves and corrected for sample dilution. The interpolated values corrected for dilution factor were plotted (mean +/− SEM, n = 5).

The histopathological changes of (a) the small intestine, (b) the liver, and (c) the lung at the terminal stage are summarized in Figure 7. Compared with the non-treated group, inflammatory cell infiltrates, epithelial changes, disrupted epithelial barrier, and alterations in the overall mucosal architecture, presence of ulcerations, granulation tissue, irregular crypts, or crypt loss were not observed in the small intestine of any of the treatment groups. In the liver, hepatocellular morphology was similar in all treatment groups compared with the untreated group. Inflammation, hemorrhage, and architectural disruption of the liver were not observed. In lung tissue sections, the surrounding lung parenchyma was similar in all groups. Only mild infiltration of immune cells was observed in the groups treated with S spores and S spores + AM + CV when compared with the control group, but no bronchial epithelial damage, tissue necrosis, hemorrhage, or lesions were observed in these two treatment groups.

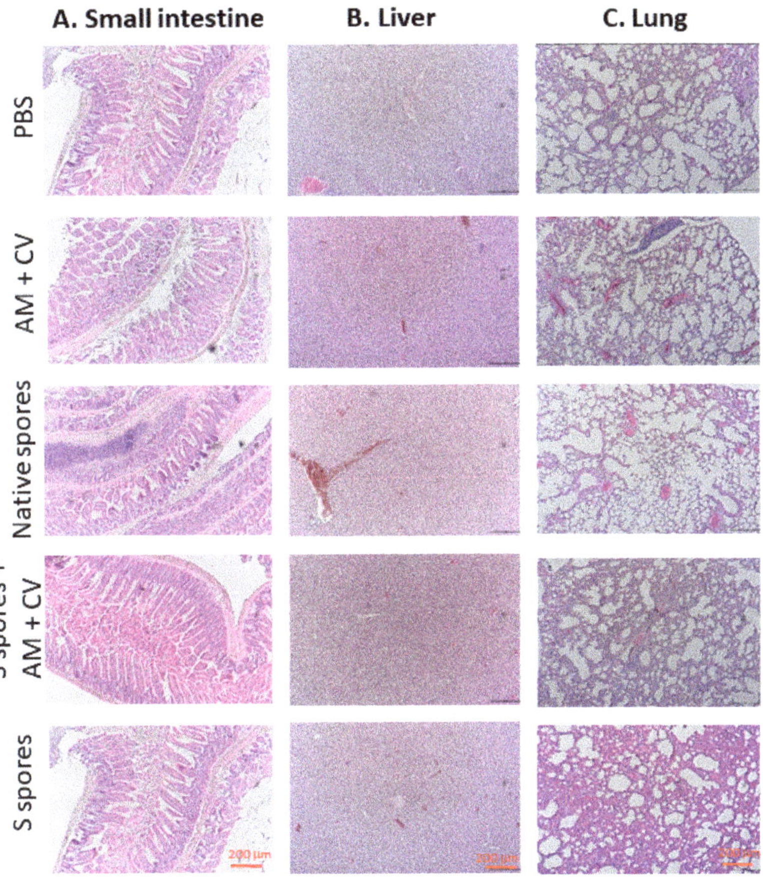

Figure 7. Representative hematoxylin and eosin (H & E) staining of (**A**) small intestine, (**B**) liver, and (**C**) lung at the terminal stage (magnification: 100×) of mice treated with native *B. subtilis* spores, S spores, AM and CV, or PBS (control) after three courses (days 1–4, days 14–17, and days 28–31).

2.5. Adjuvant Effects of B. subtilis Spores, AM/CV on CoronaVac Vaccine

The antibody production (IgA, IgM, and IgG against SARS-CoV-2 spike (S) and nucleocapsid (N) proteins) in the serum of all tested groups are shown in Figure 8. The antibody levels (IgA, IgM, and IgG against SARS-CoV-2 S and N proteins) of all groups tested were low on Day 1. After the first CoronaVac vaccination, a slight increase in IgM against both N and S proteins was observed in all vaccinated groups with no significant difference (all

$p > 0.05$). For IgG antibodies against the N protein, the activities were low compared to the sera IgG before vaccination. A significant increase in IgG production against the S protein was observed only in the S spores-fed group ($p < 0.05$), suggesting that oral administration of S spores 14 days before CoronaVac could enhance the production of antibodies against the S protein of SARS-CoV-2. In the mice fed with AM and CV, IgM and IgG antibody activities against both N and S proteins were similar to those in the group that received CoronaVac alone. Notably, one mouse in the AM/CV group produced a relatively higher level of IgA against N protein after the first and second CoronaVac vaccinations. After the second CoronaVac vaccination, higher concentrations of IgG against S protein were observed in all vaccinated groups compared to the serum activities of the corresponding groups after the first vaccination. For N protein, higher levels of IgG against N protein were observed in both the S spores and AM/CV groups compared to the CoronaVac alone group. The mice sera were further tested for neutralizing antibody activities against SARS-CoV-2 using a commercially available SARS-CoV-2 neutralizing antibody ELISA assay (Figure 8b). After the first vaccination of CoronaVac, a significantly stronger neutralization activity was observed in the S spores-fed group (% neutralization = 22.2%) ($p < 0.05$) when compared with the other two groups: CoronaVac +AM/CV and CoronaVac only (4% and 6.2%). In summary, a significantly higher concentration of neutralizing IgG against S protein could be induced in mice with CoronaVac vaccination with prior oral administration of S spores.

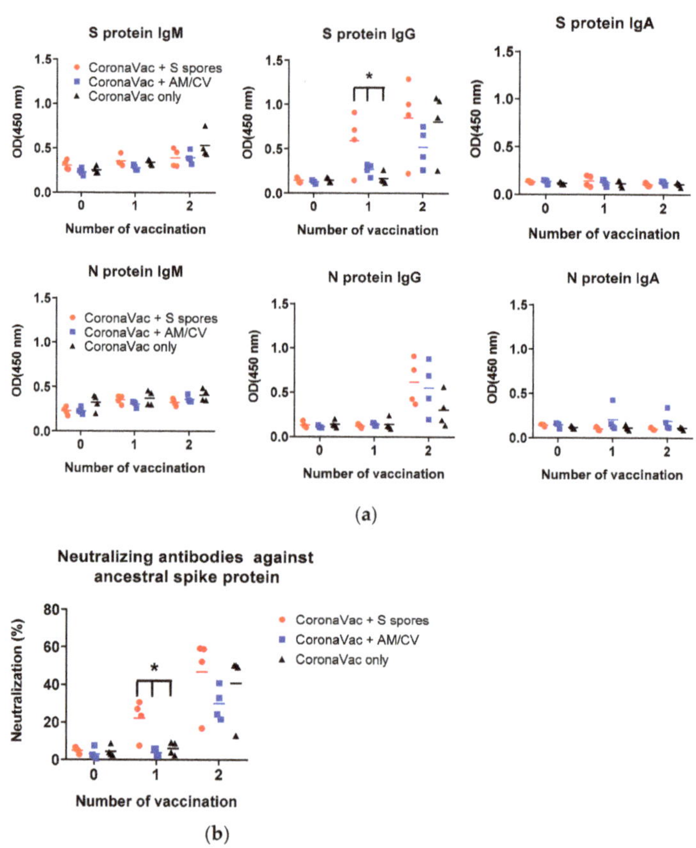

Figure 8. Summary of the CoronaVac animal study. (**a**) Serum (1:50) IgM, IgG, and IgA and (**b**) serum-neutralizing antibodies against ancestral SARS-CoV-2 spike protein from mice (n = 4) orally administered

with *Astragalus membranaceus* (Fisch.) Bge (AM) and *Coriolus versicolor* (CV), genetically engineered *B. subtilis* (S spores), or PBS (control) for 2 courses (days 1–4 and days 14–17). The % neutralization of the positive control serum was 62.8 ± 0.3%. * $p < 0.05$.

3. Discussion

In SARS-CoV-2 infection, innate immunity-related monocytes and macrophages are also responsible for early pathogen recognition, initiation and resolution of inflammation, and repair of tissue damage [21]. Previous findings in the field of immune memory have shown that B and T cell-mediated adaptive immunity after infection is enhanced by "Trained Immunity" [22]. This effect has been investigated for the tuberculosis vaccine strain *Bacillus Calmette-Guerin* (BCG) [23]. Natea developed the concept of "Trained Immunity" based on epidemiological observations that people who have received commonplace vaccinations, such as BCG against *Mycobacterium tuberculosis* infection, tend to be more resistant to other infections. Using severe combined immunodeficiency (SCID) mice infected with *Candida albicans*, it was found that they could be protected against reinfection in a monocyte-dependent manner [24]. It is suggested that this protective effect is mediated by a BCG-induced increase in innate immune cell function with several distinct features, including higher proinflammatory cytokine responses to unrelated secondary pathogens and activation of the innate immune effector cells such as macrophages and DC. This immune process is mediated by transcriptomic and epigenetic changes in myeloid cells, such as monocytes and macrophages [25]. Monocytes can develop immunological memory, a functional feature widely recognized as innate immune training, to distinguish it from memory in adaptive immune cells. Upon a secondary immune challenge, either homologous or heterologous, trained monocytes/macrophages exhibit a more robust production of proinflammatory cytokines, such as IL-1β, IL-6, and TNF-α, than untrained monocytes [26]. The functional reprogramming of monocytes leads to increased innate immunity-related cytokine and chemokine production that could be induced by the bioactive β-glucans of the fungal cell wall [27].

SARS-CoV-2 infection is closely related to the innate immune system and adaptive immune system. DCs are one of the most important cells in generating immune responses, linking innate immunity and adaptive immunity in viral diseases [28]. DC has also been shown to exhibit immune memory responses [29]. DCs are the most potent professional antigen-presenting cells involved in the antibody production process. In our study, we have shown that *B. subtilis* spores, AM, and CV could stimulate DCs, monocytes, and macrophages to produce IL-1β, IL-6, TNF-α, and other cytokines and chemokines for immunomodulation. The combination of S protein-engineered *B. subtilis*, AM, and CV may produce better outcomes in developing an oral vaccine-like supplement against COVID-19 by boosting the memory response and immune activities of both monocytes and dendritic cells in innate immunity.

Oral vaccinations with *B. subtilis* spores expressing immunogens/antigens for various infectious diseases have been shown to be effective in producing serum IgG and mucosal IgA against the pathogens in pigs [30], mice, and guinea pigs [19,31,32]. Interestingly, the antibody titers in these published studies were quite similar, and only relatively weak antibody responses were elicited (OD < 1.0). The results of our animal studies with *B. subtilis* spores expressing S proteins alone were similar to those published. Oral administration of β-glucan enriched CV and AM extracts enhanced mucosal IgA production and could not be achieved with either CV or AM alone. In our case, the antibody immune responses elicited by oral vaccination with *B. subtilis* spores with or without CV/AM extracts against COVID-19 were mild compared to the conventionally used vaccines such as CoronaVac with intramuscular injection. When mice were fed with S spores prior to CoronaVac vaccination, faster and stronger IgG responses to S protein were observed. The immunogenicity of the S spores with or without AM+CV alone may not be strong enough to induce a systemic IgG immune response against the S protein. However, the mucosal immune response induced by S spores could produce a faster and stronger neutralizing

IgG response against the S protein prior to vaccination with the commercial COVID-19 vaccine CoronaVac, thereby producing stronger immunogenicity and IgG response against the ancestral SARS-CoV-2 coronavirus.

With the increasing popularity of probiotics for health promotion, oral probiotic vaccination should be a means to reach those groups. Probiotics have great potential to be incorporated into oral vaccine delivery systems, which may facilitate the induction of mucosal immunity without latent risks of pathogenicity [33]. In addition to AM/CV, β-glucan-enriched yeasts with immunostimulatory properties have recently been tested as a vaccine vehicle and adjuvant. An extended-release vaccine delivery system (GP-diABZI-RBD) consisting of the original SARS-CoV-2 WA1 strain RBD as the antigen and diABZI STING agonist in conjunction with yeast β-glucan particles (GP-diABZI) has been tested [34]. The respiratory and gastrointestinal tracts are the main targets of COVID-19 infection, with 12 to 61% of patients reported to have various gastrointestinal symptoms such as diarrhea, direct damage to intestinal mucosal epithelial cells, malnutrition, and intestinal flora disorders [35]. The mucosal immunity enhanced by our spores, together with β-glucan CV and AM enriched, may provide additional protection against the gastrointestinal tract's infection by COVID-19.

In both in vitro and in vivo studies, we have demonstrated that *B. subtilis* S spores with or without AM and CV extracts are non-cytotoxic and safe to use. Moreover, a pilot clinical trial was conducted with 16 participants [9] who received oral *B. subtilis* spores expressing the S protein RBD of SARS-CoV-2 on the spore surface or placebo for three courses on three consecutive days. These participants were vaccinated with either BBIBP-CorV [36] or BNT162b2 [37] prior to the clinical trial. In the placebo group, the neutralizing antibody levels gradually declined, whereas the participants receiving the S protein spores showed an increase in neutralizing antibody levels against SARS-CoV-2. No observable local or systemic adverse effects were reported. Apart from the vaccine-like activities, the innate immune-boosting effects of both *B. subtilis* spores and S spores together with CV and AM may provide protection against various respiratory and intestinal infections. *Bacillus*-based probiotics have been shown to strengthen the intestinal barrier and limit inflammatory responses, which may indeed improve digestive health [38]. Recently, *Bacillus* species have been shown to be a potential source of anti-SARS-CoV-2 major protease inhibitors [39].

4. Materials and Methods

4.1. Preparation of B. subtilis *Spores (S Spores) Expressing the Ancestral SARS-CoV-2 RBD of S Protein*

Expression constructs containing a cascade of full-length CotC from *B. subtilis*, linker region (peptide sequence: GGGEAAAKGGG) and the RBD of S protein from SARS-CoV-2 were codon optimized for *B. subtilis*, synthesized by Invitrogen GeneArt gene synthesis, and further cloned into shuttle vector pHT01 for *E. coli* and *B. subtilis*. *E. coli* strain DH5α (NEB) was used for the cloning and transformed into *B. subtilis* strain WB800N (Mo Bi Tec) for protein expression. *B. subtilis* transformants were grown at 37 °C, 200 rpm, until OD600 reached 1.0 in 2× LB supplemented with chloramphenicol (5 μg/mL). The culture was induced with a final concentration of 1 mM IPTG and allowed to induce for another 12 h at 37 °C, 200 rpm. The culture was harvested and centrifuged at 4200 rpm for 15 min. The cell pellet was washed with 1× phosphate-buffered saline (PBS, pH 7.4) and resuspended in a half volume of Difco Sporulation Medium (DSM) (8 g nutrient broth No. 4, 0.1% KCl, 1 mM $MgSO_4$, and 10 μM $MnCl_2$ in 1 L of distilled water supplemented with 0.5 mM $CaCl_2$ and 1 μM $FeSO_4$). Cells were grown for 24 h at 37 °C, 200 rpm. Vegetative cells were lysed with lysozyme (0.1 mg/mL) at 37 °C for an hour and harvested with centrifugation at 10,000 rpm for 15 min, followed by washing three times with PBS.

To validate the surface expression of the S protein on the spores (S spores), immunofluorescence staining and flow cytometry analysis were performed (Figure A1). The spore was blocked with 5% normal goat serum in PBS for 1 h at room temperature and then stained with rabbit-anti-SARS-CoV-2 Spike RBD antibody (1:500 dilution) (Sino Biological,

Beijing, China, Cat. No. 40592-T62) followed by AF488 conjugated donkey anti-rabbit IgG (Invitrogen, Cat. No. R37118, 1:1000). The amount of S protein in the spores was semi-quantified by Western blotting. The spores were lysed with SDS-PAGE buffer and then denatured for 10 min at 100 °C. After centrifugation at 14,800× *g* for 15 min, spore lysate was analyzed in SDS-PAGE and then blotted with monoclonal antibodies against the RBD of S protein. The S protein RBD standard with known concentrations (0.4, 2, and 10 ng/mL) was used for semi-quantitation (Sino Biological, Beijing, China, Cat. No. 40592-V08B). Native spores of non-transformed *B. subtilis* were used as the negative control.

4.2. Preparation of the Extract of AM and CV

The dry herbs of AM and CV were purchased from a herbal supplier in Hong Kong (Figure A2). The extraction of AM and CV was performed according to the traditional custom of Chinese medicine preparation. Each of the individual herbs was extracted twice by heating under reflux at 100 °C using 10× distilled water for each extraction. The aqueous extracts for each of the individual herbs were then combined individually and filtered using cotton wool. The filtrates were concentrated under reduced pressure at 60 °C. The concentrated extracts were lyophilized. AM was authenticated using thin-layer chromatography and compared with corresponding chemical markers in accordance with the Chinese Pharmacopoeia [40] (Figure A3). CV was authenticated by quantification of polysaccharides using the phenol-sulfuric acid colorimetric method. The polysaccharide content in CV water extract was 8.02 ± 0.21% (*w*/*w*). The chemical composition of CV was further analyzed with UPLC-QTOF analysis (Figure A4). A total of 7 compounds were tentatively identified in the CV extracts by UPLC-MS. A total of 3 hydroxybenzoic acids, 2 hydroxycinnamic acids, esculetin, and quinic acid were found. The number of endotoxins present in herbal extracts was quantified with a HEK-Blue™ LPS Detection Kit 2, InvivoGen, Toulouse, France). The endotoxin levels of AM and CV were 20.1 and 19.2 EU/mL, respectively. As a small amount of endotoxins was present in the herbal extracts, the antibiotic polymyxin B sulfate (10 µg/µL) [41] was used in the in vitro studies to neutralize the effect of endotoxins in the herbal extracts.

4.3. 2.3 3-[4,5-Dimethylthiazol-2-yl]-2,5 Diphenyl Tetrazolium Bromide (MTT) Assay

An in vitro cell toxicity test of the tested compounds on primary human PBMC, purified from human buffy coat, was performed using the MTT assay. Cells were treated with AM, CV (0.1 to 10 mg/mL), native, or S spores (5 to 500 µg/mL) for 48 h. MTT (5 µg/mL) in phosphate-buffered saline (PBS) was then added to each well, and the plates were further incubated for 2 h at 37 °C. The supernatant was removed, and 100 µL of dimethyl sulfoxide (DMSO) was added to each well to dissolve the purple formazan crystals. Absorbance at a wavelength of 540 nm was measured using a microplate reader. Results were expressed as the percentage of untreated cells.

4.4. Co-Culture of Human Monocyte Derived Macrophages with Intestinal HT-29 Cells and Monocyte Derived Dendritic Cells (DC) Culture

The innate immunity-boosting effects of native and S spores (0.2–20 µg/mL), AM, and CV water extracts (20–2000 µg/mL) were evaluated. For the development of the oral vaccine-like supplement, an in vitro co-culture system of human monocytes and intestine HT-29 cells was employed to mimic the mucosal area of the intestine. Human monocytes were prepared from fresh human buffy coat obtained from healthy volunteers at the Hong Kong Red Cross Blood Transfusion Service. Peripheral blood mononuclear cells (PBMC) were isolated with Ficoll density (1.082 g/mL) centrifugation for 25 min at 1800 rpm. After RBC lysis, CD14-specific MACS beads (Miltenyi Biotec, Gaithersburg, MD, USA) were used for the enrichment of CD14+ monocytes. To induce macrophage differentiation, CD14+ monocytes were cultured in tissue-culture plates for 6 days in RPMI 1640 medium supplemented with L-glutamine, 10% FCS, 1% Penicillin-Streptomycin, 1% Sodium pyruvate, and 1% Glutamax (GIBCO) and granulocyte macrophage-colony stimulating factor (GM-CSF)

(25 ng/mL) at a density of 1.5 × 10^5/cm^2. Human intestinal epithelial HT-29 cells were grown to confluence in a 24-well culture plate and then rinsed with PBS at 37 °C to prevent cell detachment. Macrophages (5 × 10^5/mL) with different treatments were then applied to HT-29 cells in the culture plate. The epithelial cell-macrophage co-cultures were incubated with or without the tested agents for a further 18 h. The concentrations of proinflammatory cytokines/chemokines (IL-1β, IL-6, IL-8, IL-12, IFN-γ, and TNF-α) in the culture supernatant or mouse serum were quantitated with the Bio-plex human cytokine/chemokine multiplex assay using the Bio-plex 200 system (Bio-Rad, Hercules, CA, USA).

For monocyte-derived DC, monocytes were plated at 2 × 10^6 per mL per well in a 24-well plate and allowed to adhere for 45 min at 37 °C and 5% CO_2. Non-adherent cells were removed by washing the wells two to three times with a gentle stream of medium. Monocytes were then cultured in the presence of two cytokines: GM-CSF (50 ng/mL) and IL-4 (40 ng/mL) at 37 °C under 5% CO_2. On day 3, 50% of the medium was replaced with fresh medium and cytokines. The DCs were harvested and washed on day 6. Cell maturation was induced by the tested agents for 48 h. Supernatants from DC cultures were collected after cell harvest and stored at −80 °C until assayed for cytokines. The levels of IL-1β, IL-12, IL-10, TNF-α, and IFN-γ were measured using a multiplex assay.

4.5. Immunization Regimens

The experimental procedures were reviewed and approved by the Animal Experimentation Ethics Committee of the Chinese University of Hong Kong (Ref. no. 20-280-ITF). Pathogen-free BALB/c mice (aged 6–8 weeks, 15–20 g body weight) were obtained from the Laboratory Animal Services Centre, the Chinese University of Hong Kong (CUHK). All mice in this study were maintained and handled according to the CUHK Animal Experimentation Ethics Committee Guide for the Care and Use of Laboratory Animals. A total of 25 female mice were divided into 5 groups and orally administered with (1) PBS, (2) 1.0 × 10^9 S spores, (3) 1.0 × 10^9 S spores + CV (1.38 g/kg), and AM (0.74 g/kg), (4) AM (0.74 g/kg) and CV (1.38 g/kg) only, and (5) 1.0 × 10^9 native spore. Suspensions (0.5 mL aliquots) containing approximately 1 × 10^9 spores were administrated with a stainless-steel round-tip gavage cannula on days 1–4, 14–17, and 28–31 (Figure 9). Blood and fecal samples were collected 3 days before the immunization regimen and on days 8, 22, and 38. At the terminal stage, sera and tissues from the mice, including liver, lung, small and large intestine, were collected. The tissues were perfused with 1 mL of 10% neutral-buffered formalin. The tissues were dehydrated and embedded in paraffin, sectioned, and stained with hematoxylin and eosin. The stained sections were examined with light microscopy to assess the histopathological changes. To check for any toxic effect on the liver function, the sera of the mice were used to measure concentrations of the liver enzyme aspartate aminotransferase (AST) using an AST Activity Assay Kit from Abcam. Individual sera and fecal samples from each group of mice were tested for antibody response. The collected fecal materials were first lyophilized and stored at −20 °C until use. Four fecal pellets per mouse were homogenized in 200 μL of PBS containing 1% BSA and protease inhibitors. The suspensions were centrifuged (16,000× *g*, 10 min, 4 °C), and supernatants were collected. The activities of IgM, IgA, and IgG against the spike protein were measured using ELISA coated with the recombinant SARS-CoV-2 Spike RBD-His Tag protein (Sino Biological, Beijing, China, catalog number: 40592-V08B).

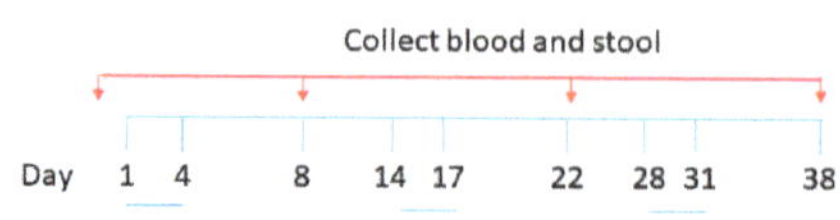

Figure 9. Schematic diagram for oral vaccine-like supplement administration and sample collection.

4.6. Adjuvant Effects of *B. subtilis* Spores, AM, and CV on CoronaVac Vaccine

CoronaVac (Sinovac Life Sciences, Beijing, China) was obtained from the Centre for Health Protection, Department of Health, Hong Kong Special Administrative Region. It is an inactivated vaccine containing inactivated SARS-CoV-2 [42–44]. Whole viral proteins of SARS-CoV-2 were used for the vaccine, and antibodies against different regions of the viral particle (e.g., nucleocapsid (N) and spike (S) proteins) can be induced by the CoronaVac vaccination. To study the adjuvant effects of *B. subtilis* spores, AM, and CV on the CoronaVac vaccine, sixteen 6-week-old female BALB/c mice were divided into 4 groups: (1) orally fed with S spores (4-day course) 14 days prior to CoronaVac vaccination (50 µL) on day 1 and 15; (2) intramuscularly immunized with CoronaVac only; (3) intramuscularly immunized with CoronaVac (50 µL) and fed with AM (0.74 g/kg) and CV (1.38 g/kg), (days 1–4 and days 14–17); and (4) immunized intramuscularly with PBS (Figure 10). Individual sera and samples of each mice group were tested for antibody response such as IgM, IgA, and IgG against N and S proteins, as described in Section 2.5.

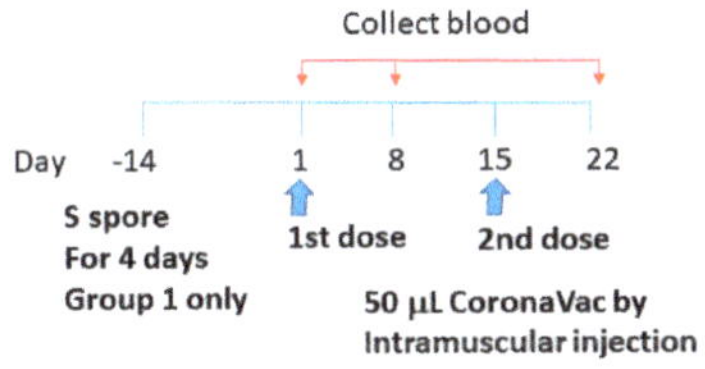

Figure 10. Schematic diagram of the vaccine administration and serum collection for studying the adjuvant effects of *B. subtilis* spores, AM, and CV on the CoronaVac vaccine.

4.7. Neutralization Assay

The sera from the CoronaVac vaccinated mice were further tested on the neutralizing antibodies activities against ancestral SARS-CoV-2 using a commercially available SARS-CoV-2 surrogate Neutralizing Antibody ELISA Kit (Thermo Fisher Scientific Inc., Waltham, MA, USA). The 96-well plate was coated with a SARS-CoV-2 Receptor Binding Domain (RBD) antigen. Samples with neutralizing antibodies competed with excessive amounts of biotinylated angiotensin-converting enzyme 2 (ACE2). ACE2 that binds to the RBD would produce signals which were inversely proportional to the level of neutralizing antibodies. The neutralizing antibody control standard included in the kit was used as a positive control. Neutralization (%) for unknown samples ≥20% was counted as positive and <20% was negative. The neutralization (%) for unknown samples was calculated by the following formula:

$$\text{Neutralization (\%)} = [1 - (\text{Absorbance of unknown sample at 450 nm}/\text{Absorbance of negative control at 450 nm})] \times 100 \quad (1)$$

4.8. Statistical Analyses

Statistical analyses and significance, as measured by the Student's *t*-test for paired samples or one-way analysis of variance (ANOVA), were performed using GraphPad PRISM software version 5.0 (GraphPad Software, San Diego, CA, USA). In all comparisons, $p < 0.05$ was considered statistically significant.

5. Conclusions

In conclusion, the combination of *B. subtilis* S spores with CV and AM may be useful as a vaccine-like supplement in COVID-19 vaccination. As with other vaccines, prolonged use of oral vaccine-like supplements may induce oral tolerance, and it is not easy to control the immune period [33,45]. Further studies, such as the in vivo functional activities of the macrophages and DC, are needed to clarify these issues before practical use. To enhance the immune response of S spores, increasing the expressions of S proteins in *B. subtilis* spores or adding multivalent SARS-CoV-2 variants S proteins [46,47] to the spores may further enhance the efficacy of the supplement against the current COVID-19 pandemic.

6. Patents

The findings from this project have been filed with Chinese (patent no. 202111143384.9) and Hong Kong (patent no. 32021042343.2) patents.

Author Contributions: Conceptualization and funding acquisition, C.-K.W., K.W.-Y.K., C.B.-S.L., and P.-C.L.; Investigation, B.C.-L.C., P.L., T.Z., S.S.-M.H., C.-P.L., W.C., F.C., and J.C.-C.S.; Methodology, M.S.-M.T., and B.C.-L.C.; Writing—original draft, B.C.-L.C.; Writing— review and editing, B.C.-L.C., M.S.-M.T., and C.-K.W. All authors have read and agreed to the published version of the manuscript.

Funding: This work was supported by the Innovation and Technology Fund (ITF) Partnership Research Programme (PRP) of the Innovation and Technology Commission, Hong Kong (Project No.: PRP/008/21FX).

Institutional Review Board Statement: The study was conducted according to the guidelines of the Declaration of Helsinki and approved by the Animal Experimentation Ethics Committee of the Chinese University of Hong Kong (Ref. no. 20-280-ITF).

Informed Consent Statement: Not applicable.

Data Availability Statement: The data presented in this study are available on request from the corresponding author.

Conflicts of Interest: The authors declare no conflict of interest. The funders had no role in the design of the study; in the collection, analyses, or interpretation of data; in the writing of the manuscript, or in the decision to publish the results.

Sample Availability: The water extracts of *Astragalus membranaceus* (Fisch.) Bge (AM) and *Coriolus versicolor* (CV) are available from the authors.

Appendix A

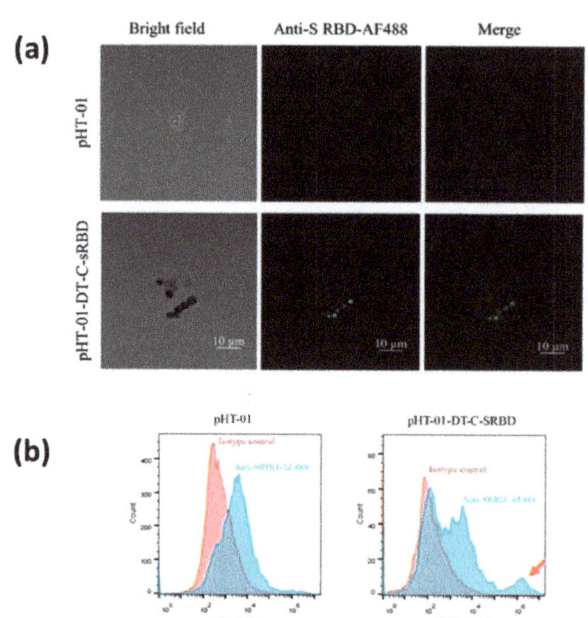

Figure A1. *Cont.*

(c)

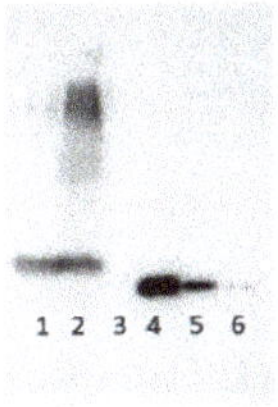

Figure A1. (**a**) Immunofluorescence staining of spike protein on recombinant *B. subtilis* spores visualized by fluorescence microscope. sRBD expression of the ancestral SARS-CoV-2 constructs linked to the coating proteins CotC were cloned and transformed into *B. subtilis* WB800N strain for protein expression. (**b**) Flow cytometry of recombinant *B. subtilis* spores transformed with empty vector pHT-01 and CotC-linked sRBD expression construct. The CotC-linked sRBD-expressing *B. subtilis* spores were immunoreactive against anti-sRBD (indicated by a red arrow). SARS-CoV-2 viral S protein (sRBD) on recombinant *B. subtilis* spores (S spores) from native *B. subtilis* spores (native spores). Immunofluorescence staining and flow cytometry analysis showed that the S protein (sRBD) is expressed on the outer surface of spores. (**c**) The spores were lysed with protein extraction buffer, and the levels of spike protein expressed on the spores were semi-quantitatively quantified by Western blot in comparison with spike protein standards of known concentrations. S proteins from 2 batches were detected (Lane 1 and 2). The amounts of S proteins were 1.7 and 2.4 ng/100 μg, respectively.

Figure A2. The sample authentication of *Astragalus membranaceus* (Fisch.) Bge (AM) and *Coriolus versicolor* (CV) was performed morphologically and chemically according to the guidelines of the Chinese Pharmacopoeia [40]. The medicinal mushroom CV belongs to the group of Basidiomycetes, in which the fruiting bodies are characterized by hymenium (fertile layer) in vertical pores on the underside of the caps. The upper surface of the cap of CV shows typical concentric zones of different colors, such as brown, white, and grey, with overlapping clusters on dead wood. The cap flat is often triangular or round, with zones of fine hairs. The pore surface is whitish or light brown, with round and twisted pores (**a**). AM was derived from the dried root of *Astragalus membranaceus* (Fisch.) Bge. of the family Fabaceae (Leguminosae). The outer surface was pale brownish-yellow or pale brown with irregular, longitudinal wrinkles or furrows. The texture was hard, tenacious, and uneasily broken. The central part of the old root was occasionally rotten-wood-shaped, blackish-brown, or hollow (**b**). The extraction of AM and CV was performed according to the traditional practice of Chinese medicine preparation. Each of the individual herbs was extracted twice by heating under reflux at 100 °C using 10× distilled water for each extraction. The aqueous extracts for each of the individual herbs were then combined individually and filtered using cotton wool. The filtrates were concentrated under reduced pressure at 60 °C. The concentrated extracts were lyophilized. All the extracts (**c**,**d**) were stored in desiccators at room temperature before use. The extraction efficiency of herbs refers to the number of water extracts extracted from Chinese medicine relative to the number of extracts in the herbs. The yield of extract (extractable components) expressed on a dry weight basis was calculated using the following equation: Yield (g/100 g) = (W1 × 100)/W2 where W1 is the weight of the extract residue obtained after water extraction. W2 is the total weight of the original plant. The extraction efficiencies of CV and AM were 9.85% and 39.27%, respectively.

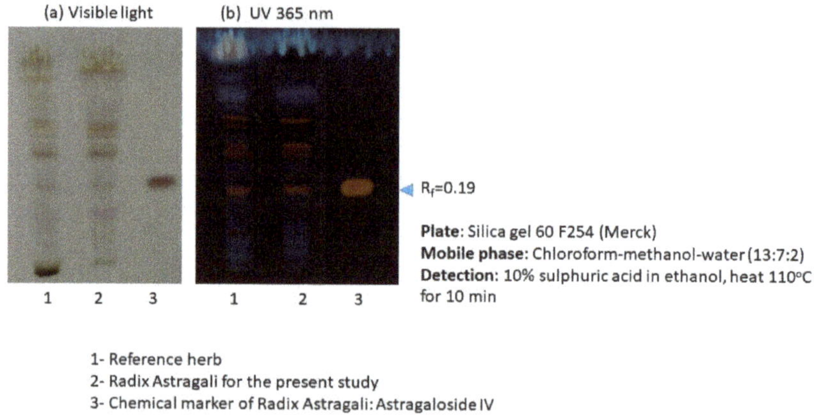

Figure A3. TLC profile of *Astragalus membranaceus* (Fisch.) Bge (AM). Two major chemical markers of RA: astragaloside IV and calycosin 7-*O*-β-D-Glucopyranoside in AM extracts, were further quantified by UPLC-QTOF analysis. The amounts of astragaloside IV and calycosin 7-*O*-β-D-Glucopyranoside were 0.0749% ± 0.0074% and 0.0050% ± 0.0004%, respectively.

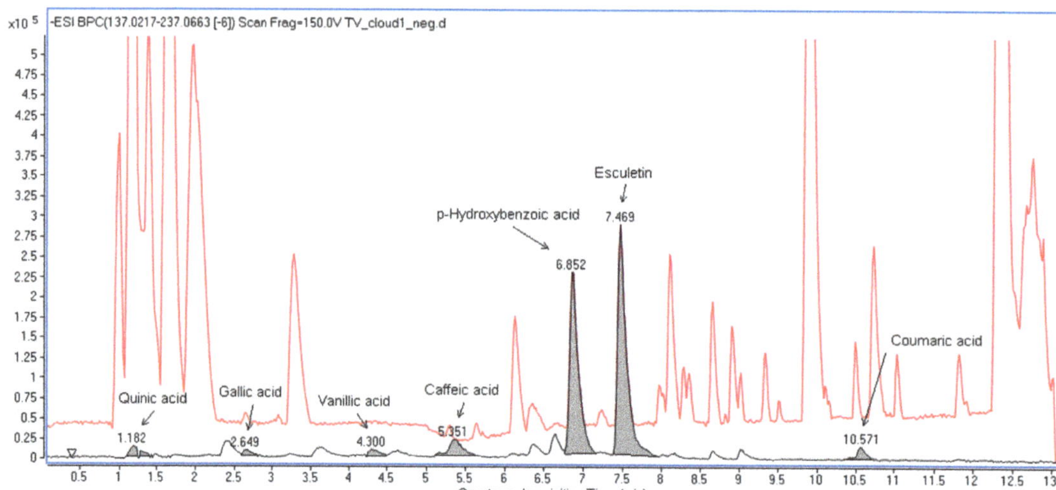

Figure A4. Base peak chromatogram illustrating the tentatively identified compounds in the *Coriolus versicolor* extracts. CV was authenticated by quantification of polysaccharides using the phenol-sulfuric acid colorimetric method. The polysaccharide content in the water extract of CV was 8.02 ± 0.21% (*w*/*w*). The chemical composition of CV was further analyzed by UPLC-QTOF. A total of 7 compounds were tentatively identified in the CV extracts by UPLC-MS. A total of 3 hydroxybenzoic acids, 2 hydroxycinnamic acids, esculetin, and quinic acid were found.

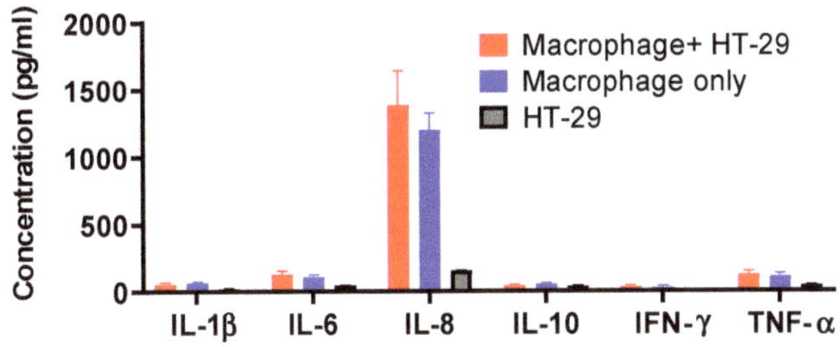

Figure A5. Cytokines production profile (IL-1β, IL-6, IL-8, IL-10, IFN-γ, and TNF-α) of macrophage and HT-29 cell co-culture, macrophage only and HT-29 cells only without treatment.

References

1. Callaway, E. COVID 'variant soup' is making winter surges hard to predict. *Nature* **2022**, *611*, 213–214. [CrossRef] [PubMed]
2. Wang, Q.; Iketani, S.; Li, Z.; Liu, L.; Guo, Y.; Huang, Y.; Bowen, A.D.; Liu, M.; Wang, M.; Yu, J.; et al. Alarming antibody evasion properties of rising SARS-CoV-2 BQ and XBB subvariants. *Cell* **2023**, *186*, 279–286. [CrossRef]
3. Hachmann, N.P.; Miller, J.; Collier, A.Y.; Ventura, J.D.; Yu, J.; Rowe, M.; Bondzie, E.A.; Powers, O.; Surve, N.; Hall, K.; et al. Neutralization Escape by SARS-CoV-2 Omicron Subvariants BA.2.12.1, BA.4, and BA.5. *N. Engl. J. Med.* **2022**, *387*, 86–88. [CrossRef]
4. Qu, P.; Faraone, J.; Evans, J.P.; Zou, X.; Zheng, Y.M.; Carlin, C.; Bednash, J.S.; Lozanski, G.; Mallampalli, R.K.; Saif, L.J.; et al. Neutralization of the SARS-CoV-2 Omicron BA.4/5 and BA.2.12.1 Subvariants. *N. Engl. J. Med.* **2022**, *386*, 2526–2528. [CrossRef] [PubMed]
5. Fonager, J.; Bennedbaek, M.; Bager, P.; Wohlfahrt, J.; Ellegaard, K.M.; Ingham, A.C.; Edslev, S.M.; Stegger, M.; Sieber, R.N.; Lassauniere, R.; et al. Molecular epidemiology of the SARS-CoV-2 variant Omicron BA.2 sub-lineage in Denmark, 29 November 2021 to 2 January 2022. *Eurosurveillance* **2022**, *27*, 2200181. [CrossRef] [PubMed]
6. Barnes, A.G.; Cerovic, V.; Hobson, P.S.; Klavinskis, L.S. *Bacillus subtilis* spores: A novel microparticle adjuvant which can instruct a balanced Th1 and Th2 immune response to specific antigen. *Eur. J. Immunol.* **2007**, *37*, 1538–1547. [CrossRef]
7. de Souza, R.D.; Batista, M.T.; Luiz, W.B.; Cavalcante, R.C.M.; Amorim, J.H.; Bizerra, R.S.P.; Martins, E.G.; Ferreira, L.C.D. *Bacillus subtilis* Spores as Vaccine Adjuvants: Further Insights into the Mechanisms of Action. *PLoS ONE* **2014**, *9*, e87454. [CrossRef]
8. Tavares Batista, M.; Souza, R.D.; Paccez, J.D.; Luiz, W.B.; Ferreira, E.L.; Cavalcante, R.C.; Ferreira, R.C.; Ferreira, L.C. Gut adhesive *Bacillus subtilis* spores as a platform for mucosal delivery of antigens. *Infect. Immun.* **2014**, *82*, 1414–1423. [CrossRef]
9. Sung, J.C.; Lai, N.C.; Wu, K.C.; Choi, M.C.; Ma, C.H.; Lin, J.; Kuok, C.N.; Leong, W.L.; Lam, W.K.; Hamied, Y.K.; et al. Safety and Immunogenicity of Inactivated *Bacillus subtilis* Spores as a Heterologous Antibody Booster for COVID-19 Vaccines. *Vaccines* **2022**, *10*, 1014. [CrossRef]
10. Vetrakova, A.; Chovanova, R.K.; Rechtorikova, R.; Krajcikova, D.; Barak, I. Bacillus subtilis spores displaying RBD domain of SARS-CoV-2 spike protein. *Comput. Struct. Biotechnol. J.* **2023**, *21*, 1550–1556. [CrossRef]
11. Hong, F.; Xiao, W.; Ragupathi, G.; Lau, C.B.; Leung, P.C.; Yeung, K.S.; George, C.; Cassileth, B.; Kennelly, E.; Livingston, P.O. The known immunologically active components of Astragalus account for only a small proportion of the immunological adjuvant activity when combined with conjugate vaccines. *Planta Med.* **2011**, *77*, 817–824. [CrossRef]
12. Ragupathi, G.; Yeung, K.S.; Leung, P.C.; Lee, M.; Lau, C.B.; Vickers, A.; Hood, C.; Deng, G.; Cheung, N.K.; Cassileth, B.; et al. Evaluation of widely consumed botanicals as immunological adjuvants. *Vaccine* **2008**, *26*, 4860–4865. [CrossRef] [PubMed]
13. Eliza, W.L.Y.; Fai, C.K.; Chung, L.P. Efficacy of Yun Zhi (*Coriolus versicolor*) on Survival in Cancer Patients: Systematic Review and Meta-Analysis. *Recent Pat. Inflamm. Allergy Drug Discov.* **2012**, *6*, 78–87. [CrossRef] [PubMed]
14. Yin, J.Y.; Chan, B.C.L.; Yu, H.; Lau, I.Y.K.; Han, X.Q.; Cheng, S.W.; Wong, C.K.; Lau, C.B.S.; Xie, M.Y.; Fung, K.P.; et al. Separation, structure characterization, conformation and immunomodulating effect of a hyperbranched heteroglycan from Radix Astragali. *Carbohydr. Polym.* **2012**, *87*, 667–675. [PubMed]
15. Han, X.Q.; Chung Lap Chan, B.; Dong, C.X.; Yang, Y.H.; Ko, C.H.; Gar-Lee Yue, G.; Chen, D.; Wong, C.K.; Bik-San Lau, C.; Tu, P.F.; et al. Isolation, structure characterization, and immunomodulating activity of a hyperbranched polysaccharide from the fruiting bodies of *Ganoderma sinense*. *J. Agric. Food Chem.* **2012**, *60*, 4276–4281. [CrossRef]
16. Han, X.Q.; Chan, B.C.; Yu, H.; Yang, Y.H.; Hu, S.Q.; Ko, C.H.; Dong, C.X.; Wong, C.K.; Shaw, P.C.; Fung, K.P.; et al. Structural characterization and immuno-modulating activities of a polysaccharide from *Ganoderma sinense*. *Int. J. Biol. Macromol.* **2012**, *51*, 597–603. [CrossRef]
17. Han, X.Q.; Yue, G.L.; Yue, R.Q.; Dong, C.X.; Chan, C.L.; Ko, C.H.; Cheung, W.S.; Luo, K.W.; Dai, H.; Wong, C.K.; et al. Structure elucidation and immunomodulatory activity of a beta glucan from the fruiting bodies of *Ganoderma sinense*. *PLoS ONE* **2014**, *9*, e100380.
18. Topfer, E.; Boraschi, D.; Italiani, P. Innate Immune Memory: The Latest Frontier of Adjuvanticity. *J. Immunol. Res.* **2015**, *2015*, 478408. [CrossRef]

19. Zhou, Z.; Dong, H.; Huang, Y.; Yao, S.; Liang, B.; Xie, Y.; Long, Y.; Mai, J.; Gong, S. Recombinant *Bacillus subtilis* spores expressing cholera toxin B subunit and *Helicobacter pylori* urease B confer protection against *H. pylori* in mice. *J. Med. Microbiol.* **2017**, *66*, 83–89. [CrossRef]
20. Potocki, W.; Negri, A.; Peszynska-Sularz, G.; Hinc, K.; Obuchowski, M.; Iwanicki, A. IL-1 Fragment Modulates Immune Response Elicited by Recombinant *Bacillus subtilis* Spores Presenting an Antigen/Adjuvant Chimeric Protein. *Mol. Biotechnol.* **2018**, *60*, 810–819. [CrossRef]
21. Knoll, R.; Schultze, J.L.; Schulte-Schrepping, J. Monocytes and Macrophages in COVID-19. *Front. Immunol.* **2021**, *12*, 720109.
22. van der Meer, J.W.M.; Joosten, L.A.B.; Riksen, N.; Netea, M.G. Trained immunity: A smart way to enhance innate immune defence. *Mol. Immunol.* **2015**, *68*, 40–44.
23. Gyssens, I.C.; Netea, M.G. Heterologous effects of vaccination and trained immunity. *Clin. Microbiol. Infect.* **2019**, *25*, 1457–1458. [PubMed]
24. Jensen, J.; Warner, T.; Balish, E. Resistance of SCID mice to *Candida albicans* administered intravenously or colonizing the gut: Role of polymorphonuclear leukocytes and macrophages. *J. Infect. Dis.* **1993**, *167*, 912–919. [CrossRef] [PubMed]
25. Cirovic, B.; de Bree, L.C.J.; Groh, L.; Blok, B.A.; Chan, J.; van der Velden, W.; Bremmers, M.E.J.; van Crevel, R.; Handler, K.; Picelli, S.; et al. BCG Vaccination in Humans Elicits Trained Immunity via the Hematopoietic Progenitor Compartment. *Cell Host Microbe* **2020**, *28*, 322–334.
26. Perez-Hernandez, C.A.; Kern, C.C.; Butkeviciute, E.; McCarthy, E.; Dockrell, H.M.; Moreno-Altamirano, M.M.B.; Aguilar-Lopez, B.A.; Bhosale, G.; Wang, H.; Gems, D.; et al. Mitochondrial Signature in Human Monocytes and Resistance to Infection in *C. elegans* During Fumarate-Induced Innate Immune Training. *Front. Immunol.* **2020**, *11*, 1715. [CrossRef]
27. Pellon, A.; Nasab, S.D.S.; Moyes, D.L. New Insights in *Candida albicans* Innate Immunity at the Mucosa: Toxins, Epithelium, Metabolism, and Beyond. *Front. Cell. Infect. Microbiol.* **2020**, *10*, 81.
28. Wang, X.; Guan, F.; Miller, H.; Byazrova, M.G.; Cndotti, F.; Benlagha, K.; Camara, N.O.S.; Lei, J.; Filatov, A.; Liu, C. The role of dendritic cells in COVID-19 infection. *Emerg. Microbes Infect.* **2023**, *12*, 2195019. [CrossRef]
29. Netea, M.G.; Dominguez-Andres, J.; Barreiro, L.B.; Chavakis, T.; Divangahi, M.; Fuchs, E.; Joosten, L.A.B.; van der Meer, J.W.M.; Mhlanga, M.M.; Mulder, W.J.M.; et al. Defining trained immunity and its role in health and disease. *Nat. Rev. Immunol.* **2020**, *20*, 375–388. [PubMed]
30. Mou, C.; Zhu, L.; Xing, X.; Lin, J.; Yang, Q. Immune responses induced by recombinant *Bacillus subtilis* expressing the spike protein of transmissible gastroenteritis virus in pigs. *Antivir. Res.* **2016**, *131*, 74–84.
31. Hu, B.; Li, C.; Lu, H.; Zhu, Z.; Du, S.; Ye, M.; Tan, L.; Ren, D.; Han, J.; Kan, S.; et al. Immune responses to the oral administration of recombinant *Bacillus subtilis* expressing multi-epitopes of foot-and-mouth disease virus and a cholera toxin B subunit. *J. Virol. Methods* **2011**, *171*, 272–279. [CrossRef] [PubMed]
32. Wang, Y.; Miao, Y.; Hu, L.P.; Kai, W.; Zhu, R. Immunization of mice against alpha, beta, and epsilon toxins of *Clostridium perfringens* using recombinant rCpa-b-x expressed by *Bacillus subtilis*. *Mol. Immunol.* **2020**, *123*, 88–96.
33. Jiang, B.; Li, Z.; Ou, B.; Duan, Q.; Zhu, G. Targeting ideal oral vaccine vectors based on probiotics: A systematical view. *Appl. Microbiol. Biotechnol.* **2019**, *103*, 3941–3953. [PubMed]
34. Lei, Z.; Zhu, L.; Pan, P.; Ruan, Z.; Gu, Y.; Xia, X.; Wang, S.; Ge, W.; Yao, Y.; Luo, F.; et al. A vaccine delivery system promotes strong immune responses against SARS-CoV-2 variants. *J. Med. Virol.* **2023**, *95*, e28475. [CrossRef] [PubMed]
35. Yao, Y.; Liu, Z.J.; Zhang, Y.K.; Sun, H.J. Mechanism and potential treatments for gastrointestinal dysfunction in patients with COVID-19. *World J. Gastroenterol.* **2022**, *28*, 6811–6826. [CrossRef] [PubMed]
36. Xia, S.; Zhang, Y.; Wang, Y.; Wang, H.; Yang, Y.; Gao, G.F.; Tan, W.; Wu, G.; Xu, M.; Lou, Z.; et al. Safety and immunogenicity of an inactivated SARS-CoV-2 vaccine, BBIBP-CorV: A randomised, double-blind, placebo-controlled, phase 1/2 trial. *Lancet Infect. Dis.* **2021**, *21*, 39–51. [CrossRef] [PubMed]
37. Polack, F.P.; Thomas, S.J.; Kitchin, N.; Absalon, J.; Gurtman, A.; Lockhart, S.; Perez, J.L.; Perez Marc, G.; Moreira, E.D.; Zerbini, C.; et al. Safety and Efficacy of the BNT162b2 mRNA Covid-19 Vaccine. *N. Engl. J. Med.* **2020**, *383*, 2603–2615. [PubMed]
38. Rhayat, L.; Maresca, M.; Nicoletti, C.; Perrier, J.; Brinch, K.S.; Christian, S.; Devillard, E.; Eckhardt, E. Effect of *Bacillus subtilis* Strains on Intestinal Barrier Function and Inflammatory Response. *Front. Immunol.* **2019**, *10*, 564. [CrossRef]
39. Alam, S.; Sadiqi, S.; Sabir, M.; Nisa, S.; Ahmad, S.; Abbasi, S.W. Bacillus species; a potential source of anti-SARS-CoV-2 main protease inhibitors. *J. Biomol. Struct. Dyn.* **2022**, *40*, 5748–5758.
40. Commercial Press. *Chinese Pharmacopoeia*; Commercial Press: Shanghai, China, 2015.
41. Tani, T.; Shimizu, T.; Tani, M.; Shoji, H.; Endo, Y. Anti-endotoxin Properties of Polymyxin B-immobilized Fibers. *Adv. Exp. Med. Biol.* **2019**, *1145*, 321–341.
42. Wu, Z.; Hu, Y.; Xu, M.; Chen, Z.; Yang, W.; Jiang, Z.; Li, M.; Jin, H.; Cui, G.; Chen, P.; et al. Safety, tolerability, and immunogenicity of an inactivated SARS-CoV-2 vaccine (CoronaVac) in healthy adults aged 60 years and older: A randomised, double-blind, placebo-controlled, phase 1/2 clinical trial. *Lancet Infect. Dis.* **2021**, *21*, 803–812. [PubMed]
43. Zhang, Y.; Zeng, G.; Pan, H.; Li, C.; Hu, Y.; Chu, K.; Han, W.; Chen, Z.; Tang, R.; Yin, W.; et al. Safety, tolerability, and immunogenicity of an inactivated SARS-CoV-2 vaccine in healthy adults aged 18–59 years: A randomised, double-blind, placebo-controlled, phase 1/2 clinical trial. *Lancet Infect. Dis.* **2021**, *21*, 181–192. [CrossRef] [PubMed]
44. Peng, Q.; Zhou, R.; Wang, Y.; Zhao, M.; Liu, N.; Li, S.; Huang, H.; Yang, D.; Au, K.K.; Wang, H.; et al. Waning immune responses against SARS-CoV-2 variants of concern among vaccinees in Hong Kong. *EBioMedicine* **2022**, *77*, 103904. [PubMed]

45. Ribeiro, L.A.; Azevedo, V.; Le Loir, Y.; Oliveira, S.C.; Dieye, Y.; Piard, J.C.; Gruss, A.; Langella, P. Production and targeting of the *Brucella abortus* antigen L7/L12 in *Lactococcus lactis*: A first step towards food-grade live vaccines against brucellosis. *Appl. Environ. Microbiol.* **2002**, *68*, 910–916. [CrossRef]
46. Wang, R.; Sun, C.; Ma, J.; Yu, C.; Kong, D.; Chen, M.; Liu, X.; Zhao, D.; Gao, S.; Kou, S.; et al. A Bivalent COVID-19 Vaccine Based on Alpha and Beta Variants Elicits Potent and Broad Immune Responses in Mice against SARS-CoV-2 Variants. *Vaccines* **2022**, *10*, 702. [CrossRef]
47. Castro, J.T.; Azevedo, P.; Fumagalli, M.J.; Hojo-Souza, N.S.; Salazar, N.; Almeida, G.G.; Oliveira, L.I.; Faustino, L.; Antonelli, L.R.; Marcal, T.G.; et al. Promotion of neutralizing antibody-independent immunity to wild-type and SARS-CoV-2 variants of concern using an RBD-Nucleocapsid fusion protein. *Nat. Commun.* **2022**, *13*, 4831.

Disclaimer/Publisher's Note: The statements, opinions and data contained in all publications are solely those of the individual author(s) and contributor(s) and not of MDPI and/or the editor(s). MDPI and/or the editor(s) disclaim responsibility for any injury to people or property resulting from any ideas, methods, instructions or products referred to in the content.

molecules

MDPI

Communication

Identification of Flavonoids from *Scutellaria barbata* D. Don as Inhibitors of HIV-1 and Cathepsin L Proteases and Their Structure–Activity Relationships

Ting-Ting Tang [1,†], Su-Mei Li [2,†], Bo-Wen Pan [1,†], Jun-Wei Xiao [1,†], Yu-Xin Pang [1], Shou-Xia Xie [2], Ying Zhou [1,*], Jian Yang [3] and Ying Wei [1,*]

[1] College of Pharmacy, Guizhou University of Traditional Chinese Medicine, Guiyang 550025, China; tangingtt@163.com (T.-T.T.); pandawater@163.com (B.-W.P.); sethverlo77@hotmail.com (J.-W.X.); pyxmarx@126.com (Y.-X.P.)

[2] Department of Pharmacology, Shenzhen People's Hospital (The Second Clinical Medical College, Jinan University, The First Affiliated Hospital, Southern University of Science and Technology), Shenzhen 518020, China; li.sumei@szhospital.com (S.-M.L.); szshouxia@163.com (S.-X.X.)

[3] College of Pharmacy and Nutrition, University of Saskatchewan, 107 Wiggins Road, Saskatoon, SK S7N 5E5, Canada; jian.yang@usask.ca

* Correspondence: yingzhou71@126.com (Y.Z.); weiying1969@126.com (Y.W.); Tel.: +86-0851-88233090 (Y.Z.)

† These authors contributed equally to this work.

Abstract: *Scutellaria barbata* D. Don (**SB**, Chinese: Ban Zhi Lian), a well-known medicinal plant used in traditional Chinese medicine, is rich in flavonoids. It possesses antitumor, anti-inflammatory, and antiviral activities. In this study, we evaluated the inhibitory activities of **SB** extracts and its active components against HIV-1 protease (HIV-1 PR) and SARS-CoV2 viral cathepsin L protease (Cat L PR). UPLC/HRMS was used to identify and quantify the major active flavonoids in different **SB** extracts, and fluorescence resonance energy transfer (FRET) assays were used to determine HIV-1 PR and Cat L PR inhibitions and identify structure–activity relationships. Molecular docking was also performed, to explore the diversification in bonding patterns of the active flavonoids upon binding to the two PRs. Three **SB** extracts (SBW, SB30, and SB60) and nine flavonoids inhibited HIV-1 PR with an IC_{50} range from 0.006 to 0.83 mg/mL. Six of the flavonoids showed 10~37.6% inhibition of Cat L PR at a concentration of 0.1 mg/mL. The results showed that the introduction of the 4′-hydroxyl and 6-hydroxyl/methoxy groups was essential in the 5,6,7-trihydroxyl and 5,7,4′-trihydroxyl flavones, respectively, to enhance their dual anti-PR activities. Hence, the 5,6,7,4′-tetrahydroxyl flavone scutellarein (HIV-1 PR, IC_{50} = 0.068 mg/mL; Cat L PR, IC_{50} = 0.43 mg/mL) may serve as a lead compound to develop more effective dual protease inhibitors. The 5,7,3′,4′-tetrahydroxyl flavone luteolin also showed a potent and selective inhibition of HIV-1 PR (IC_{50} = 0.039 mg/mL).

Keywords: *Scutellaria barbata* D. Don; flavonoids; HIV-1 protease; cathepsin L protease; structure–activity relationships

check for updates

Citation: Tang, T.-T.; Li, S.-M.; Pan, B.-W.; Xiao, J.-W.; Pang, Y.-X.; Xie, S.-X.; Zhou, Y.; Yang, J.; Wei, Y. Identification of Flavonoids from *Scutellaria barbata* D. Don as Inhibitors of HIV-1 and Cathepsin L Proteases and Their Structure–Activity Relationships. *Molecules* **2023**, *28*, 4476. https://doi.org/10.3390/molecules28114476

Academic Editor: H. P. Vasantha Rupasinghe

Received: 13 April 2023
Revised: 26 May 2023
Accepted: 28 May 2023
Published: 31 May 2023

Copyright: © 2023 by the authors. Licensee MDPI, Basel, Switzerland. This article is an open access article distributed under the terms and conditions of the Creative Commons Attribution (CC BY) license (https://creativecommons.org/licenses/by/4.0/).

1. Introduction

Globally, 766 million confirmed cases of coronavirus disease 2019 (COVID-19), including 6.93 million deaths, have been reported to the World Health Organization (WHO) [1]. Approximately 10–20% of people infected with severe acute respiratory syndrome coronavirus 2 (SARS-CoV-2) develop long COVID-19, with sequelae such as fatigue, shortness of breath, persistent cough, depression and anxiety, brain fog, myocardial inflammation, and myocardial infarction. In particular, cardiovascular health and mortality are emerging as a new epidemic, substantially changing the lives of millions of people globally [2]. To date, 1.33 trillion COVID-19 vaccines have been administered [1]. However, SARS-CoV-2 mutates quickly and is highly infectious, disparities in vaccinations exist, and effective treatments for long COVID-19 are lacking. Current challenges include the development of more effective vaccines and drugs against COVID-19 [2,3]. During experiments of antiviral drug discovery, two types of anti-SARS-CoV-2 agents, virus-protein targeted agents (blocked virus life cycle) and host protein targeted agents (involved in the viral life cycle), need to be developed to tackle COVID-19 and long COVID-19 [4]. Since both SARS-CoV-2 and HIV are RNA viruses relying on proteases for maturation, HIV-1 protease (HIV-1 PR), an aspartic protease, and cathepsin L protease (Cat L PR), a cysteine protease, have been shown to play a vital role in the lifecycle of HIV and SARS-CoV-2, and are considered key antiviral drug targets [5,6].

In spite of SARS-CoV-2 Cat L PR being a different class of protease (cellular protease) [4], several HIV-1 PR inhibitors have exhibited potent activities towards Cat L PR [7]. Combinations with other therapeutics are currently under clinical trial for the treatment of COVID-19 [7]. Furthermore, the US Food and Drug Administration (FDA) advisory committee has recommended full approval of Paxlovid (Nirmatrelvir/Ritonavir) for treatment of COVID-19 [8]. Although Nirmatrelvir is a substrate of cytochrome P450 (CYP) 3A4, co-administration of Ritonavir can significantly reduce the CYP3A4 metabolism of Nirmatrelvir and result in a high serum level [8]. Other promising drug combinations including Nelfinavir/Cepharanthine and Lopinavir/Ritonavir have also been investigated for COVID-19 treatment [9,10].

Over the past decade, seven Cat L inhibitors have been developed as anti-coronavirus agents, with K111777 and oxocarbazate being the most promising candidates [6]. Furthermore, recent studies have shown that ten FDA-approved drugs exhibited Cat L inhibitory activity [6]. These drugs could be repositioned for COVID-19 treatment, especially for patients in the early stages of infection or who are asymptomatic. Cat L inhibitors can block viral entry on the surface of host cells, without affecting the adaptive immunity [6]. However, some serious adverse reactions may arise from drug–drug interactions [9]. Owing to these drug–drug interactions and rapid development of viral resistance, a continued search for HIV-1 PR or/and Cat L PRs inhibitors is critical for the prevention and treatment of SARS-CoV-2 and other viral infections [6,8].

Plant-derived natural products remain a rich source of therapeutic agents for human illnesses [11]. Some natural products and their derivatives have exhibited antiviral activities and are under clinical trial as potential therapeutic agents [12]. *Scutellaria barbata* D. Don (**SB**, Chinese: Ban Zhi Lian) is a well-known traditional medicinal plant widely distributed in Japan, Korea, and China. It has been used in a wide range of applications in folk medicine, including treatment of snakebites, tumors, hepatitis, sore throat, pulmonary abscess, and hemoptysis. In the 2020 edition of the Chinese Pharmacopoeia (volume one), **SB** (tablet or extract) is listed for medicinal use in treating pneumonia, bronchitis, pharyngitis, and pulmonary abscess [13,14]. More than 200 components, predominately flavonoids and neoclerodane diterpenoids, have been isolated and characterized from methanol or ethanol extracts of **SB** whole grass or aerial parts [15]. The potential therapeutic effects of **SB** extracts and major components have been extensively studied. Huang et al. reported that **SB** aqueous extracts and six major flavonoids prevented SARS-CoV-2 infection via inhibition of Mpro and TMPRSS2 proteases [16]. Zhou et al. screened nine flavonoids obtained from **SB** for their anti-HBV activity [17]. Some flavones in **SB** were shown to prevent parainfluenza viral infection [18]. Moreover, anti-cardiovascular disease activity was observed for **SB** extracts and some flavones during both in vivo and in vitro studies [19]. However, the inhibitory effects of **SB** extracts and its flavonoids against HIV-1 and Cat L PRs have not been investigated. Thus, in the current study, we evaluated the anti-HIV-1 PR and anti-Cat L PR activities of four extracts and nine flavonoids from **SB**.

2. Results and Discussion

2.1. Inhibition of HIV-1, Cat L, and Renin PRs by SB Extracts

SB extract was prepared from the whole plant according to a previously described method [20]. Using a high-throughput screening approach [20,21], the safety of four **SB** extracts (SBW, SB30, SB60, and SB85: prepared from water or MeOH-H_2O) was initially evaluated using renin protease (Table 1) (Please refer to the Supplementary Materials for details). SBW and SB30 had anti-human renin PR activities with IC_{50} values of 0.59 and 0.70 mg/mL, respectively, suggesting that these two extracts may present minor toxicity to humans. Three **SB** extracts (SBW, SB30, and SB60) showed potent activities against HIV-1 PR, with respective IC_{50} of 0.006, 0.028 and 0.03 mg/mL. However, no inhibition was observed for any **SB** extract against Cat L PR. SB85 did not show any activity towards the three PRs. These results indicated that polar **SB** extracts could potently inhibit HIV-1 PR.

Table 1. The IC_{50} values of *SG* extracts against HIV-1, Cat L, and renin PR (n = 3).

Name	$IC_{50} \pm$ SD (mg/mL)		
	Cat L PR	**HIV-1 PR**	**Renin PR**
SGW	>100	0.006 ± 2.79	0.59 ± 4.76
SG30	>100	0.028 ± 6.72	0.70 ± 0.87
SG60	>100	0.03 ± 1.71	>100
SG85	>100	>100	>100
PC1	-	$1.7 \times 10^{-4} \pm 1.91$	-
PC2	$6.8 \times 10^{-7} \pm 0.99$	-	-
PC3	-	-	$9.0 \times 10^{-4} \pm 0.80$

-: no test. PC1: pepstatin A, positive control for HIV-1 protease. PC2: cathepsin L inhibitor, positive control for cathepsin L protease. PC3: positive control for renin protease.

2.2. *Identification and Quantitation of Ingredients in* **SB** *Extracts Using UPLC-HRMS*

In this study, we identified nine compounds inhibiting HIV-1 PR and six compounds inhibiting Cat L PR from **SB**. The five strongest inhibitors (scutellarin, scutellarein, luteolin, hispidulin, and apigenin) were quantitated (Figures 1 and 2 and Table 2). Calibration curves of the five compounds followed a linear regression (R^2 = 0.9994–0.9998) (Please refer to the Supplementary Materials for details). Scutellarin was the most abundant flavonoid in the **SB** extracts, with contents of 25.18 μg/mg, 23.53 μg/mg, 30.35 μg/mg, and 22.49 μg/mg in SBW, SB30, SB60, and SB85, respectively. The contents of hispidulin and scutellarein ranged from 1.04 to 1.97 μg/mg, except for scutellarein in SB60 (0.74 μg/mg). The contents of luteolin and apigenin were less than 0.3 μg/mg in the **SB** extracts.

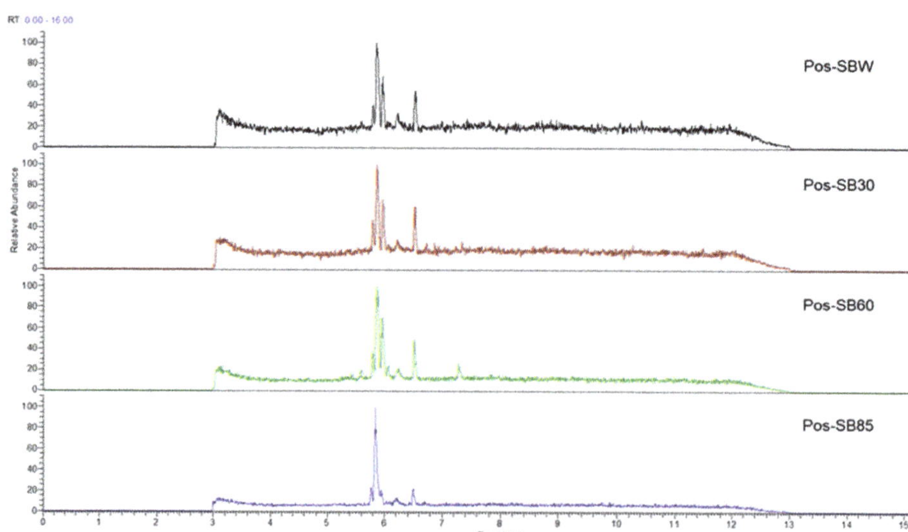

Figure 1. Bioactive flavonoids in **SB** extracts analyzed by UPLC-MS (positive).

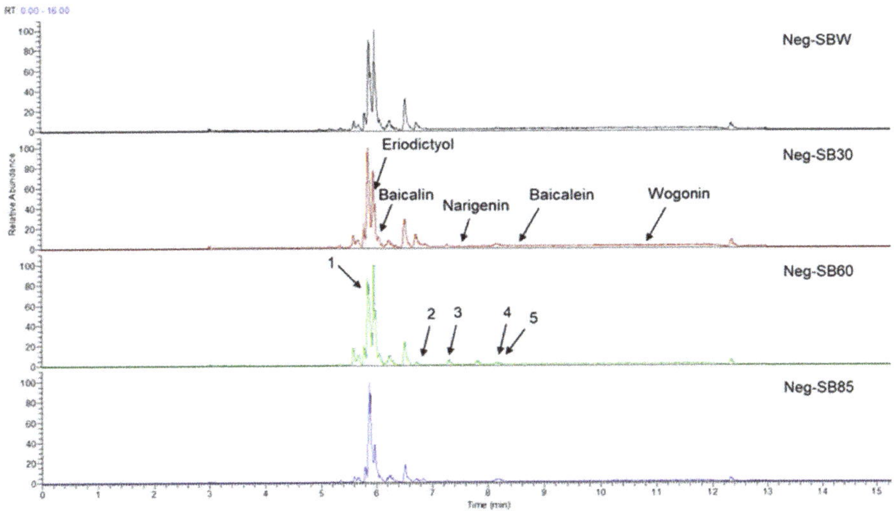

Figure 2. Bioactive flavonoids in **SB** extracts analyzed by UPLC-MS (negative).

Table 2. Contents of the five strongest inhibitors in **SB** extracts.

No.	Rt (min)	Name	SBW	SB30	SB60	SB85	Regression Equation
			Content (μg/mg)				
1	5.86	Scutellarin	25.18	23.53	30.35	22.49	$Y = 5.76 \times 10^3 X - 8.353 \times 10^4; R^2 = 0.9996$
2	6.73	Scutellarein	1.17	1.97	0.74	1.04	$Y = 5.05 \times 10^3 X - 6.36 \times 10^5; R^2 = 0.9994$
3	7.28	Luteolin	0.04	0.12	0.09	0.14	$Y = 9.496 \times 10^3 X + 3.015 \times 10^5; R^2 = 0.9977$
4	8.23	Apigenin	0.10	0.15	0.16	0.29	$Y = 9.979 \times 10^3 X + 1.689 \times 10^5; R^2 = 0.9998$
5	8.32	Hispidulin	1.04	1.72	1.71	1.78	$Y = 8.976 \times 10^2 X - 1.397 \times 10^5; R^2 = 0.9996$

2.3. Inhibitory Activity of Nine Flavonoids Obtained from **SB** *against Cat L and HIV-1 PRs, and Their Respective Structure–Activity Relationships*

The anti-Cat L and anti-HIV-1 PR activities of the nine flavonoids isolated from **SB** were screened using SensoLyte® 520 Cathepsin L and HIV-1 Assay Kits *Fluorimetric*, respectively, according to a previously reported procedure [20,21]. The nine flavonoid compounds were scutellarin, baicalin, scutellarein, hispidulin, apigenin, luteolin, narigenin, eriodictyol, and wogonin. As shown in Table 3 (Please refer to the Supplementary Materials for details), all nine flavonoids exhibited inhibitory activities against HIV-1 PR (IC_{50} = 0.039–0.83 mg/mL), with luteolin as the most potent inhibitor. Towards Cat L PR, apigenin, hispidulin, scutellarin, scutellarein, eriodictyol, narigenin, and baicalin gave an inhibition of 4.95–37.6% at a concentration of 0.1 mg/mL. However, only the IC_{50} for Scutellarein was able to be determined, at 0.43 mg/mL. We are developing other experimental protocols to confirm the current results and figure out whether there are any behavioral differences between Scutellarein and the other flavonoids in the experimental system. Nevertheless, these flavonoids had a preference for HIV-1 PR and were much weaker inhibitors than the positive controls.

Table 3. The inhibition of nine flavonoids from **SB** extracts against Cat L and HIV-1 PRs (n = 3).

Name	Structures	Cat L PR		HIV-1 PR	
		% Inhibition ± SD at Concentration 0.1 (mg/mL)	IC_{50} ± SD (mg/mL)	% Inhibition ± SD at Concentration 1.0 (mg/mL)	IC_{50} ± SD (mg/mL)
5,6,7-Trihydroxyl flavonoid glycoside (A type)					
Scutellarin	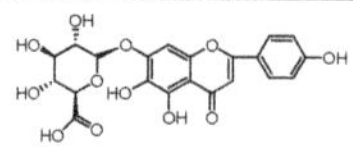	18.3 ± 2.84	>100	56.2 ± 5.30	0.60 ± 6.81
Baicalin	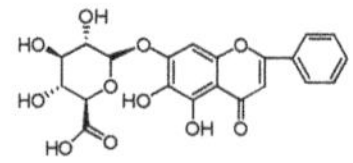	4.95 ± 3.53	>100	53.5 ± 6.45	0.83 ± 6.15
5,7,4′-Trihydroxyl flavone (B type)					
Scutellarein	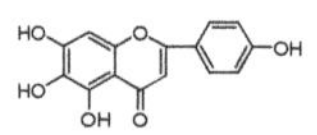	15.6 ± 0.06	0.43 ± 3.17	100.0 ± 0.45	0.068 ± 2.75
Hispidulin	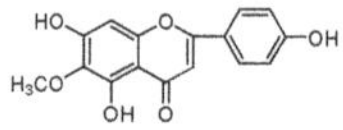	19.8 ± 3.23	>100	86.0 ± 3.76	0.048 ± 2.95
Apigenin	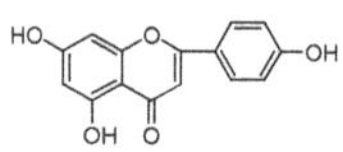	37.6 ± 1.5	>100	91.3 ± 0.87	0.13 ± 3.78
5,7,3′,4′-Tetrahydroxyl flavone (C type)					
Luteolin	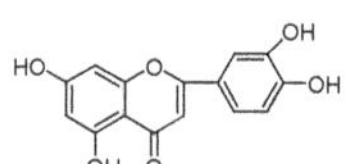	−14.6 ± 10.9	>100	100.0 ± 0.31	0.039 ± 2.42
5,7,4′-Trihydroxyl flavanone (D type)					
Narigenin	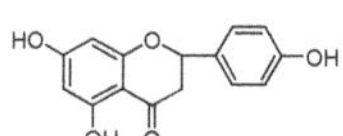	11.2 ± 1.5	>100	89.3 ± 0.00	0.18 ± 1.95
* Eriodictyol	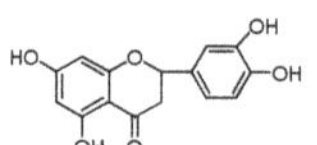	14.3 ± 2.32	>100	100.0 ± 4.98	0.19 ± 5.23
5,7-dihydroxyl-8-methoxy flavone (E type)					
Wogonin	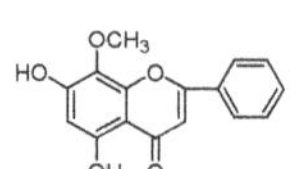	−5.03 ± 2.01	>100	64.8 ± 5.07	0.40 ± 5.29
PC1			4.5×10^{-7} ± 2.30		
PC2					1.2×10^{-4} ± 1.39

* Reprinted with permission from Ref. [20].

We further evaluated the flavonoids and activities in their respective categories (Table 3 and Figure 3). For the 5,6,7-trihydroxyl flavone glycosides (A type), scutellarin exhibited a slightly higher activity against HIV-1 PR than baicalin. However, towards Cat L PR, the inhibitory activity of scutellarin was about three-fold stronger than that of baicalin. This implies that 4′-hydroxyl is critical for inhibition of Cat L PR.

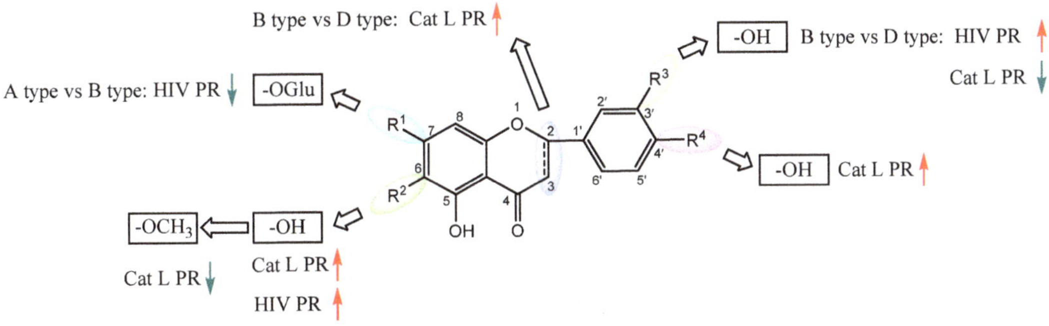

Figure 3. Structure–activity relationships of flavonoids from **SB** against HIV-1 and Cat L PRs.

Within the 5,7,4′-trihydroxyl flavone (type B) family, all three compounds, scutellarein, hispidulin, and apigenin, showed potent activities against HIV-1 PR. The stronger activities for scutellarein and hispidulin suggests that C6-subsitution potentiates the inhibitory function. Towards Cat L PR, a moderate inhibition was observed for the compounds. The higher activity of Apigenin implies that C6-substitution is unfavorable for inhibition of Cat L PR. This also illustrates the structural variance and requirements between HIV-1 PR and Cat L PR for binding their respective substrates and inhibitors. Further analysis between type A and type B flavonoids indicated that C7-glycosylation is detrimental to HIV-1 PR inhibition but may be beneficial for Cat L PR inhibition.

For the 5,7,3′,4′-tetrahydroxyl flavones (type C) and 5,7-dihydroxyl-8-methoxy flavones (type E), only one compound was examined in each category. The type C compound Luteolin and type E compound Wogonin showed good inhibition of HIV-1 PR but no inhibition on Cat L PR. Comparison of these two compounds implied that hydroxylation of the C-ring provides better inhibition of HIV-1 PR; however, further studies are needed to evaluate the contribution of C8-methoxylation to HIV-1 PR inhibition. Further comparison of Luteolin with Apigenin suggested that C3′-hydroxylation provides better inhibition of HIV-1 PR for flavone compounds.

In the 5,7,4′-trihydroxyl flavanones (type D), both compounds, Eriodictyol and Narigenin, exhibited comparable effects towards both HIV-1 PR and Cat L PR, although their inhibitions of HIV-1 PR were much stronger. This implies that C3′-hydroxylation has limited effects on PR inhibition for flavanones. Upon comparing these two flavanone compounds with structurally corresponding flavone compounds, Luteolin and Apigenein, we could conclude that the flavone structure with flat conformation of the B-ring is favorable for HIV-1 PR inhibition. However, we could not reach a conclusion about which type of flavonoid (flavones vs. flavanones) is more potent for Cat L PR inhibition.

2.4. Molecular Docking Results

Molecular docking was performed for the top four inhibitors (Scutellarein, Hispidulin, Apigenin, and Luteolin) against both HIV-1 PR and Cat L PR [21]. As shown in Table 4, the binding energies of Scutellarein, Hispidulin, Apigenin, and Luteolin were −46.959 to −38.811 kcal/mol towards Cat L PR, and −56.377 to −51.378 kcal/mol towards HIV-1 PR. The docking scores were −8.887 to −6.722 towards HIV-1 PR; however, only Scutellarein showed good docking to Cat L PR (−7.332). These docking results were consistent with the experimental results that the four compounds were potent inhibitors of HIV-1 PR but not Cat L PR.

Table 4. The docking scores of active flavonoids from **SB** against Cat L and HIV-1 PRs.

Name	Cat L PR (PDB ID:3OF9)			HIV-1 PR (PDB ID: 1QBS)		
	Docking Score	Glide Gscore	Glide Emodel (kcal/mol)	Docking Score	Glide Gscore	Glide Emodel (kcal/mol)
Scutellarein	−7.332	−7.380	−41.662	−6.722	−6.770	−52.931
Luteolin	−3.565	−3.437	−38.811	−8.887	−8.927	−54.263
Hispidulin	−4.172	−4.212	−46.959	−7.543	−7.583	−56.377
Apigenin	−4.919	−4.959	−41.526	−7.220	−7.260	−51.378
Cat L inhibitor	−7.822	−7.823	−93.170	-	-	-
Pepstatin A	-	-	-	−10.940	−10.941	−131.591

"-": not tested.

Docking conformations of the three compounds (Scutellarein, Apigenin, and Hispidulin) to Cat L PR are shown in Figure 4. Three H-bonds and one salt bridge were formed between Scutellarein and Cat L PR (Figure 4a). The H-bonds were formed between 7-OH and Met162 (1.99 Å), the ketonic functional group at position 4 and Met71 (2.06 Å), and 4′-OH and Asp115 (2.10 Å). For Apigenin (Figure 4b), two H-bonds were formed between 7-OH and Gly21 (2.03 Å) and between 4′-OH and Asp163 (1.86 Å). Three π-π interactions were also formed between A and C rings of Apigenin and Trp190 of Cat L PR. For Hispidulin (Figure 4c), two H-bonds were formed between 7-OH and Asp163 (1.91 Å) and between 4′-OH and Glu160 (2.18 Å).

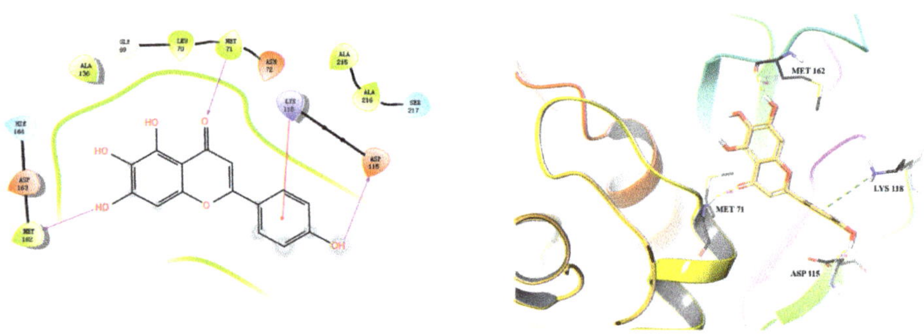

(a)

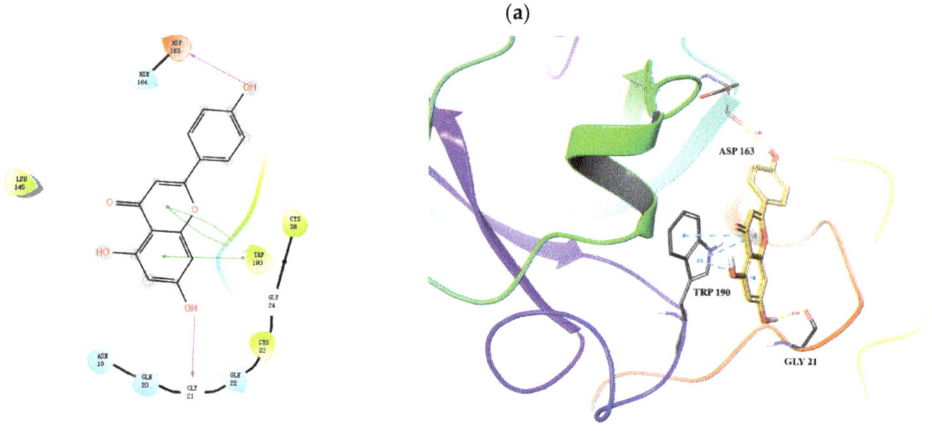

(b)

Figure 4. *Cont.*

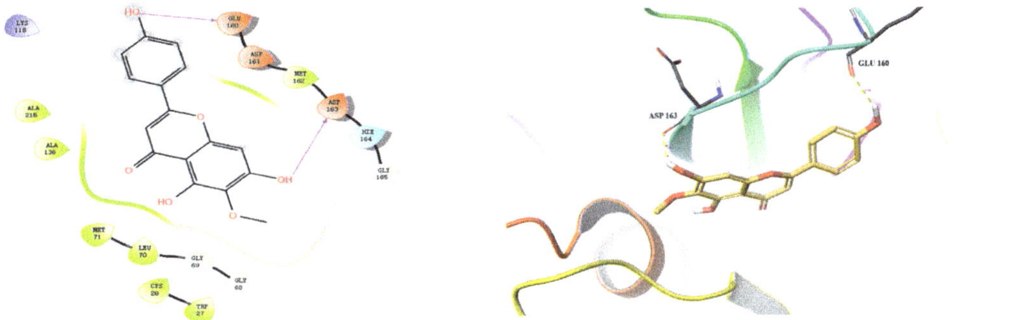

(**c**)

Figure 4. Predicted binding of scutellarein (**a**), apigenin (**b**), and hispidulin (**c**) to Cat L PR. Carbon is colored earth-yellow, while oxygen, hydrogen, nitrogen, and sulfur atoms are colored red, white, blue, and yellow, respectively. The blue dotted lines represent π-π interactions, the green dotted line shows a π-cation interaction and the hydrogen bond is depicted as a yellow dotted line.

Docking conformations of three compounds (Scutellarein, Luteolin, and Hispidulin) to HIIV-1 PR are shown in Figure 5. For Scutellarein (Figure 5a), 6,7-di-OH formed two H-bonds with Asp29(B) (1.66 Å and 1.92 Å). For Luteolin (Figure 5b); 7-OH formed one H-bond with Gly27(B) (2.07 Å); the 4-ketonic functional group formed two H-bonds with Ile50(A) (1.94 Å) and Ile50 (B) (2.28 Å), respectively; and 3′-OH formed one H-bond with Gly48(A) (2.02 Å). For Hispidulin (Figure 5c), the 4-ketonic functional group formed two H-bonds with the Ile50(A) (1.94 Å) and Ile50(B) (2.19 Å), respectively. Taken together, the present docking analysis supports the studies of enzyme inhibition and structure–activity relationships, and may provide modification guidance for the development of more effective PR inhibitors.

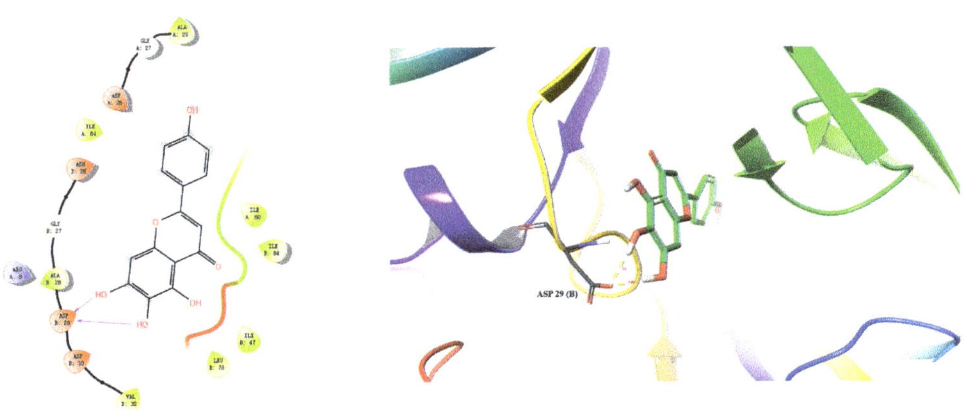

(**a**)

Figure 5. *Cont.*

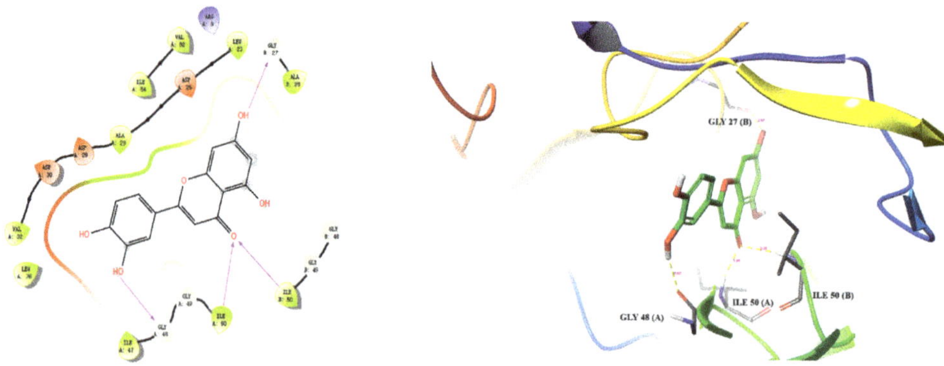

(**b**)

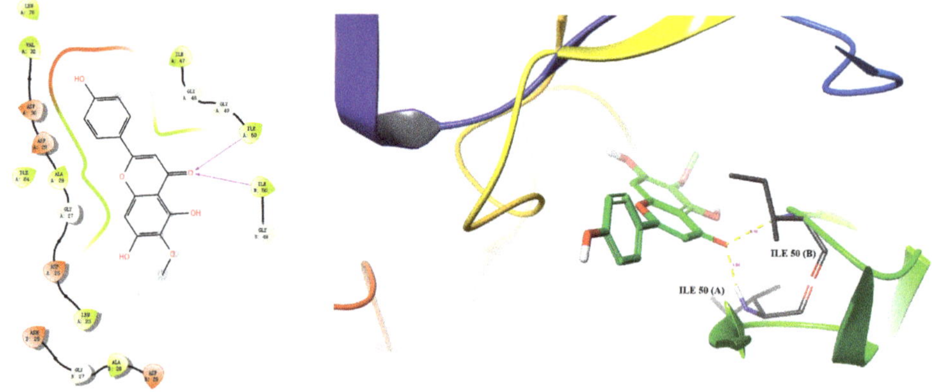

(**c**)

Figure 5. Predicted binding of scutellarein (**a**), luteolin (**b**), and hispidulin (**c**) to HIV-1 PR. Carbon is colored green, while oxygen, hydrogen, nitrogen, and sulfur atoms are colored red, white, blue, and yellow, respectively. The blue dotted lines represent π-π interactions, the green dotted line shows a π-cation interaction and the hydrogen bond is depicted as a yellow dotted line.

3. Conclusions

In this study, we determined the inhibitory activities of nine flavonoids and four **SB** extracts against HIV-1 PR and Cat L PR. Three **SB** extracts (SBW, SB30, and SB60) and all nine flavonoids (Luteolin, Hispidulin, Scutellarein, Apigenin, Narigenin, Eriodictyol, Scutellarin, Baicalin, and Wogonin) inhibited HIV-1 PR, with IC_{50} in the range of 0.006 to 0.83 mg/mL. Six flavonoids (Scutellarein, Apigenin, Hispidulin, Narigenin, Eriodictyol and Scutellarin) showed 10–37.6% inhibition of Cat L PR at a concentration of 0.1 mg/mL.

As the most abundant flavonoid in the **SB** extracts, scutellarin exhibited only moderate inhibition of both PRs. Interestingly, as an aglycone and one of the metabolites of scutellarin [22], scutellarein showed potent inhibition of HIV-1 and Cat L PRs. This indicates that free 5,6,7-hydroxyl groups are likely critical for the inhibitions and provide better bonding affinity within the active sites of both PRs. The introduction of 4′-hydroxyl and 6-hydroxyl/methoxy groups was shown to be essential in 5,6,7-trihydroxyl and 5,7,4′-trihydroxyl flavones, respectively, to enhance dual anti-PR activities. Hence, among the 5,6,7,4′-tetrahydroxyl flavones, scutellarein may serve as a leading molecule for developing more effective dual PR inhibitors. The 5,7,3′,4′-tetrahydroxyl flavone luteolin was a selective and potent inhibitor of HIV-1 PR. Quite a few flavonoids, such as quercetin and kaempferol, demonstrated potent inhibitions of HIV-1 PR and deserve further studies to develop more potent PR inhibitors [23,24]. Our current study also revealed that three **SB** extracts (SBW, SB30, and SB60) and their respective active components, including luteolin and scutellarein, exhibited potent inhibitory activities against HIV-1 PR. However, the selected phytochemical samples are poorly soluble in a neutral assay buffer, and this may lower the bio-availability towards Cat L PR. To develop more dual- and even pan-viral protease inhibitors from medicinal plants, further studies are warranted, to explore potential protease inhibitors based on these active components, for the prevention and/or treatment of COVID-19.

Supplementary Materials: The following are available online at https://www.mdpi.com/article/10.3390/molecules28114476/s1, 1. The chemical profile of **SB** extracts analyzed by UPLC-MS (positive). 2. Predicted binding models of Cat L PR with scutellarein (4a), apigenin (4b), and hispidulin (4c). 3. Predicted binding models of HIV-1 PR with scutellarein (5a), luteolin (5b), and hispidulin (5c). 4. The quantitative curves and maps of scutellarin, scutellarein, apigenin, luteolin, and hispidulin. 5. Four extracts and the main ingredients of anti-HIV and cathepsin L protease inhibition.

Author Contributions: Y.W., J.-W.X. and T.-T.T. carried out the bioactivity assay, analyzed the data, and wrote the draft of the manuscript; B.-W.P. carried out the molecular docking study; S.-M.L. and S.-X.X. carried out the UPLC-MS analysis; and Y.Z., Y.-X.P. and J.Y. revised the manuscript. All authors have read and agreed to the published version of the manuscript.

Funding: This work was supported by an Innovation Group Project of Guizhou Province (grant no. Qian Jiao He KY (2021)018); Guizhou Provincial Science and Technology Projects (grant no. Qian Ke He Zi Cheng (2020)4Y213, ZK (2023) general404 and ZK (2022) general480); Research Center of Molecular Bioactivity from Traditional Chinese Medicine and Ethnomedicine (grant no. 3411-4110000520364); Supported by Shenzhen Fund for Guangdong Provincial High-level Clinical Key Specialties (grant no. SZXK059), the Youth Science and Technology Talent Growth Project of Guizhou Province (grant no. Qian Jiao He KY (2022) 253), and the Project of Guizhou Provincial Health Commission (grant no. gzwkj2022-467).

Institutional Review Board Statement: Not applicable.

Informed Consent Statement: Not applicable.

Data Availability Statement: The data presented in this study are available in the article and on request from the corresponding author.

Conflicts of Interest: The authors declare no conflict of interest.

Sample Availability: The abovementioned samples are available from the authors.

References

1. WHO: Coronavirus Disease 2019 (COVID-19): Situation Report: COVID-19 Partners Platform. Available online: https://covid19.who.int/ (accessed on 17 May 2023).
2. Raman, B.; Bluemke, D.A.; Lüscher, T.F.; Neubauer, S. Long COVID: Post-acute sequelae of COVID-19 with a cardiovascular focus. *Eur. Heart J.* **2022**, *43*, 1157–1172. [CrossRef] [PubMed]
3. Liu, Y.; Eggo, R.M.; Kucharski, A.J. Secondary attack rate and super spreading events for SARS-CoV-2. *Lancet* **2020**, *39*, e47. [CrossRef] [PubMed]
4. Li, G.D.; Hilgenfeld, R.; Whitley, R.; Clercq, E.D. Therapeutic Strategies for COVID-19: Progress and Lessons Learned, Nature Reviews Drug Discovery. Available online: https://doi.org/10.1038/s41573-023-00672-y (accessed on 25 May 2023).
5. Ghosh, A.K.; Osswald, H.L.; Prato, G. Recent progress in the development of HIV-1 protease inhibitors for the treatment of HIV/AIDS. *J. Med. Chem.* **2016**, *59*, 5172–5208. [CrossRef] [PubMed]
6. Liu, T.; Luo, S.; Libby, P.; Shi, G.P. Cathepsin L-selective inhibitors: A potentially promising treatment for COVID-19 patients. *Pharmacol. Ther.* **2020**, *213*, 107587–107601. [CrossRef] [PubMed]
7. Rameswari, C.; Saurabh, A. COVID-19: Characteristics and Therapeutics. *Cell* **2021**, *10*, 206–234.
8. NIH: Ritonavir-Boosted Nirmatrelvir (Paxlovid). Available online: https://www.covid19treatmentguidelines.nih.gov/therapies/antivirals-including-antibody-products/ritonavir-boosted-nirmatrelvir{-}{-}paxlovid-/ (accessed on 20 April 2023).
9. Ohashi, H.; Watashi, K.; Saso, W.; Shionoya, K.; Iwanami, S.; Hirokawa, T.; Shirai, T.; Kanaya, S.; Ito, Y.; Kim, K.S.; et al. Potential anti-COVID-19 agents, cepharanthine and nelfifinavir, and their usage for combination treatment. *iScience* **2021**, *24*, 102367. [CrossRef] [PubMed]

10. Nutho, B.; Mahalapbutr, P.; Hengphasatporn, K.; Pattaranggoon, N.C.; Simanon, N.; Shigeta, Y.; Hannongbua, S.; Rungrotmongkol, T. Why are Lopinavir and Ritonavir effective against the newly emerged Coronavirus 2019? atomistic insights into the inhibitory mechanisms. *Biochemistry* **2020**, *59*, 1769–1779. [CrossRef] [PubMed]
11. Yang, Y.; Islam, M.S.; Wang, J.; Li, Y.; Chen, X. Traditional Chinese Medicine in the treatment of patients infected with 2019-new Coronavirus (SARS-CoV-2): A review and perspective. *Int. J. Biol. Sci.* **2020**, *16*, 1708–1717. [CrossRef] [PubMed]
12. Damle, M. *Glycyrrhiza glabra* (Liquorice)-a potent medicinal herb. *Int. J. Herb. Med.* **2014**, *2*, 132–136.
13. Chinese Pharmacopoeia Commission. *Chinese Pharmacopoeia [S]*; China Medical Science and Technology Press: Beijing, China, 2020; Volume 1, p. 122.
14. Jiang, Y.H.; Lyu, J.J.; Zhao, X.Y. Study on the intervention effect of extract of Chinese medicine for unblocking the collaterals on A549 cells infected by mycoplasma pneumoniae. *Mod. J. Integr. Tradit. Chin. West. Med.* **2021**, *30*, 115–119.
15. Wang, L.; Chen, W.; Li, M.M.; Zhang, F.; Chen, K.X.; Chen, W.S. A review of the ethnopharmacology, phytochemistry, pharmacology, and quality control of *Scutellaria barbata* D. Don. *J. Ethnopharmacol.* **2020**, *254*, 112260–112272. [CrossRef] [PubMed]
16. Huang, S.T.; Chen, Y.; Chang, W.C.; Chen, H.F.; Lai, H.C.; Lin, Y.C.; Wang, W.J.; Wang, Y.C.; Yang, C.S.; Wang, S.C.; et al. *Scutellaria barbata* D. Don inhibits the Main Proteases (Mpro and TMPRSS2) of Severe Acute Respiratory Syndrome Coronavirus 2 (SARS-CoV-2) Infection. *Viruses* **2021**, *13*, 826. [CrossRef] [PubMed]
17. Zhou, L.L.; Di, X.H.; Ding, X. Screening of anti-HBV extracts of *Scutellaria barbata* D. Don in vitro. *Tradit. Chinses Med. Mater.* **2015**, *38*, 1042–1045.
18. Guo, S.S.; Shi, Y.J.; Gao, Y.J.; Su, D.; Cui, X.L. The cytology mechanism of anti-parainfluenza virus infection of total flavone of *Scutellaria barbata*. *Yao Xue Xue Bao Acta Pharm. Sin.* **2009**, *44*, 1348–1352.
19. Chen, Q.; Rahman, K.; Wang, S.J.; Zhou, S.; Zhang, H. *Scutellaria barbata*: A review on chemical constituents, pharmacological activities and clinical applications, *Curr. Pharm. Des.* **2020**, *26*, 160–175. [CrossRef] [PubMed]
20. Pan, B.W.; Xiao, J.W.; Li, S.M.; Yang, X.; Zhou, X.; Sun, Q.W.; Chen, M.; Xie, S.X.; Sakharkar, M.K.; Yang, J.; et al. Inhibitors of HIV-1 and Cathepsin L proteases identified from the insect gall of *Hypericum kouytchense*. *Pharmaceuticals* **2022**, *15*, 1499. [CrossRef] [PubMed]
21. Pan, B.W.; Li, S.M.; Xiao, J.W.; Yang, X.; Xie, S.X.; Zhou, Y.; Yang, J.; Wei, Y. Dual inhibition of HIV-1 and Cathepsin L proteases by *Sarcandra glabra*. *Molecules* **2022**, *27*, 5552. [CrossRef] [PubMed]
22. Xing, J.F.; You, H.S.; Dong, Y.L.; Lu, J.; Chen, S.Y.; Zhu, H.F.; Dong, Q.; Wang, M.Y.; Dong, W.H. Metabolic and pharmacokinetic studies of scutellarin in rat plasma, urine, and feces. *Acta Pharmacol. Sin.* **2011**, *32*, 655–663. [CrossRef] [PubMed]
23. Xu, H.X.; Wan, M.; Dong, H.; But, P.P.H.; Foo, L.Y. Inhibitory activity of flavonoids and tannins against HIV-1 protease. *Bio. Pharm. Bull.* **2000**, *23*, 1072–1076. [CrossRef] [PubMed]
24. Wang, L.; Song, J.; Liu, A.L.; Xiao, B.; Li, S.; Wen, Z.; Lu, Y.; Du, G.H. Research progress of the antiviral bioactivities of natural flavonoids. *Nat. Prod. Biopro.* **2020**, *10*, 271–283. [CrossRef] [PubMed]

Disclaimer/Publisher's Note: The statements, opinions and data contained in all publications are solely those of the individual author(s) and contributor(s) and not of MDPI and/or the editor(s). MDPI and/or the editor(s) disclaim responsibility for any injury to people or property resulting from any ideas, methods, instructions or products referred to in the content.

molecules

MDPI

Review

Fusarium-Derived Secondary Metabolites with Antimicrobial Effects

Meijie Xu [1], Ziwei Huang [1], Wangjie Zhu [1], Yuanyuan Liu [1], Xuelian Bai [2,*] and Huawei Zhang [1,*]

[1] School of Pharmaceutical Sciences, Zhejiang University of Technology, Hangzhou 310014, China
[2] College of Life and Environmental Sciences, Hangzhou Normal University, Hangzhou 311121, China
* Correspondence: baixl2013@163.com (X.B.); hwzhang@zjut.edu.cn (H.Z.)

Abstract: Fungal microbes are important in the creation of new drugs, given their unique genetic and metabolic diversity. As one of the most commonly found fungi in nature, *Fusarium* spp. has been well regarded as a prolific source of secondary metabolites (SMs) with diverse chemical structures and a broad spectrum of biological properties. However, little information is available concerning their derived SMs with antimicrobial effects. By extensive literature search and data analysis, as many as 185 antimicrobial natural products as SMs had been discovered from *Fusarium* strains by the end of 2022. This review first provides a comprehensive analysis of these substances in terms of various antimicrobial effects, including antibacterial, antifungal, antiviral, and antiparasitic. Future prospects for the efficient discovery of new bioactive SMs from *Fusarium* strains are also proposed.

Keywords: *Fusarium*; secondary metabolite; antimicrobial effect; antibacterial; antifungal; antiviral; antiparasitic

check for updates

Citation: Xu, M.; Huang, Z.; Zhu, W.; Liu, Y.; Bai, X.; Zhang, H. *Fusarium*-Derived Secondary Metabolites with Antimicrobial Effects. *Molecules* **2023**, *28*, 3424. https://doi.org/10.3390/molecules28083424

Academic Editors: Xun Song, Chenyang Li and Yifu Guan

Received: 24 March 2023
Revised: 7 April 2023
Accepted: 11 April 2023
Published: 13 April 2023

Copyright: © 2023 by the authors. Licensee MDPI, Basel, Switzerland. This article is an open access article distributed under the terms and conditions of the Creative Commons Attribution (CC BY) license (https://creativecommons.org/licenses/by/4.0/).

1. Introduction

Antimicrobial agents play a significant role in the treatment of infectious diseases caused by pathogenic microorganisms with various modes of action. Since the fortuitous discovery of penicillin in 1928, hundreds of antibiotics have been approved for clinical use. However, some of these drugs have become less efficacy or unavailability simultaneously owing to the development of antimicrobial resistance (AMR), in which a pathogenic microbe evolves a survival mechanism that protects the drug target by modification or replacement, or degradation or modification of the antibiotic to render it harmless, such as MRSA (methicillin-resistant *Staphylococcus aureus*), multidrug-resistant *S. aureus* (MDRS), VREF (vancomycin-resistant *Enterococcus faecium*), CRKP (cephalosporin-resistant *Klebsiella pneumoniae*) [1]. Antimicrobial resistance has become an increasing threat to human health and is widely considered to be the next global pandemic [2]. Therefore, it is an urgent need for the discovery of new antimicrobial drugs with novel structural scaffolds and new modes of action.

Microorganisms are well recognized as a prolific source of biomolecules with diverse chemical structures and various biological properties. Microbial natural products have been, to date, our most successful defense against infectious disease. As one of the most commonly isolated filamentous fungi in terrestrial and marine environments, *Fusarium* spp. possess the potential capability to biosynthesize structurally diverse secondary metabolites (SMs), including alkaloids, peptides, amides, terpenoids, quinones, pyranones, and miscellaneous compounds [3]. Up to now, however, no document highlighting *Fusarium*-derived SMs with antimicrobial effects has been reported. With the aim to enrich our knowledge, this review comprehensively summarizes the occurrence of these antimicrobial substances, including antibacterials, antifungals, antivirals, and antiparasitics.

As of December 2022, the Dictionary of Natural Products (DNP) database listed 783 *Fusarium*-derived SMs, many of them also occurring in other microbial genera. By extensive literature search, as many as 185 antimicrobial SMs (**1–185**) had been discovered from *Fusarium* strains and are, respectively, introduced in terms of various antimicrobial activities, including antibacterial, antifungal, antiviral, and antiparasitic. Their detailed information is supplied in the Supplementary Materials.

2. Antibacterial Secondary Metabolites

Bacterial infection is a common clinical disease that can affect a variety of organs and tissues. *Fusarium*-derived antibacterial SMs have a wide array of structural motifs, most of which are polyketides, followed by alkaloids, terpenoids, and cyclopeptides. According to antibacterial properties, these chemicals are divided into three groups, including anti-Gram-positive bacterial SMs (**1–50**, Figure 1), anti-Gram-negative bacterial SMs (**51–64**, Figure 2) and both anti-Gram-positive and anti-Gram-negative bacterial SMs (**65–81**, Figure 3).

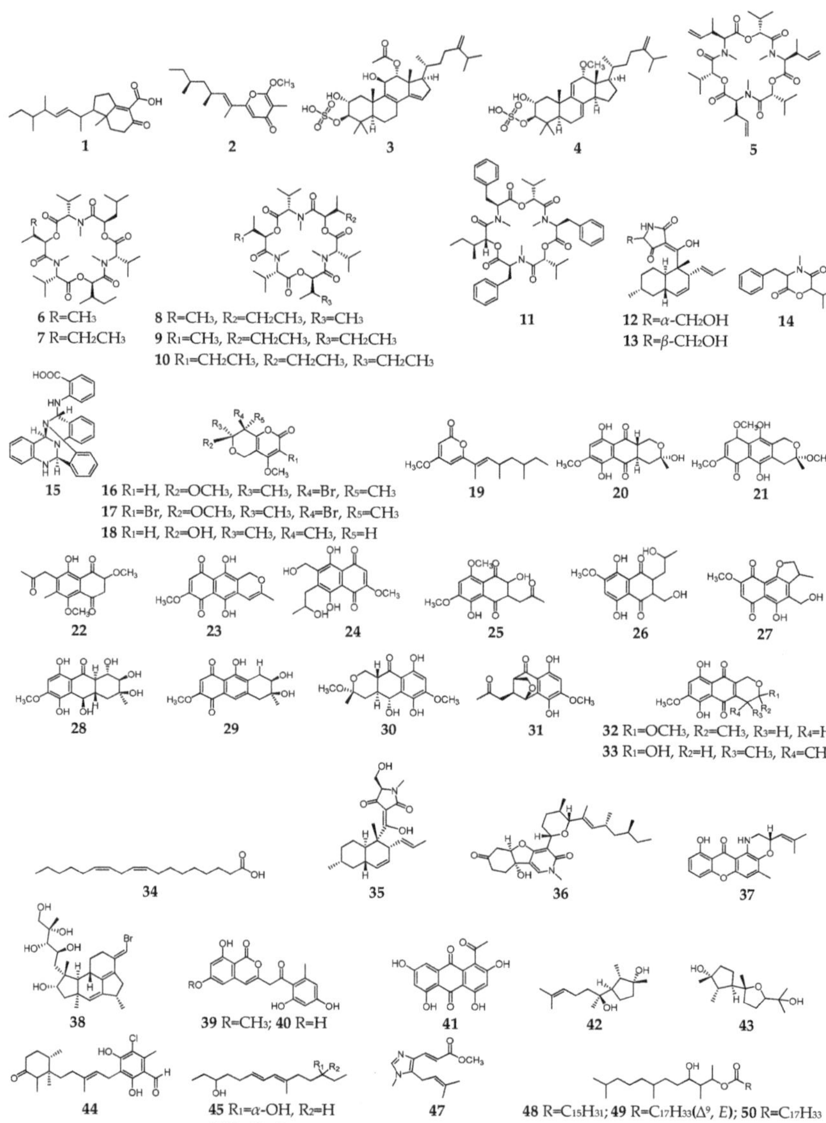

Figure 1. *Fusarium*-derived anti-Gram-positive bacterial SMs (**1–50**).

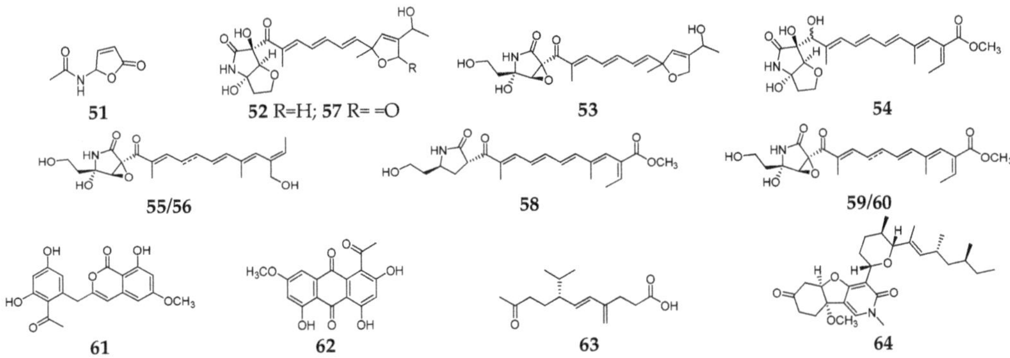

Figure 2. *Fusarium*-derived anti-Gram-negative bacterial SMs (**51–64**).

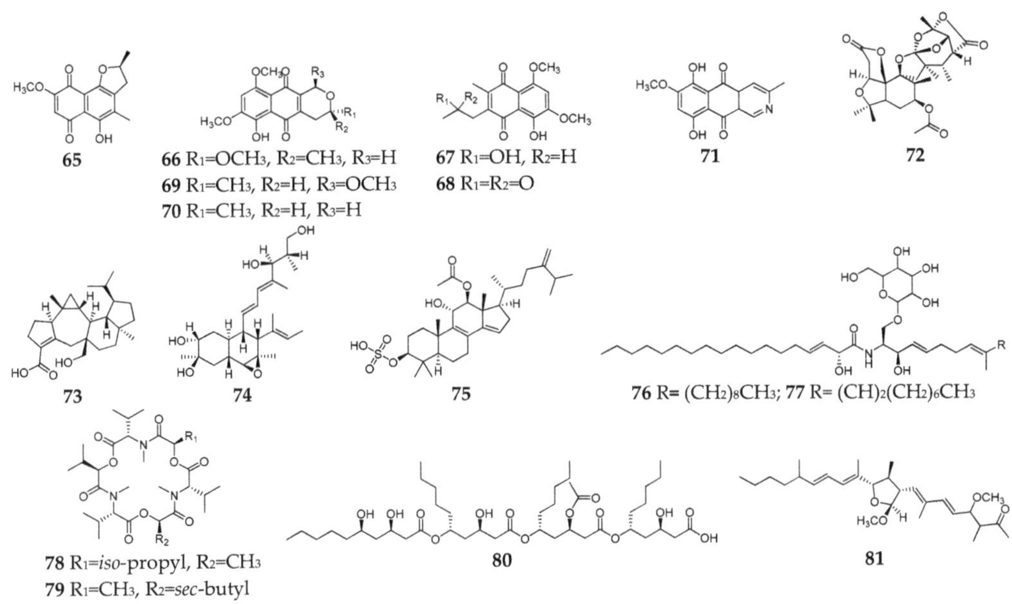

Figure 3. *Fusarium*-derived anti-Gram-positive and anti-Gram-negative bacterial SMs (**65–81**).

2.1. Anti-Gram-Positive Bacterial SMs

Fifty *Fusarium*-derived SMs (**1–50**, Figure 1) had been characterized and displayed various bactericidal effects on Gram-positive strains, such as *Staphylococcus aureus*, methicillin-resistant *Staphylococcus aureus*, multidrug-resistant *S. aureus*, *Mycobacterium tuberculosis*, *Bacillus subtilis*, etc. Fusariumins C (**1**) and D (**2**) are two new polyketides produced by an endophytic strain *F. oxysporum* ZZP-R1 from coastal plant *Rumex midair* Makino displayed medium effect on *S. aureus* with MIC (minimum inhibitory concentration) values of 6.25 and 25.0 μM, respectively [4]. Two triterpene sulfates (**3** and **4**) isolated from *F. compactum* exhibited weak activity toward *S. aureus* and *Streptococcus* strains in the range of 6–50 μg/mL [5]. Enniatins (**5–10**), a group of antibiotics commonly synthesized by various *Fusarium* strains, are six-membered cyclic depsipeptides formed by the union of three molecules of D-α-hydroxyisovaleric acid and three *N*-methyl-*L*-amino acids [6]. Three enniatins (**8–10**), beauvericin A (**11**) and trichosetin (**12**) were obtained from an endophytic fungus, *Fusarium* sp. TP-G1 and showed moderate anti-*S. aureus* and anti-methicillin-resistant *S. aureus* effects with MIC values in the range of 2–16 μg/mL [7]. Two enantiomers

(**12** and **13**) were separated from the culture broth of *F. oxysporum* FKI-4553 and found to have an inhibitory effect on the undecaprenyl pyrophosphate synthase activity of *S. aureus* with IC_{50} values of 83 and 30 μM, respectively [8].

Lateritin (**14**) derived from *Fusarium* sp. 2TnP1–2 showed anti-*S. aureus* activity at 2 μg per disc with 7 mm of inhibition zone [9]. A new polycyclic quinazoline alkaloid (**15**) displayed moderate antibacterial activity against methicillin-resistant *S. aureus* and multidrug-resistant *S. aureus*, with the same MIC value of 6.25 μg/mL [10]. Three pyranopyranones (**16–18**) showed weak inhibitory activities against *S. aureus*, methicillin-resistant *S. aureus*, and multidrug-resistant *S. aureus* [11]. Compound **19** was a new pyran-2-one with weak activity against methicillin-resistant *S. aureus* and was shown to be the inhibitor of the quorum-sensing mechanism of *S. aureus* and *Pseudomonas aeruginosa* [12]. Trans-dihydrofusarubin (**20**) and seven analogs (**21–27**) had significant antibiotic activity against *S. aureus* (MIC values < 4 μg/mL), and compounds **26** and **27** exhibited potent activity against *S. pyogenes* [13]. Five naphthoquinones **28–32** showed anti-*Mycobacterium tuberculosis* activity with MICs ranging from 25 to 50 μg/mL [14]. Compounds **32** and **33** displayed moderate antibacterial activity against *S. aureus* and potent activities against *B. cereus* and *S. pyogenes* with MIC values of <1 μg/mL as compared to ciprofloxacin, whose MIC value was 0.15 and 10 μg/mL, respectively [15].

Linoleic acid (**34**) and *epi*-equisetin (**35**) had certain inhibitory activity against *S. aureus* and multidrug-resistant *S. aureus* [16]. (−)-4,6′-anhydrooxysporidinone (**36**) was obtained from *F. oxysporum* and showed weak anti-multidrug-resistant *S. aureus* and moderate anti-*B. subtilis* effects [17]. Fusaroxazin (**37**), a novel antimicrobial xanthone derivative from *F. oxysporum*, possessed significant antibacterial activity towards *S. aureus* and *B. cereus*, with MIC values of 5.3 and 3.7 μg/mL, respectively [18]. Neomangicol B (**38**) isolated from the mycelial extract of a marine *Fusarium* strain was found to inhibit *B. subtilis* growth with a potency similar to that of the antibiotic gentamycin [19]. Three aromatic polyketides (**39–41**) were produced by strain *F. proliferatum* ZS07 and possessed potent antibacterial activity against *B. subtilis* with the same MIC values of 6.25 μg/mL [20]. Two sesterterpenes (**42** and **43**) produced by *F. avenaceum* SF-1502 displayed stronger antibacterial activity against *B. megaterium* than positive controls (ampicillin, erythromycin, and streptomycin) [21]. 4,5-Dihydroascochlorin (**44**) had strong antibacterial activity towards *Bacillus megaterium* [22]. Fusariumnols A (**45**) and B (**46**) were two novel anti-*S. epidermidis* aliphatic unsaturated alcohols isolated from *F. proliferatum* 13,294 [23]. Fungerin (**47**) displayed weak antibacterial activity against *S. aureus* and *S. pneumoniae* [24]. Compounds **48–50** were purified from *F. oxysporum* YP9B and showed a potent inhibitory effect on *S. aureus, E.faecalis, S. mutans, B. cereus*, and *M. smegmatis* with MICs of less than 4.5 μg/mL [25].

2.2. Anti-Gram-Negative Bacterial SMs

Butenolide (**51**) was a fusarium mycotoxin from unknown origin strain *Fusaium* sp. and showed selective inhibitory activity against *E. coli* [26]. Extensive chemical investigation of the endophytic fungus *F. solani* JK10 afforded nine 2-pyrrolidone derivatives (**52–60**), which displayed antibacterial activity against *E. coli* with MIC values of 5–10 μg/mL. Particularly, three lucilactaene analogs (**52–54**) had strong inhibitory effects on *Acinetobacter* sp., comparable to the positive control streptomycin [27]. One new aromatic polyketide, karimunones B (**61**), together with compounds **62** and **63**, was obtained from sponge-associated *Fusarium* sp. KJMT.FP.4.3 and exhibited anti-multidrug resistant *Salmonella enterica* ser. Typhi activity with a MIC of 125 μg/mL [28]. Fusapyridon A (**64**) is produced by an endophytic strain, *Fusarium* sp. YG-45 demonstrated moderate antibacterial activity against *Pseudomonas aeruginosa* with a MIC value of 6.25 μg/mL [29].

2.3. Both Anti-Gram-Positive and Anti-Gram-Negative Bacterial SMs

Seventeen *Fusarium*-derived SMs (**65–81**, Figure 3) were shown to have both anti-Gram-positive and anti-Gram-negative activity. Seven naphthoquinones (**65–71**) demonstrated moderate activities against an array of Gram-positive and Gram-negative bacteria, such as

B. megaterium, *B. subtilis*, *C. perfringens*, *E. coli*, methicillin-resistant *S. aureus*, *P. aeruginosa*, *S. aureus*, and *S. pyogenes* [13,21,30,31]. The mechanism of action (MoA) study indicated that compounds **66** and **71** could stimulate the oxygen consumption of bacterial cells and induce cyanide-insensitive oxygen consumption, which results in the generation of superoxide anion and hydrogen peroxide [32]. Compounds **72–75** were polycyclic terpenoids, respectively, produced by three *Fusarium* strains [33–35]. Compound **72** had significant activity against *S. aureus* and *P. aeruginosa* with a MIC value of 6.3 μg/mL, and **73** showed moderate activities against *Salmonella enteritidis* and *Micrococcus luteus* with MIC values of 6.3 and 25.2 μg/mL, respectively, while **74** showed a broad spectrum of antibacterial activity and **75** exhibited moderate antibacterial activities against *S. aureus* and *E. coli* with the same MIC value of 16 μg/mL. Two xanthine oxidase inhibitory cerebrosides (**76** and **77**) were identified and purified from the culture broth of *Fusarium* sp. IFB-121 and showed strong antibacterial activities against *B. subtilis*, *E. coli*, and *P. fluorescens* with MICs of less than 7.8 μg/mL [36]. Enniatins J_1 (**78**) and J_3 (**79**) were two hexadepsipeptides with an array of antibacterial activity toward *C. perfringens*, *E. faecium*, *E. coli*, *S. dysenteriae*, *S. aureus*, *Y. enterocolitica*, and lactic acid bacteria except for *B. adolescentis* [37]. Halymecin A (**80**) was produced by a marine-derived *Fusarium* sp. FE-71-1 and exhibited a moderate inhibitory effect on *E. faecium*, *K. pneumoniae*, and *P. vulgaris* with the MIC value of 10 μg/mL [38]. Fusaequisin A (**81**) was isolated from rice cultures of *F. equiseti* SF-3-17 and found to have moderate antimicrobial activity against *S. aureus* NBRC 13,276 and *P. aeruginosa* ATCC 15,442 [39].

3. Antifungal Secondary Metabolites

Invasive fungal infections are very common in immunocompromised patients (such as acquired immune deficiency syndrome and organ transplantation) and have become a global problem resulting in 1.7 million deaths every year [40–42]. Furthermore, the overuse of antifungal agents increases opportunistic pathogen resistance, which had been listed as one of the dominant threats by the World Health Organization in 2019. Therefore, the urgent need for new antimycotics with novel targets is undeniable. Till the end of 2022, twenty-seven antifungal SMs (**82–108**, Figure 4) had been discovered from *Fusarium* strains. Compounds **82–84** are three anti-*C. albicans* glycosides belong to the papulacandin class [43,44]. The MoA study suggested that compound **82** is an inhibitor of glutamine synthetase (GS) enzyme for (1,3)-β-glucan biosynthesis [43]. CR377 (**85**) was a new α-furanone derivative from an endophytic *Fusarium* sp. CR377 and showed a similar antifungal effect on *C. albicans* with nystatin [45]. Compounds **86** and **87** were two zearalenone analogs and exhibited weak activity against *Cryptococcus neoformans* [46]. Neofusapyrone (**88**) produced by a marine-derived *Fusarium* sp. FH-146 displayed moderate activity against *A. clavatus* F318a with a MIC value of 6.25 μg/mL [47]. Six cyclic depsipeptides **89–94** had been isolated from several *Fusarium* strains and found to have significant inhibitory activities against pathogenic fungi, such as *C. albicans* [48], *C. glabrata*, *C. krusei*, *V. ceratosperma*, and *A. fumigates* [49]. Cyclosporin A (**91**) has long been recognized as an immunosuppressant agent and could inhibit the growth of sensitive fungi after their germination [50,51]. Parnafungins A-D (**95–98**) were isoxazolidinone-containing natural products and demonstrated broad-spectrum antifungal activity with no observed activity against bacteria. The targeted pathway of these alkaloids was determined to be the mRNA 3′-cleavage and polyadenylation process [52,53]. One *N*-hydroxypyridine derivative (**99**) showed antifungal activity against *C. albicans* and *Penicillium chrysogenum* with MICs of 16 and 8 μg/mL, respectively [54]. Indole acetic acid (**100**) exhibited activity against the fluconazole-resistant *C. albicans* (MIC = 125 μg/mL) [55].

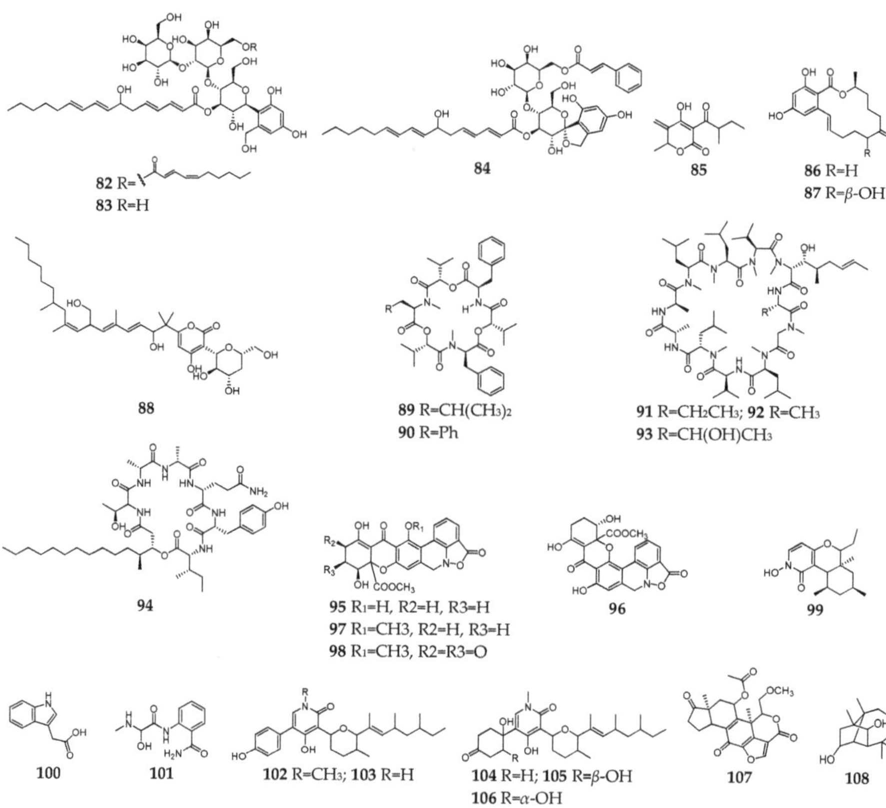

Figure 4. *Fusarium*-derived antifungal SMs (**82–108**).

Fusaribenzamide A (**101**) possessed a significant anti-*C. albicans* activity with MIC of 11.9 μg/disc compared to nystatin (MIC = 4.9 μg/disc) [56]. Three pyridone derivatives (**102–104**) displayed significant activities against multidrug-sensitive *S. cerevisiae* 12geneΔ0HSR-iERG6, and the MoA study indicated that these substances have a potent inhibitory effect on NADH-cytochrome C oxidoreductase [57]. Compounds **105–107** were derived from strain *F. oxysporum* N17B, and the former (**105** and **106**) showed selective fungistatic activity against *Aspergillus fumigatus*, and the latter (**107**) had selective potent activity against *C. albicans* through inhibition of phosphatidylinositol 3-kinase [58]. Culmorin (**108**) displayed remarkable antifungal activity against both marine (*S. marina, M. pelagica*) and medically relevant fungi (*A. fumigatus, A. niger, C. albicans, T. mentagrophytes*) [59,60].

4. Both Antibacterial and Antifungal Secondary Metabolites

Till the end of 2022, forty-one SMs (**109–149**, Figure 5) with both antibacterial and antifungal effects had been discovered from *Fusarium* spp. Among these *Fusarium*-derived 1,4-naphthoquinone analogs (**109–115**), compound **109** showed potent anti-Gram-positive bacteria activity against *B. cereus* and *S. pyogenes* with MIC of <1 μg/mL and anti-*C. albicans* activity with IC_{50} (the half maximal inhibitory concentration) of 6.16 μg/mL [14], and **110–115** demonstrated moderate inhibitory effects on *S. aureus, C. albicans*, and *B. subtilis* [61]. Bikaverin (**116**) was found to have anti-*E. coli* and antifungal (*P. notatum, Alternaria humicola*, and *A. flavus*) activity [48,62,63]. Lateropyrone (**117**) was the same SM as *F. acuminatum, F. lateritium*, and *F. tricinctum* and displayed good antibacterial activity against *B. subtilis, S. aureus, S. pneumoniae*, methicillin-resistant *S. aureus, Mycobacterium tuberculosis*, and vancomycin-resistant of *E. faecalis* and significant inhibitory activity towards the growth of *C. albicans* [64–67]. BE-29,602 (**118**) was a novel antibiotic of the papulacandin family, showing good activity against *C. albicans*, *S. cerevisiae*, *S. pombe* with MIC values < 1 μg/mL and moderate activity against *B. subtilis* and *P. chrysogenum* with the MIC values < 8 μg/mL [44,68]. Fusarielin A (**119**) was a meroterpenoid with moderate antifungal activities against *A. fumigatus* and *F. nivale* and weak antibacterial effect on *S. aureus*, methicillin-resistant *S. aureus*, and multidrug-resistant *S. aureus* [11,69]. Three helvolic acid derivatives (**120–122**) displayed potent antifungal and antibacterial activities against *B. subtilis*, *S. aureus*, *E. coli*, *B. cinerea*, *F. Graminearum*, and *P. capsica* [70]. Fusartricin (**123**) had moderate antimicrobial activity against *E. aerogenes*, *M. tetragenu*, and *C. albicans* with the same MIC value of 19 μM [34].

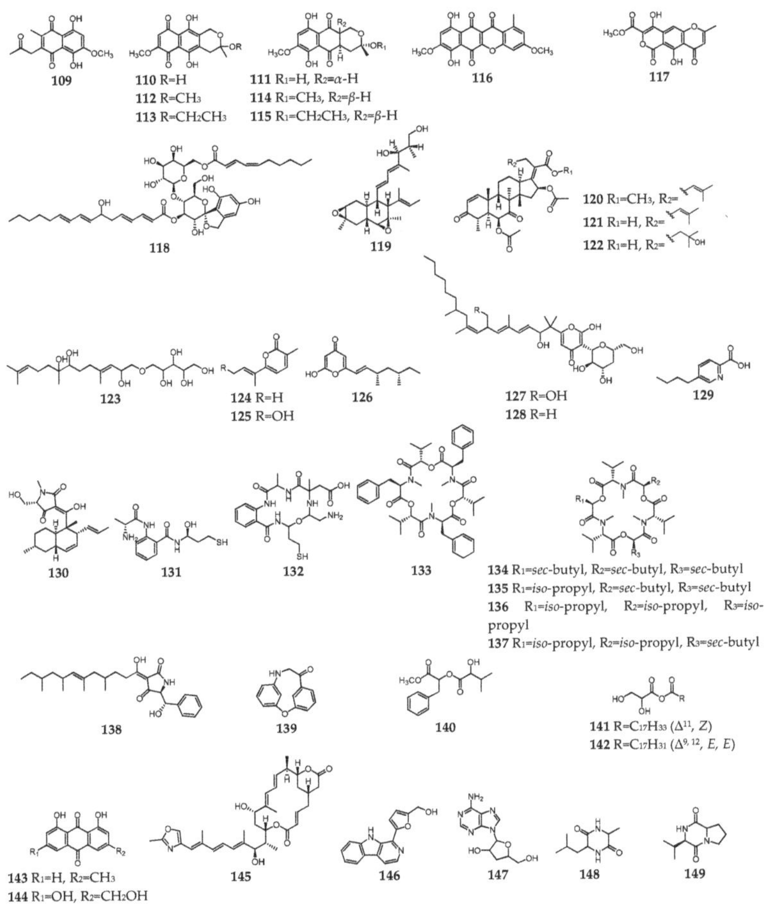

Figure 5. *Fusarium*-derived antibacterial and antifungal SMs (**109–149**).

Compounds **124–128** are pyrone family members and showed antimicrobial activity against bacteria (such as *B. subtilis*, *S. aureus*, *Vibrio parahaemolyticus*, *C. kefyr*, and *P. aeruginosa*) and fungi (such as *A. clavatus*, *Geotrichum candidum*, *C. albicans*, *M. albican*, and *S. cerevisiae*) [47,71–74]. Fusaric acid (**129**), one of the most significant mycotoxins from *Fusarium* strains, displayed a broad spectrum of moderate antimicrobial activity against *Bacillus* species, *Acinetobacter baumannii*, *Phytophthora infestans*, etc. [75–77]. Equisetin (**130**) was shown to be active against several strains of Gram-positive bacteria (*B. subtilis*, *Mycobacterium phlei*, *S. aureus*, *methicillin-resistant S. aureus*, and *S. erythraea*) and the Gram-negative bacteria *Neisseria perflava* at concentrations of 0.5–4.0 μg/mL, as well as antifungal activity toward *P. syringae* and *R. cerealis* [78,79]. Fusarithioamides A (**131**) and B (**132**) demonstrated antibacterial potential towards *B. cereus*, *S. aureus*, and *E. coli* compared to ciprofloxacin and selective antifungal activity towards *C. albicans* compared to clotrimazole [80,81]. Beauvericin (**133**) and enniatins A, A1, B and B1 (**134–137**) are cyclic hexadepsipeptides with a wide array of highly antimicrobial activities against bacteria (such as *B. subtilis*, *S. aureus*, *methicillin-resistant S. aureus*, etc.) and fungi (such as *C. albicans*, *B. bassiana*, *T. harzianum*, etc.) [82–86]. Unlike most antibiotics, cell organelles or enzyme systems are the targets of the antibiotic **133** [87]. As a drug efflux pump modulator, furthermore, compound **133** had the capability to reverse the multi-drug resistant phenotype of *C. albicans* by blocking the ATP-binding cassette transporters and to repress the expression of many filament-specific genes, including the transcription factor BRG1, global regulator TORC1 kinase [88]. Fusaramin (**138**) displayed anti-Gram-positive and anti-Gram-negative bacterial activity and could inhibit the growth of *S. cerevisiae* 12geneΔ0HSR-iERG6 [57]. Compounds **139–142** were isolated from *F. oxysporum* YP9B and exhibited a significant antimicrobial effect against bacterial and fungi at concentrations of 0.8–6.3 μg/mL [25]. Seven SMs (**143–149**) were separated from an endophytic fungus *F. equiseti*, and showed antibacterial (such as *B. subtilis*, *S. aureus*, *B. megaterium*) and anti-*C. albicans* activities [89].

5. Antiviral Secondary Metabolites

The infections by viruses in humans resulted in millions of deaths globally and are accountable for viral diseases, including HIV/AIDS, hepatitis, influenza, herpes simplex, common cold, etc. [90]. The emergence of new viruses like Ebola and coronaviruses (SARS-CoV, SARS-CoV-2) emphasizes the need for more innovative strategies to develop better antiviral drugs. Twenty-three *Fusarium*-derived SMs (**64, 99, 105, 135–137, 140–142, 144–147, 149–158**, Figure 6) had been shown to have antiviral effects. The isolation of fusaricide (**99**) was guided by the Rev (regulation of virion expression) binding assay [54]. Fusapyridon A (**64**) and oxysporidinone (**105**) displayed antiviral activity against the coronavirus (HCoV-OC43) with IC_{50} values of 13.33 and 6.65 μM, respectively [91]. Their enniatins (**135–137**) were found to protect human lymphoblastoid cells from HIV-1 infection with an in vitro "therapeutic index" of approximately 200 (IC_{50} = 1.9, EC_{50} = 0.01 μg/ mL, respectively) [92]. The antiviral activity against HSV type-1 was determined to be 0.312 μM for compound **140** and 1.25 μM for **141** and **142** [25]. Three indole alkaloids (**150–152**) were obtained from a marine-derived *Fusarium* sp. L1 and exhibited inhibitory activity against the Zika virus (ZIKV) with EC_{50} values of 7.5, 4.2, and 5.0 μM, respectively [93]. A chemical study of an endophytic fungus *F. equiseti* led to the isolation of compounds **144–147** and **153–157**, of which **149** and **157** showed good potency against hepatitis C virus NS3/4A protease, while **144** and **155** were the most potent hepatitis C virus NS3/4A protease inhibitors [89]. Coculnol (**158**) was a penicillic acid from a coculture of *F. solani* FKI-6853 and *Talaromyces* sp. FKA-65 displayed an inhibitory effect on A/PR/8/34 (H1N1) with an IC_{50} value of 283 μg/mL [94].

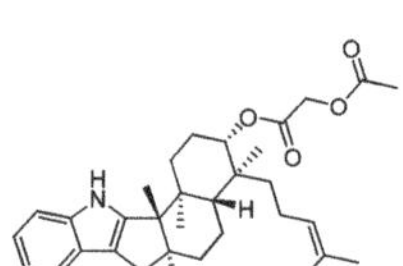
150

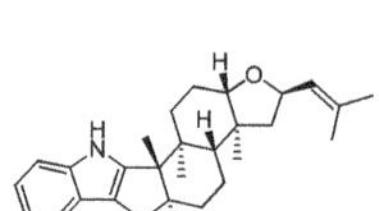
151

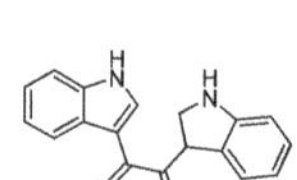
152

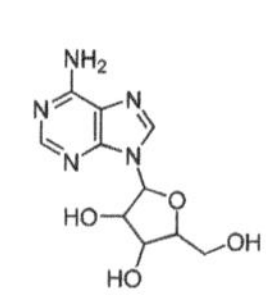
153

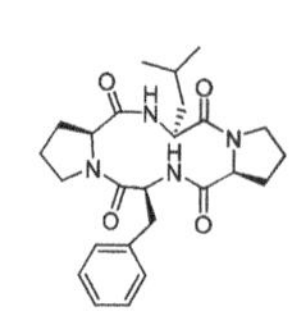
154

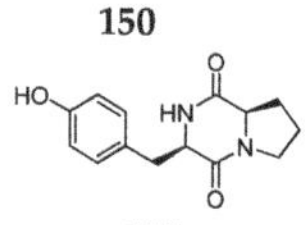
155

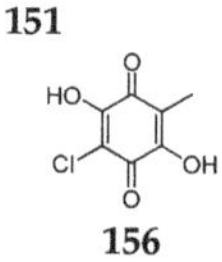
156

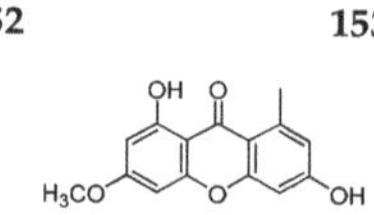
157

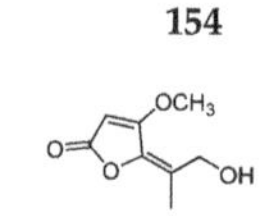
158

Figure 6. *Fusarium*-derived antiviral SMs (**150–158**).

6. Antiparasitic Secondary Metabolites

Parasitic diseases caused by protozoa, helminths and ectoparasites affect millions of people each year and result in substantial morbidity and mortality, particularly in tropical regions [95]. Therefore, new antiparasitic agents are urgently needed to treat and control these diseases. A total of 39 antiparasitic SMs (**23, 28, 29, 59, 108, 93, 116, 133–137, 159–185**, Figure 7) had been isolated and characterized from *Fusarium* strains. Five naphthoquinones (**23, 29, 30, 109,** and **159**) and one anthraquinone (**160**) showed weak inhibitory activity toward the most deadly malaria parasite *Plasmodium falciparum* K1 with IC_{50} values in the range 9.8–26.1 μM [96]. However, compound **93** displayed significant antiplasmodial activity toward *P. falciparum* (D6 clone) with an IC_{50} value of 0.34 μM [49]. Bikaverin (**116**) was specifically effective against *Leishmania brasiliensis*, which is one of the main causes of cutaneous leishmaniasis in the Americas [97]. Beauvericin (**133**) was reported to inhibit *Trypanosoma cruzi* with an IC_{50} value of 2.43 μM and *L. braziliensis* with an EC_{50} value of 1.86 μM [98,99]. In addition to antibacterial and antifungal effects, enniatins (**134–137**) exhibited mild anti-leishmanial activity by inhibition of the activity of thioredoxin reductase enzyme of *P. falciparum* [6]. Integracides F, G, H, and J (**161–164**) were also shown to have stronger anti-leishmanial activity towards *L. donovani* than the positive control pentamidine (IC_{50} = 6.35 μM) [100]. Among twelve lucilactaene derivatives (**165–176**), compounds **166–168** showed very potent antimalarial activity toward *P. falciparum* (IC_{50} = 0.0015, 0.15, and 0.68 μM, respectively) [101–103]. Structure−activity relationship study suggested that epoxide is extremely detrimental, and demethylation of the lucilactaene methyl ester and formation of the free carboxylic acid group resulted in a 300-fold decrease in activity. Nine cyclic tetrapeptides (**177–185**) are apicomplexan histone deacetylase (HDA) inhibitors [104–106]. Particularly, compound **177** was an excellent inhibitory agent (IC_{50} < 2 nM) and showed in vivo high efficacy against *P. berghei* malaria in mice at less than 10 mg/kg.

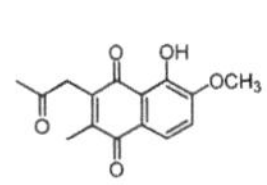

159

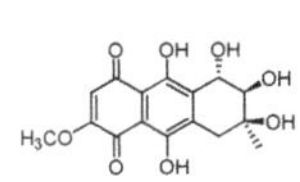

160

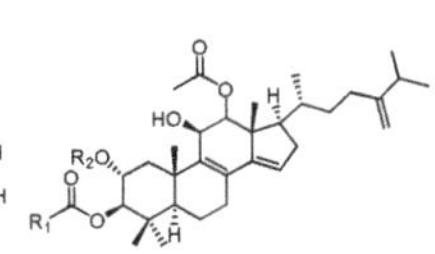

161 R_1=CH_3, R_2=H
162 R_1=$(CH_2)_9OH$, R_2=H
163 R_1=CH_3, R_2=$COCH_3$
164 R_1=*p*-hydroxyphenyl, R_2=H

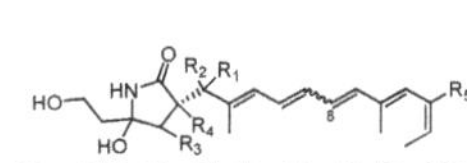

165 R_1=OH, R_2=H, R_3=R_4=O, R_5=$COOCH_3$
166 R_1=R_2=O, R_3=R_4=H, R_5=$COOCH_3$
171 R_1=R_2=O, R_3=R_4=O, R_5=CH_2OH
172 R_1=R_2=O, R_3=R_4=O, R_5=CH_2OH (Δ^8, *Z*)
175 R_1=R_2=O, R_3=R_4=O, R_5=CH_3

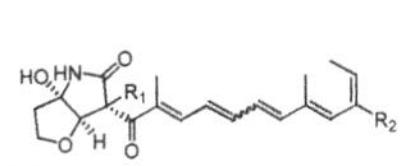

167 R_1=H, R_2=$COOCH_3$
168 R_1=OH, R_2=$COOCH_3$
169 R_1=H, R_2=COOH
170 R_1=H, R_2=COOH (Δ^8, *Z*)
176 R_1=H, R_2=CH_3

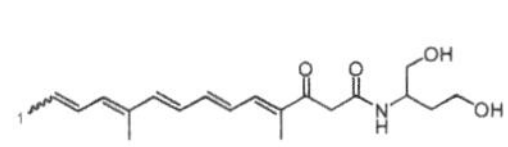

173 (Δ^1, *Z*)
174 (Δ^1, *E*)

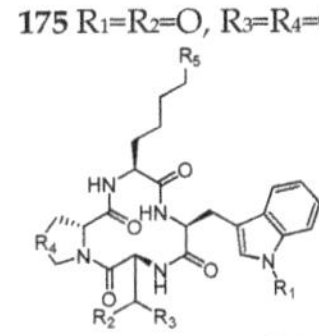

177 R_1=OCH_3, R_2=CH_2CH_3, R_3=CH_3, R_4=CH_2CH_2, R_5=$CH_2COCH_2CH_3$
178 R_1=H, R_2=CH_2CH_3, R_3=CH_3, R_4=CH_2CH_2, R_5=$CH_2COCH_2CH_3$
179 R_1=OCH_3, R_2=CH_2CH_3, R_3=CH_3, R_4=CH_2, R_5=$CH_2COCH_2CH_3$
180 R_1=OCH_3, R_2=CH_3, R_3=CH_3, R_4=CH_2CH_2, R_5=$CH_2COCH_2CH_3$
181 R_1=OCH_3, R_2=CH_2CH_3, R_3=CH_3, R_4=CH_2CH_2, R_5=$CH_2COCH(OH)CH_3$
182 R_1=OCH_3, R_2=CH_2CH_3, R_3=CH_3, R_4=CH_2CH_2, R_5=CH_2CH(*S*-OH)CH_2CH_3
183 R_1=OCH_3, R_2=CH_2CH_3, R_3=CH_3, R_4=CH_2CH_2, R_5=$CH_2CH_2CH(OH)CH_3$
184 R_1=OCH_3, R_2=CH_2CH_3, R_3=CH_3, R_4=CH_2CH_2, R_5=$CH_2CH_2CH_2CH_3$
185 R_1=OCH_3, R_2=phenyl, R_3=H, R_4=CH_2CH_2, R_5=CH_2COOH

Figure 7. *Fusarium*-derived antiparasitic SMs (**159–185**).

7. Conclusions

In summary, the genus *Fusarium* is one of the excellent producers of antimicrobial SMs, some of which have great potential in new drug development, such as anti-Gram-positive bacterial terpenes **38**, **42** and **43**, anti-Gram-negative lucilactaenes **52–54**, antifungal papulacandin **82** and pyridones **102–104**, antiviral enniatins **135–137**, and antiparasitic integracides **161–164**, etc. In the past two decades, however, the rate of discovery of novel SMs from *Fusarium* has constantly been decreasing [3]. Fortunately, a growing number of evidence suggest that the potential of *Fusarium* spp. to make novel SMs is still immense since most of their SM biosynthetic gene clusters (BGCs) are inactive or un-awakened under traditional fermentation and culture conditions [107]. More and more cryptic BGCs responsible for the biosynthesis of novel SMs have been disclosed by various bio-informative tools and approaches and efficiently activated using genome mining strategies, such as BGC heterogeneous expression [108], promoter engineering [109] and gene transcriptional regulation [110]. In addition, more efforts should be made to analyze and interpret the action mechanisms of *Fusarium*-derived leading compounds, which have similar or more potent antimicrobial effects compared to positive controls.

Supplementary Materials: The following supporting information can be downloaded at: https://www.mdpi.com/article/10.3390/molecules28083424/s1, Table S1: Detail information for *Fusarium*-derived anti-Gram-positive bacterial SMs; Table S2: Detail information for *Fusarium*-derived anti-Gram-negative bacterial SMs; Table S3: Detail information for *Fusarium*-derived both anti-Gram-positive and anti-Gram-negative bacterial SMs; Table S4: Detail information for *Fusarium*-derived antifungal SMs; Table S5: Detail information for *Fusarium*-derived both antibacterial and antifungal SMs; Table S6: Detail information for *Fusarium*-derived antiviral SMs; Table S7: Detail information for *Fusarium*-derived antiparasitic SMs.

Author Contributions: Conceptualization, Funding acquisition and Project administration, H.Z.; Writing—original draft, M.X., Z.H., W.Z. and Y.L.; Writing—review & editing, X.B. and H.Z. All authors have read and agreed to the published version of the manuscript.

Funding: This work was co-financially supported by the National Key Research and Development Program of China (2022YFC2804203 and 2018YFC0311004) and the National Natural Science Foundation of China (41776139).

Institutional Review Board Statement: Not applicable.

Informed Consent Statement: Not applicable.

Data Availability Statement: The data presented in this study are available in the Supplementary Material.

Conflicts of Interest: The authors declare no conflict of interest.

References

1. Denissen, J.; Reyneke, B.; Waso-Reyneke, M.; Havenga, B.; Barnard, T.; Khan, S.; Khan, W. Prevalence of ESKAPE pathogens in the environment: Antibiotic resistance status, community-acquired infection and risk to human health. *Int. J. Hyg. Environ. Health* **2022**, *244*, 114006. [CrossRef] [PubMed]
2. Nadimpalli, M.L.; Chan, C.W.; Doron, S. Antibiotic resistance: A call to action to prevent the next epidemic of inequality. *Nat. Med.* **2021**, *27*, 187–188. [CrossRef]
3. Li, M.; Yu, R.; Bai, X.; Wang, H.; Zhang, H. *Fusarium*: A treasure trove of bioactive secondary metabolites. *Nat. Prod. Rep.* **2020**, *37*, 1568–1588. [CrossRef] [PubMed]
4. Chen, J.; Bai, X.; Hua, Y.; Zhang, H.; Wang, H. Fusariumins C and D, two novel antimicrobial agents from *Fusarium oxysporum* ZZP-R1 symbiotic on Rumex madaio Makino. *Fitoterapia* **2019**, *134*, 1–4. [CrossRef] [PubMed]
5. Brill, G.M.; Kati, W.M.; Montgomery, D.; Karwowski, J.P.; Humphrey, P.E.; Jackson, M.; Clement, J.J.; Kadam, S.; Chen, R.H.; McAlpine, J.B. Novel triterpene sulfates from *Fusarium compactum* using a rhinovirus 3C protease inhibitor screen. *J. Antibiot.* **1996**, *49*, 541–546. [CrossRef]
6. Zaher, A.M.; Makboul, M.A.; Moharram, A.M.; Tekwani, B.L.; Calderon, A.I. A new enniatin antibiotic from the endophyte *Fusarium tricinctum* Corda. *J. Antibiot.* **2015**, *68*, 197–200. [CrossRef]

7. Shi, S.; Li, Y.; Ming, Y.; Li, C.; Li, Z.; Chen, J.; Luo, M. Biological activity and chemical composition of the endophytic fungus *Fusarium* sp. TP-G1 obtained from the root of *Dendrobium officinale* Kimura et Migo. *Rec. Nat. Prod.* **2018**, *12*, 549–556. [CrossRef]
8. Inokoshi, J.; Shigeta, N.; Fukuda, T.; Uchida, R.; Nonaka, K.; Masuma, R.; Tomoda, H. *Epi*-trichosetin, a new undecaprenyl pyrophosphate synthase inhibitor, produced by *Fusarium oxysporum* FKI-4553. *J. Antibiot.* **2013**, *66*, 549–554. [CrossRef]
9. Du, Z.; Song, C.; Yu, B.; Luo, X. Secondary metabolites produced by *Fusarium* sp. 2TnP1-2, an endophytic fungus from *Trewia nudiflora*. *Chin. J. Med. Chem.* **2008**, *18*, 452–456.
10. Nenkep, V.; Yun, K.; Son, B.W. Oxysporizoline, an antibacterial polycyclic quinazoline alkaloid from the marine-mudflat-derived fungus *Fusarium oxysporum*. *J. Antibiot.* **2016**, *69*, 709–711. [CrossRef]
11. Nenkep, V.; Yun, K.; Zhang, D.; Choi, H.D.; Kang, J.S.; Son, B.W. Induced production of bromomethylchlamydosporols A and B from the marine-derived fungus *Fusarium tricinctum*. *J. Nat. Prod.* **2010**, *73*, 2061–2063. [CrossRef]
12. Alfattani, A.; Marcourt, L.; Hofstetter, V.; Queiroz, E.F.; Leoni, S.; Allard, P.M.; Gindro, K.; Stien, D.; Perron, K.; Wolfender, J.L. Combination of pseudo-LC-NMR and HRMS/MS-based molecular networking for the rapid identification of antimicrobial metabolites from *Fusarium petroliphilum*. *Front. Mol. Biosci.* **2021**, *8*, 725691. [CrossRef] [PubMed]
13. Baker, R.A.; Tatum, J.H.; Nemec, S., Jr. Antimicrobial activity of naphthoquinones from Fusaria. *Mycopathologia* **1990**, *111*, 9–15. [CrossRef]
14. Kornsakulkarn, J.; Dolsophon, K.; Boonyuen, N.; Boonruangprapa, T.; Rachtawee, P.; Prabpai, S.; Kongsaeree, P.; Thongpanchang, C. Dihydronaphthalenones from endophytic fungus *Fusarium* sp. BCC14842. *Tetrahedron* **2011**, *67*, 7540–7547. [CrossRef]
15. Shah, A.; Rather, M.A.; Hassan, Q.P.; Aga, M.A.; Mushtaq, S.; Shah, A.M.; Hussain, A.; Baba, S.A.; Ahmad, Z. Discovery of anti-microbial and anti-tubercular molecules from *Fusarium solani*: An endophyte of *Glycyrrhiza glabra*. *J. Appl. Microbiol.* **2017**, *122*, 1168–1176. [CrossRef]
16. Chen, C.; Luo, X.; Li, K.; Guo, C.; Li, J.; Lin, X. Antibacterial secondary metabolites from a marine sponge-derived fungus *Fusarium equiseti* SCSIO 41019. *Chin. J. Antibiot.* **2019**, *44*, 1035–1040. [CrossRef]
17. Wang, Q.X.; Li, S.F.; Zhao, F.; Dai, H.Q.; Bao, L.; Ding, R.; Gao, H.; Zhang, L.X.; Wen, H.A.; Liu, H.W. Chemical constituents from endophytic fungus *Fusarium oxysporum*. *Fitoterapia* **2011**, *82*, 777–781. [CrossRef]
18. Mohamed, G.A.; Ibrahim, S.R.M.; Alhakamy, N.A.; Aljohani, O.S. Fusaroxazin, a novel cytotoxic and antimicrobial xanthone derivative from *Fusarium oxysporum*. *Nat. Prod. Res.* **2022**, *36*, 952–960. [CrossRef] [PubMed]
19. Renner, M.K.; Jensen, P.R.; Fenical, W. Neomangicols: Structures and absolute stereochemistries of unprecedented halogenated sesterterpenes from a marine fungus of the genus *Fusarium*. *J. Org. Chem.* **1998**, *63*, 8346–8354. [CrossRef]
20. Li, S.; Shao, M.-W.; Lu, Y.-H.; Kong, L.-C.; Jiang, D.-H.; Zhang, Y.-L. Phytotoxic and antibacterial metabolites from *Fusarium proliferatum* ZS07 Isolated from the gut of long-horned grasshoppers. *J. Agric. Food Chem.* **2014**, *62*, 8997–9001. [CrossRef]
21. Jiang, C.X.; Li, J.; Zhang, J.M.; Jin, X.J.; Yu, B.; Fang, J.G.; Wu, Q.X. Isolation, identification, and activity evaluation of chemical constituents from soil fungus *Fusarium avenaceum* SF-1502 and endophytic fungus *Fusarium proliferatum* AF-04. *J. Agric. Food Chem.* **2019**, *67*, 1839–1846. [CrossRef] [PubMed]
22. Hussain, H.; Drogies, K.-H.; Al-Harrasi, A.; Hassan, Z.; Shah, A.; Rana, U.A.; Green, I.R.; Draeger, S.; Schulz, B.; Krohn, K. Antimicrobial constituents from endophytic fungus *Fusarium* sp. *Asian Pac. J. Trop. Dis.* **2015**, *5*, 186–189. [CrossRef]
23. Lu, W.; Zhu, G.; Yuan, W.; Han, Z.; Dai, H.; Basiony, M.; Zhang, L.; Liu, X.; Hsiang, T.; Zhang, J. Two novel aliphatic unsaturated alcohols isolated from a pathogenic fungus *Fusarium proliferatum*. *Synth. Syst. Biotechnol.* **2021**, *6*, 446–451. [CrossRef] [PubMed]
24. Wen, H.; Li, Y.; Liu, X.; Ye, W.; Yao, X.; Che, Y. Fusagerins A-F, new alkaloids from the fungus *Fusarium* sp. *Nat. Prod. Bioprospect.* **2015**, *5*, 195–203. [CrossRef]
25. Kılıç, G.; Tosun, G.; Bozdeveci, A.; Erik, İ.; Öztürk, E.; Reis, R.; Sipahi, H.; Cora, M.; Karaoğlu, Ş.A.; Yaylı, N. Antimicrobial, cytotoxic, antiviral wffects, and apectroscopic characterization of metabolites produced by *fusarium oxysporum* YP9B. *Rec. Nat. Prod.* **2021**, *15*, 547–567. [CrossRef]
26. Valla, A.; Giraud, M.; Labia, R.; Morand, A. In vitro inhibitory activity against bacteria of a fusarium mycotoxin and new synthetic derivatives. *Bull. Soc. Chim. Fr.* **1997**, *6*, 601–603. [CrossRef]
27. Kyekyeku, J.O.; Kusari, S.; Adosraku, R.K.; Bullach, A.; Golz, C.; Strohmann, C.; Spiteller, M. Antibacterial secondary metabolites from an endophytic fungus, *Fusarium solani* JK10. *Fitoterapia* **2017**, *119*, 108–114. [CrossRef]
28. Sibero, M.T.; Zhou, T.; Fukaya, K.; Urabe, D.; Radjasa, O.K.K.; Sabdono, A.; Trianto, A.; Igarashi, Y. Two new aromatic polyketides from a sponge-derived *Fusarium*. *Beilstein. J. Org. Chem.* **2019**, *15*, 2941–2947. [CrossRef]
29. Tsuchinari, M.; Shimanuki, K.; Hiramatsu, F.; Murayama, T.; Koseki, T.; Shiono, Y. Fusapyridons A and B, novel pyridone alkaloids from an endophytic fungus, *Fusarium* sp. YG-45. *Z. Naturforsch. B.* **2007**, *62*, 1203–1207. [CrossRef]
30. Supratman, U.; Hirai, N.; Sato, S.; Watanabe, K.; Malik, A.; Annas, S.; Harneti, D.; Maharani, R.; Koseki, T.; Shiono, Y. New naphthoquinone derivatives from *Fusarium napiforme* of a mangrove plant. *Nat. Prod. Res.* **2021**, *35*, 1406–1412. [CrossRef]
31. Khan, N.; Afroz, F.; Begum, M.N.; Roy Rony, S.; Sharmin, S.; Moni, F.; Mahmood Hasan, C.; Shaha, K.; Sohrab, M.H. Endophytic *Fusarium solani*: A rich source of cytotoxic and antimicrobial napthaquinone and aza-anthraquinone derivatives. *Toxicol. Rep.* **2018**, *5*, 970–976. [CrossRef]
32. Haraguchi, H.; Yokoyama, K.; Oike, S.; Ito, M.; Nozaki, H. Respiratory stimulation and generation of superoxide radicals in *Pseudomonas aeruginosa* by fungal naphthoquinones. *Arch. Microbiol.* **1997**, *167*, 6–10. [CrossRef]

33. Yan, C.; Liu, W.; Li, J.; Deng, Y.; Chen, S.; Liu, H. Bioactive terpenoids from Santalum album derived endophytic fungus *Fusarium* sp. YD-2. *RSC Adv.* **2018**, *8*, 14823–14828. [CrossRef] [PubMed]
34. Zhang, J.; Liu, D.; Wang, H.; Liu, T.; Xin, Z. Fusartricin, a sesquiterpenoid ether produced by an endophytic fungus *Fusarium tricinctum* Salicorn 19. *Eur. Food Res. Technol.* **2014**, *240*, 805–814. [CrossRef]
35. Dong, J.W.; Cai, L.; Li, X.J.; Duan, R.T.; Shu, Y.; Chen, F.Y.; Wang, J.P.; Zhou, H.; Ding, Z.T. Production of a new tetracyclic triterpene sulfate metabolite sambacide by solid-state cultivated *Fusarium sambucinum* B10.2 using potato as substrate. *Bioresour. Technol.* **2016**, *218*, 1266–1270. [CrossRef] [PubMed]
36. Shu, R.; Wang, F.; Yang, Y.; Liu, Y.; Tan, R. Antibacterial and xanthine oxidase inhibitory cerebrosides from *Fusarium* sp. IFB-121, and endophytic fungus in Quercus variabilis. *Lipids* **2004**, *39*, 667–673. [CrossRef]
37. Sebastià, N.; Meca, G.; Soriano, J.M.; Mañes, J. Antibacterial effects of enniatins J(1) and J(3) on pathogenic and lactic acid bacteria. *Food Chem. Toxicol.* **2011**, *49*, 2710–2717. [CrossRef] [PubMed]
38. Chen, C.; Imamura, N.; Nishijima, M.; Adachi, K.; Sakai, M.; Sano, H. Halymecins, new antimicroalgal substances produced by fungi isolated from marine algae. *J. Antibiot.* **1996**, *49*, 998–1005. [CrossRef]
39. Shiono, Y.; Shibuya, F.; Murayama, T.; Koseki, T.; Poumale, H.M.P.; Ngadjui, B.T. A polyketide metabolite from an endophytic *Fusarium equiseti* in a medicinal plant. *Z. Naturforsch. B.* **2013**, *68*, 289–292. [CrossRef]
40. Ivanov, M.; Ćirić, A.; Stojković, D. Emerging antifungal targets and strategies. *Int. J. Mol. Sci.* **2022**, *23*, 2756. [CrossRef]
41. Van Daele, R.; Spriet, I.; Wauters, J.; Maertens, J.; Mercier, T.; Van Hecke, S.; Brüggemann, R. Antifungal drugs: What brings the future? *Med. Mycol.* **2019**, *57*, S328–S343. [CrossRef]
42. Campoy, S.; Adrio, J.L. Antifungals. *Biochem. Pharmacol.* **2017**, *133*, 86–96. [CrossRef] [PubMed]
43. Jackson, M.; Frost, D.J.; Karwowski, J.P.; Humphrey, P.E.; Dahod, S.K.; Choi, W.S.; Brandt, K.; Malmberg, L.-H.; Rasmussen, R.R.; Scherr, M.H. Fusacandins A and B; Novel Antifungal Antibiotics of the Papulacandin Class from *Fusarium sambucinum* I. Identity of the Producing Organism, Fermentation and Biological Activity. *J. Antibiot.* **1995**, *48*, 608–613. [CrossRef] [PubMed]
44. Chen, R.H.; Tennant, S.; Frost, D.; O'Beirne, M.J.; Karwowski, J.P.; Humphrey, P.E.; Malmberg, L.-H.; Choi, W.; Brandt, K.D.; West, P. Discovery of saricandin, a novel papulacandin, from a *Fusarium* species. *J. Antibiot.* **1996**, *49*, 596–598. [CrossRef]
45. Brady, S.F.; Clardy, J. CR377, a new pentaketide antifungal agent isolated from an endophytic fungus. *J. Nat. Prod.* **2000**, *63*, 1447–1448. [CrossRef]
46. Saetang, P.; Rukachaisirikul, V.; Phongpaichit, S.; Sakayaroj, J.; Shi, X.; Chen, J.; Shen, X. β-Resorcylic macrolide and octahydronaphthalene derivatives from a seagrass-derived fungus *Fusarium* sp. PSU-ES123. *Tetrahedron* **2016**, *72*, 6421–6427. [CrossRef]
47. Hiramatsu, F.; Miyajima, T.; Murayama, T.; Takahashi, K.; Koseki, T.; Shiono, Y. Isolation and structure elucidation of neofusapyrone from a marine-derived *Fusarium* species, and structural revision of fusapyrone and deoxyfusapyrone. *J. Antibiot.* **2006**, *59*, 704–709. [CrossRef]
48. Xu, X.; Zhao, S.; Yu, Y.; Chen, Z.; Shen, H.; Zhou, L. Beauvericin K, a new antifungal beauvericin analogue from a marine-derived *Fusarium* sp. *Nat. Prod. Commun.* **2016**, *11*, 1825–1826. [CrossRef]
49. Ibrahim, S.R.M.; Abdallah, H.M.; Elkhayat, E.S.; Al Musayeib, N.M.; Asfour, H.Z.; Zayed, M.F.; Mohamed, G.A. Fusaripeptide A: New antifungal and anti-malarial cyclodepsipeptide from the endophytic fungus *Fusarium* sp. *J. Asian Nat. Prod. Res.* **2018**, *20*, 75–85. [CrossRef] [PubMed]
50. Dreyfuss, M.; Härri, E.; Hofmann, H.e.a.; Kobel, H.; Pache, W.; Tscherter, H. Cyclosporin A and C: New metabolites from *Trichoderma polysporum* (Link ex Pers.) *Rifai. Appl. Microbiol. Biot.* **1976**, *3*, 125–133. [CrossRef]
51. Baráth, Z.; Baráthová, H.; Betina, V.; Nemec, P. Ramihyphins—Antifungal and morphogenic antibiotics from *Fusarium* sp. S-435. *Folia. Microbiol.* **1974**, *19*, 507–511. [CrossRef] [PubMed]
52. Parish, C.A.; Smith, S.K.; Calati, K.; Zink, D.; Wilson, K.; Roemer, T.; Jiang, B.; Xu, D.; Bills, G.; Platas, G. Isolation and structure elucidation of parnafungins, antifungal natural products that inhibit mRNA polyadenylation. *J. Am. Chem. Soc.* **2008**, *130*, 7060–7066. [CrossRef] [PubMed]
53. Overy, D.; Calati, K.; Kahn, J.N.; Hsu, M.J.; Martin, J.; Collado, J.; Roemer, T.; Harris, G.; Parish, C.A. Isolation and structure elucidation of parnafungins C and D, isoxazolidinone-containing antifungal natural products. *Bioorg. Med. Chem. Lett.* **2009**, *19*, 1224–1227. [CrossRef] [PubMed]
54. McBrien, K.D.; Gao, Q.; Huang, S.; Klohr, S.E.; Wang, R.R.; Pirnik, D.M.; Neddermann, K.M.; Bursuker, I.; Kadow, K.F.; Leet, J.E. Fusaricide, a new cytotoxic *N*-hydroxypyridone from *Fusarium* sp. *J. Nat. Prod.* **1996**, *59*, 1151–1153. [CrossRef]
55. Hilário, F.; Chapla, V.; Araujo, A.; Sano, P.; Bauab, T.; dos Santos, L. Antimicrobial Screening of Endophytic Fungi Isolated from the Aerial Parts of *Paepalanthus chiquitensis* (Eriocaulaceae) Led to the Isolation of Secondary Metabolites Produced by *Fusarium fujikuroi*. *J. Braz. Chem. Soc.* **2016**, *28*, 1389–1395. [CrossRef]
56. Ibrahim, S.M.; Mohamed, G.; Khayat, M.; Al Haidari, R.; El-Kholy, A.; Zayed, M. A new antifungal aminobenzamide derivative from the endophytic fungus *Fusarium* sp. *Pharmacogn. Mag.* **2019**, *15*, 204–207. [CrossRef]
57. Sakai, K.; Unten, Y.; Iwatsuki, M.; Matsuo, H.; Fukasawa, W.; Hirose, T.; Chinen, T.; Nonaka, K.; Nakashima, T.; Sunazuka, T.; et al. Fusaramin, an antimitochondrial compound produced by *Fusarium* sp., discovered using multidrug-sensitive *Saccharomyces cerevisiae*. *J. Antibiot.* **2019**, *72*, 645–652. [CrossRef]
58. Woscholski, R.; Kodaki, T.; McKinnon, M.; Waterfield, M.D.; Parker, P.J. A comparison of demethoxyviridin and wortmannin as inhibitors of phosphatidylinositol 3-kinase. *FEBS Lett.* **1994**, *342*, 109–114. [CrossRef]
59. Pedersen, P.B.; Miller, J.D. The fungal metabolite culmorin and related compounds. *Nat. Toxins* **1999**, *7*, 305–309. [CrossRef]

60. Strongman, D.; Miller, J.; Calhoun, L.; Findlay, J.; Whitney, N. The biochemical basis for interference competition among some lignicolous marine fungi. *Bot. Mar.* **1987**, *30*, 21–26. [CrossRef]
61. Kurobane, I.; Zaita, N.; Fukuda, A. New metabolites of *Fusarium martii* related to dihydrofusarubin. *J. Antibiot.* **1986**, *39*, 205–214. [CrossRef]
62. Limón, M.C.; Rodríguez-Ortiz, R.; Avalos, J. Bikaverin production and applications. *Appl. Microbiol. Biotechnol.* **2010**, *87*, 21–29. [CrossRef]
63. Deshmukh, R.; Mathew, A.; Purohit, H.J. Characterization of antibacterial activity of bikaverin from *Fusarium* sp. HKF15. *J. Biosci. Bioeng.* **2014**, *117*, 443–448. [CrossRef] [PubMed]
64. Bushnell, G.W.; Li, Y.-L.; Poulton, G.A. Pyrones. X. Lateropyrone, a new antibiotic from the fungus *Fusarium lateritium* Nees. *Can. J. Chem.* **1984**, *62*, 2101–2106. [CrossRef]
65. Clark, T.N.; Carroll, M.; Ellsworth, K.; Guerrette, R.; Robichaud, G.A.; Johnson, J.A.; Gray, C.A. Antibiotic mycotoxins from an endophytic *Fusarium acuminatum* isolated from the medicinal plant *Geum macrophyllum*. *Nat. Prod. Commun.* **2018**, *13*, 1934578X1801301017. [CrossRef]
66. Ariantari, N.P.; Frank, M.; Gao, Y.; Stuhldreier, F.; Kiffe-Delf, A.-L.; Hartmann, R.; Höfert, S.-P.; Janiak, C.; Wesselborg, S.; Müller, W.E.G.; et al. Fusaristatins D–F and (7*S*,8*R*)-(−)-chlamydospordiol from *Fusarium* sp. BZCB-CA, an endophyte of *Bothriospermum chinense*. *Tetrahedron* **2021**, *85*, 132065–132071. [CrossRef]
67. Ola, A.R.B.; Thomy, D.; Lai, D.; Brötz-Oesterhelt, H.; Proksch, P. Inducing secondary metabolite production by the endophytic fungus *Fusarium tricinctum* through coculture with *Bacillus subtilis*. *J. Nat. Prod.* **2013**, *76*, 2094–2099. [CrossRef]
68. Okada, H.; Nagashima, M.; Suzuki, H.; Nakajima, S.; Kojiri, K.; Suda, H. BE-29602, a new member of the papulacandin family. *J. Antibiot.* **1996**, *49*, 103–106. [CrossRef]
69. Kobayashi, H.; Sunaga, R.; Furihata, K.; Morisaki, N.; IWasaki, S. Isolation and structures of an antifungal antibiotic, fusarielin A, and related compounds produced by a *Fusarium* sp. *J. Antibiot.* **1995**, *48*, 42–52. [CrossRef] [PubMed]
70. Liang, X.A.; Ma, Y.M.; Zhang, H.C.; Liu, R. A new helvolic acid derivative from an endophytic *Fusarium* sp. of Ficus carica. *Nat. Prod. Res.* **2016**, *30*, 2407–2412. [CrossRef] [PubMed]
71. Janevska, S.; Arndt, B.; Niehaus, E.-M.; Burkhardt, I.; Rösler, S.M.; Brock, N.L.; Humpf, H.-U.; Dickschat, J.S.; Tudzynski, B. Gibepyrone biosynthesis in the rice pathogen *Fusarium fujikuroi* is facilitated by a small polyketide synthase gene cluster. *J. Biol. Chem.* **2016**, *291*, 27403–27420. [CrossRef]
72. Zhou, G.; Qiao, L.; Zhang, X.; Sun, C.; Che, Q.; Zhang, G.; Zhu, T.; Gu, Q.; Li, D. Fusaricates H-K and fusolanones A-B from a mangrove endophytic fungus *Fusarium solani* HDN15-410. *Phytochemistry* **2019**, *158*, 13–19. [CrossRef]
73. Evidente, A.; Conti, L.; Altomare, C.; Bottalico, A.; Sindona, G.; Segre, A.L.; Logrieco, A. Fusapyrone and deoxyfusapyrone, two antifungal α-pyrones from *Fusarium semitectum*. *Nat. Toxins* **1994**, *2*, 4–13. [CrossRef] [PubMed]
74. Altomare, C.; Perrone, G.; Zonno, M.C.; Evidente, A.; Pengue, R.; Fanti, F.; Polonelli, L. Biological characterization of fusapyrone and deoxyfusapyrone, two bioactive secondary metabolites of *Fusarium semitectum*. *J. Nat. Prod.* **2000**, *63*, 1131–1135. [CrossRef] [PubMed]
75. Son, S.; Kim, H.; Choi, G.; Lim, H.; Jang, K.; Lee, S.; Lee, S.; Sung, N.; Kim, J.C. Bikaverin and fusaric acid from *Fusarium oxysporum* show antioomycete activity against *Phytophthora infestans*. *J. Appl. Microbiol.* **2008**, *104*, 692–698. [CrossRef] [PubMed]
76. Bacon, C.W.; Hinton, D.M.; Hinton, A., Jr. Growth-inhibiting effects of concentrations of fusaric acid on the growth of *Bacillus mojavensis* and other biocontrol *Bacillus* species. *J. Appl. Microbiol.* **2006**, *100*, 185–194. [CrossRef] [PubMed]
77. Poleto, L.; da Rosa, L.O.; Fontana, R.C.; Rodrigues, E.; Poletto, E.; Baldo, G.; Paesi, S.; Sales-Campos, C.; Camassola, M. Production of antimicrobial metabolites against pathogenic bacteria and yeasts by *Fusarium oxysporum* in submerged culture processes. *Bioproc. Biosyst. Eng.* **2021**, *44*, 1321–1332. [CrossRef] [PubMed]
78. Vesonder, R.F.; Tjarks, L.W.; Rohwedder, W.K.; Burmeister, H.R.; Laugal, J.A. Equisetin, an antibiotic from *Fusarium equisetin* NRRL 5537, identified as a derivative of *N*-methyl-2, 4-pyrollidone. *J. Antibiot.* **1979**, *32*, 759–761. [CrossRef]
79. Ratnaweera, P.B.; de Silva, E.D.; Williams, D.E.; Andersen, R.J. Antimicrobial activities of endophytic fungi obtained from the arid zone invasive plant Opuntia dillenii and the isolation of equisetin, from endophytic *Fusarium* sp. *BMC Complement. Altern. Med.* **2015**, *15*, 220. [CrossRef]
80. Ibrahim, S.R.M.; Elkhayat, E.S.; Mohamed, G.A.A.; Fat'hi, S.M.; Ross, S.A. Fusarithioamide A, a new antimicrobial and cytotoxic benzamide derivative from the endophytic fungus *Fusarium chlamydosporium*. *Biochem. Biophys. Res. Commun.* **2016**, *479*, 211–216. [CrossRef]
81. Ibrahim, S.R.M.; Mohamed, G.A.; Al Haidari, R.A.; Zayed, M.F.; El-Kholy, A.A.; Elkhayat, E.S.; Ross, S.A. Fusarithioamide B, a new benzamide derivative from the endophytic fungus *Fusarium chlamydosporium* with potent cytotoxic and antimicrobial activities. *Bioorg. Med. Chem.* **2018**, *26*, 786–790. [CrossRef] [PubMed]
82. Jiang, Z.; Barret, M.-O.; Boyd, K.G.; Adams, D.R.; Boyd, A.S.; Burgess, J.G. JM47, a cyclic tetrapeptide HC-toxin analogue from a marine *Fusarium* species. *Phytochemistry* **2002**, *60*, 33–38. [CrossRef] [PubMed]
83. Roig, M.; Meca, G.; Marin, R.; Ferrer, E.; Manes, J. Antibacterial activity of the emerging *Fusarium* mycotoxins enniatins A, A(1), A(2), B, B(1), and B(4) on probiotic microorganisms. *Toxicon* **2014**, *85*, 1–4. [CrossRef]

84. Meca, G.; Sospedra, I.; Valero, M.A.; Manes, J.; Font, G.; Ruiz, M.J. Antibacterial activity of the enniatin B, produced by *Fusarium tricinctum* in liquid culture, and cytotoxic effects on Caco-2 cells. *Toxicol. Mech. Method.* **2011**, *21*, 503–512. [CrossRef]
85. Meca, G.; Soriano, J.M.; Gaspari, A.; Ritieni, A.; Moretti, A.; Manes, J. Antifungal effects of the bioactive compounds enniatins A, A(1), B, B(1). *Toxicon* **2010**, *56*, 480–485. [CrossRef]
86. Tsantrizos, Y.S.; Xu, X.-J.; Sauriol, F.; Hynes, R.C. Novel quinazolinones and enniatins from *Fusarium lateritium* Nees. *Can. J. Chem.* **1993**, *71*, 1362–1367. [CrossRef]
87. Meca, G.; Sospedra, I.; Soriano, J.M.; Ritieni, A.; Moretti, A.; Manes, J. Antibacterial effect of the bioactive compound beauvericin produced by *Fusarium proliferatum* on solid medium of wheat. *Toxicon* **2010**, *56*, 349–354. [CrossRef]
88. Wu, Q.; Patocka, J.; Nepovimova, E.; Kuca, K. A Review on the Synthesis and Bioactivity Aspects of Beauvericin, a *Fusarium* Mycotoxin. *Front. Pharmacol.* **2018**, *9*, 1338. [CrossRef]
89. Hawas, U.W.; Al-Farawati, R.; Abou El-Kassem, L.T.; Turki, A.J. Different Culture Metabolites of the Red Sea Fungus *Fusarium equiseti* Optimize the Inhibition of Hepatitis C Virus NS3/4A Protease (HCV PR). *Mar. Drugs* **2016**, *14*, 190. [CrossRef]
90. Tompa, D.R.; Immanuel, A.; Srikanth, S.; Kadhirvel, S. Trends and strategies to combat viral infections: A review on FDA approved antiviral drugs. *Int. J. Biol. Macromol.* **2021**, *172*, 524–541. [CrossRef]
91. Chang, S.; Yan, B.; Chen, Y.; Zhao, W.; Gao, R.; Li, Y.; Yu, L.; Xie, Y.; Si, S.; Chen, M. Cytotoxic hexadepsipeptides and anti-coronaviral 4-hydroxy-2-pyridones from an endophytic *Fusarium* sp. *Front. Chem.* **2022**, *10*, 1106869. [CrossRef] [PubMed]
92. McKee, T.C.; Bokesch, H.R.; McCormick, J.L.; Rashid, M.A.; Spielvogel, D.; Gustafson, K.R.; Alavanja, M.M.; Cardelline, J.H., 2nd; Boyd, M.R. Isolation and characterization of new anti-HIV and cytotoxic leads from plants, marine, and microbial organisms. *J. Nat. Prod.* **1997**, *60*, 431–438. [CrossRef] [PubMed]
93. Guo, Y.W.; Liu, X.J.; Yuan, J.; Li, H.J.; Mahmud, T.; Hong, M.J.; Yu, J.C.; Lan, W.J. l-Tryptophan induces a marine-derived *Fusarium* sp. to produce indole alkaloids with activity against the Zika virus. *J. Nat. Prod.* **2020**, *83*, 3372–3380. [CrossRef] [PubMed]
94. Nonaka, K.; Chiba, T.; Suga, T.; Asami, Y.; Iwatsuki, M.; Masuma, R.; Ōmura, S.; Shiomi, K. Coculnol, a new penicillic acid produced by a coculture of *Fusarium solani* FKI-6853 and *Talaromyces* sp. FKA-65. *J. Antibiot.* **2015**, *68*, 530–532. [CrossRef]
95. Lee, S.M.; Kim, M.S.; Hayat, F.; Shin, D. Recent Advances in the Discovery of Novel Antiprotozoal Agents. *Molecules* **2019**, *24*, 3386. [CrossRef]
96. Trisuwan, K.; Khamthong, N.; Rukachaisirikul, V.; Phongpaichit, S.; Preedanon, S.; Sakayaroj, J. Anthraquinone, cyclopentanone, and naphthoquinone derivatives from the sea fan-derived fungi *Fusarium* spp. PSU-F14 and PSU-F135. *J. Nat. Prod.* **2010**, *73*, 1507–1511. [CrossRef]
97. Balan, J.; Fuska, J.; Kuhr, I.; Kuhrová, V. Bikaverin, an antibiotic from *Gibberella fujikuroi*, effective against *Leishmania brasiliensis*. *Folia Microbiol.* **1970**, *15*, 479–484. [CrossRef]
98. Nascimento, A.M.d.; Conti, R.; Turatti, I.C.; Cavalcanti, B.C.; Costa-Lotufo, L.V.; Pessoa, C.; Moraes, M.O.d.; Manfrim, V.; Toledo, J.S.; Cruz, A.K. Bioactive extracts and chemical constituents of two endophytic strains of *Fusarium oxysporum*. *Rev. Bras. Farmacogn.* **2012**, *22*, 1276–1281. [CrossRef]
99. Campos, F.F.; Sales Junior, P.A.; Romanha, A.J.; Araújo, M.S.; Siqueira, E.P.; Resende, J.M.; Alves, T.; Martins-Filho, O.A.; Santos, V.L.d.; Rosa, C.A. Bioactive endophytic fungi isolated from *Caesalpinia echinata* Lam. (Brazilwood) and identification of beauvericin as a trypanocidal metabolite from *Fusarium* sp. *Mem. Inst. Oswaldo Cruz* **2015**, *110*, 65–74. [CrossRef]
100. Ibrahim, S.R.; Abdallah, H.M.; Mohamed, G.A.; Ross, S.A. Integracides H-J: New tetracyclic triterpenoids from the endophytic fungus *Fusarium* sp. *Fitoterapia* **2016**, *112*, 161–167. [CrossRef]
101. Abdelhakim, I.; Bin Mahmud, F.; Motoyama, T.; Futamura, Y.; Takahashi, S.; Osada, H. Dihydrolucilactaene, a potent antimalarial compound from *Fusarium* sp. RK97-94. *J. Nat. Prod.* **2021**, *85*, 63–69. [CrossRef] [PubMed]
102. Kato, S.; Motoyama, T.; Futamura, Y.; Uramoto, M.; Nogawa, T.; Hayashi, T.; Hirota, H.; Tanaka, A.; Takahashi-Ando, N.; Kamakura, T. Biosynthetic gene cluster identification and biological activity of lucilactaene from *Fusarium* sp. RK97-94. *Biosci. Biotechnol. Biochem.* **2020**, *84*, 1303–1307. [CrossRef] [PubMed]
103. Abdelhakim, I.A.; Motoyama, T.; Nogawa, T.; Mahmud, F.B.; Futamura, Y.; Takahashi, S.; Osada, H. Isolation of new lucilactaene derivatives from P450 monooxygenase and aldehyde dehydrogenase knockout *Fusarium* sp. RK97-94 strains and their biological activities. *J. Antibiot.* **2022**, *75*, 361–374. [CrossRef] [PubMed]
104. Singh, S.B.; Zink, D.L.; Polishook, J.D.; Dombrowski, A.W.; Darkin-Rattray, S.J.; Schmatz, D.M.; Goetz, M.A. Apicidins: Novel cyclic tetrapeptides as coccidiostats and antimalarial agents from *Fusarium pallidoroseum*. *Tetrahedron Lett.* **1996**, *37*, 8077–8080. [CrossRef]
105. Singh, S.B.; Zink, D.L.; Liesch, J.M.; Dombrowski, A.W.; Darkin-Rattray, S.J.; Schmatz, D.M.; Goetz, M.A. Structure, histone deacetylase, and antiprotozoal activities of apicidins B and C, congeners of apicidin with proline and valine substitutions. *Org. Lett.* **2001**, *3*, 2815–2818. [CrossRef]
106. Von Bargen, K.W.; Niehaus, E.-M.; Bergander, K.; Brun, R.; Tudzynski, B.; Humpf, H.-U. Structure elucidation and antimalarial activity of apicidin F: An apicidin-like compound produced by *Fusarium fujikuroi*. *J. Nat. Prod.* **2013**, *76*, 2136–2140. [CrossRef]
107. Liu, Y.; Xu, M.; Tang, Y.; Shao, Y.; Wang, H.; Zhang, H. Genome features and antiSMASH analysis of an endophytic strain *Fusarium* sp. R1. *Metabolites* **2022**, *12*, 521. [CrossRef] [PubMed]
108. Zhang, H.; Zhang, C.; Li, Q.; Ma, J.; Ju, J. Metabolic blockade-based genome mining reveals lipochain-linked dihydro-β-alanine synthetases involved in autucedine biosynthesis. *Org. Lett.* **2022**, *24*, 5535–5540. [CrossRef]

109. Kang, H.S.; Charlop-Powers, Z.; Brady, S.F. Multiplexed CRISPR/Cas9- and TAR-mediated promoter engineering of natural product biosynthetic gene clusters in yeast. *ACS Synth. Biol.* **2016**, *5*, 1002–1010. [CrossRef]
110. Keller, N.P. Fungal secondary metabolism: Regulation, function and drug discovery. *Nat. Rev. Microbiol.* **2019**, *17*, 167–180. [CrossRef]

Disclaimer/Publisher's Note: The statements, opinions and data contained in all publications are solely those of the individual author(s) and contributor(s) and not of MDPI and/or the editor(s). MDPI and/or the editor(s) disclaim responsibility for any injury to people or property resulting from any ideas, methods, instructions or products referred to in the content.

molecules

MDPI

Review

Compilation of the Antimicrobial Compounds Produced by *Burkholderia* Sensu Stricto

Mariana Rodríguez-Cisneros [1,†], Leslie Mariana Morales-Ruíz [1,†], Anuar Salazar-Gómez [2], Fernando Uriel Rojas-Rojas [3,4] and Paulina Estrada-de los Santos [1,*]

[1] Departamento de Microbiología, Escuela Nacional de Ciencias Biológicas, Instituto Politécnico Nacional, Prol. de Carpio y Plan de Ayala S/N Col. Santo Tomás Alc. Miguel Hidalgo, Ciudad de México 11340, Mexico

[2] Escuela Nacional de Estudios Superiores Unidad León, Universidad Nacional Autónoma de México (ENES-León UNAM), Blvd. UNAM 2011, León, Guanajuato 37684, Mexico

[3] Laboratorio de Ciencias AgroGenómicas, Escuela Nacional de Estudios Superiores Unidad León, Universidad Nacional Autónoma de México (ENES-León UNAM), Blvd. UNAM 2011, León, Guanajuato 37684, Mexico

[4] Laboratorio Nacional PlanTECC, Escuela Nacional de Estudios Superiores Unidad León, Universidad Nacional Autónoma de México (ENES-León UNAM), Blvd. UNAM 2011, León, Guanajuato 37684, Mexico

* Correspondence: pestrada@ipn.mx or pestradadelossantos@gmail.com; Tel.: +52-557296000 (ext. 62352)

† These authors contributed equally to this work.

Abstract: Due to the increase in multidrug-resistant microorganisms, the investigation of novel or more efficient antimicrobial compounds is essential. The World Health Organization issued a list of priority multidrug-resistant bacteria whose eradication will require new antibiotics. Among them, *Acinetobacter baumannii, Pseudomonas aeruginosa*, and Enterobacteriaceae are in the "critical" (most urgent) category. As a result, major investigations are ongoing worldwide to discover new antimicrobial compounds. *Burkholderia*, specifically *Burkholderia* sensu stricto, is recognized as an antimicrobial-producing group of species. Highly dissimilar compounds are among the molecules produced by this genus, such as those that are unique to a particular strain (like compound CF66I produced by *Burkholderia cepacia* CF-66) or antimicrobials found in a number of species, e.g., phenazines or ornibactins. The compounds produced by *Burkholderia* include N-containing heterocycles, volatile organic compounds, polyenes, polyynes, siderophores, macrolides, bacteriocins, quinolones, and other not classified antimicrobials. Some of them might be candidates not only for antimicrobials for both bacteria and fungi, but also as anticancer or antitumor agents. Therefore, in this review, the wide range of antimicrobial compounds produced by *Burkholderia* is explored, focusing especially on those compounds that were tested in vitro for antimicrobial activity. In addition, information was gathered regarding novel compounds discovered by genome-guided approaches.

Keywords: *Burkholderia* sensu stricto; antimicrobials; non-ribosomal peptides

check for updates

Citation: Rodríguez-Cisneros, M.; Morales-Ruíz, L.M.; Salazar-Gómez, A.; Rojas-Rojas, F.U.; Estrada-de los Santos, P. Compilation of the Antimicrobial Compounds Produced by *Burkholderia* Sensu Stricto. *Molecules* **2023**, *28*, 1646. https://doi.org/10.3390/molecules28041646

Academic Editors: Xun Song, Chenyang Li and Yifu Guan

Received: 5 January 2023
Revised: 26 January 2023
Accepted: 28 January 2023
Published: 8 February 2023

Copyright: © 2023 by the authors. Licensee MDPI, Basel, Switzerland. This article is an open access article distributed under the terms and conditions of the Creative Commons Attribution (CC BY) license (https://creativecommons.org/licenses/by/4.0/).

1. Introduction

Burkholderia sensu lato comprises more than 100 species, which were gradually discovered during 30 years of research. In recent years, using comparative genomics, this large group was divided into seven genera, namely *Burkholderia* sensu stricto (s.s.), *Paraburkholderia, Caballeronia, Robbsia, Mycetohabitans, Trinickia*, and *Pararobbsia* [1–5]. The species contained in these genera thrive in soil, water, rhizosphere, plant nodules, fungi, and in animal and human infections. *Burkholderia* s.s. is formed by three groups of species: (a) the *Burkholderia pseudomallei* group (composed of 8 species), (b) the *Burkholderia* species that are mostly plant pathogenic bacteria (containing 4 species), and (c) the *Burkholderia cepacia* complex (Bcc) (composed of 25 species). The *B. pseudomallei* group is of worldwide importance because the species *B. pseudomallei* and *Burkholderia* mallei cause the mortal (if not treated) melioidosis diseases in humans and animals, and glanders, specifically in equines, respectively [6,7]. Recently, "*Burkholderia mayonis*" and "*Burkholderia savannae*" were described within the *B. pseudomallei* group [8]. Although some species from the Bcc are plant pathogens, there is a small group where *Burkholderia plantarii, Burkholderia gladioli,* and *Burkholderia glumae* are included; however, they are not part of the Bcc. Recently, "*Burkholderia perseverans*" was added to this group; this species produces volatile compounds that inhibit plant pathogens but has not been described as a pathogen per se [9]. The Bcc species are best known as opportunistic pathogens, mainly in cystic fibrosis (CF) and immunocompromised patients [10]. Within the Bcc, the last species described was *Burkholderia orbicola* [11]. Another important feature within the Bcc is their resistance to many antibiotics [12], which especially endangers the lives of CF and immunocompromised patients.

Bcc is also known for their phenotypic and genotypic diversity [13], which includes features/functions for biotechnological uses. This functional ability has been shown through biopesticidal activity in the rhizosphere [14] and by the bioremediation of xenobiotics compounds [15,16]. The Bcc are also able to produce a large array of compounds involved in the inhibition of pathogenic bacteria, fungi, and yeasts, which is important for tackling multidrug-resistant microorganisms [17]. Interestingly, the *B. pseudomallei* group encodes the largest capacity for secondary metabolite biosynthesis (>11% of their genomes) [18]. Moreover, the Bcc account for significant antibiotic biosynthetic capacity, e.g., *Burkholderia ambifaria* involves 9% of its genome in secondary metabolism, and *B. gladioli* and *B. glumae* dedicate 10% or more of their genome to antibiotic biosynthesis. Therefore, this review aims to enumerate in detail all antimicrobial compounds produced by *Burkholderia* s.s., detailing activities demonstrated in vitro and reviewing the novel compounds discovered by genome-guided approaches. The compounds are grouped and discussed according to common chemical features and shown as a list in Supplementary Table S1 with the numbers given in bold.

2. N-Containing Heterocycles

The analogs of nitrogen-based heterocycles occupy an exclusive position as a valuable source of therapeutic agents in medicinal chemistry [19]. Many of these compounds are volatile organic compounds (VOCs) or volatile nitrogen compounds [20]. **Pyrazine**-derived compounds (VOC) produced by *Burkholderia seminalis* JRBHU6 have been named PPDH and identified as (pyrrolo (1,2-a) pyrazine-1,4-dione, hexahydro) (**1**) and PPDHMP identified as $C_{11}H_{18}N_2O_2$ (pyrrolo (1,2-a) pyrazine-1,4-dione, hexahydro-3(2-methyl-propyl)) (**2**) [21]. The pyrrolo [1,2-a] pyrazine core occurs in nature and is frequently used in drug design. Pyrrole has therapeutic significance as an anticancer, antimicrobial, and antiviral agent [22]. Compounds **1** and **2** produced by *B. seminalis* JRBHU6 inhibit the fungi genera *Fusarium, Aspergillus, Microsporum, Trichophyton*, and *Trichoderma*, and the bacterial genera *Staphyloccous, Pseudomonas, Escherichia, Shigella*, and *Klebsiella* [21]. Molecular docking with bioactive compounds **1** and **2** was carried out to identify protein targets. According to the analysis, **2** showed full fitness to human proteins cell division protein kinase 7 and mitogen-activated protein kinase 8, suggesting a putative role in the inhibition of protein kinase activity in sensitive microorganisms. Good binding affinity and full fitness were also found with bacterial proteins such as choloylglycine hydrolase, camphor t-monooxygenase, chitinase B, and tyrosine phenol-lyase, while **1** showed good strong full fitness only with chitinase B. Other N-containing antimicrobial compounds are **iminopyrrolidines** produced by *B. plantarii* 9424. These compounds are 2-imino-3-methylene-5-L(carboxy-L-valyl)-pyrrolidine (**3**) and 2-imino-3-methylene-5-L(carboxy-L-threoninyl)-pyrrolidine (**4**). These are amino acid conjugates and have high in vitro inhibitory activity against the bacteria *Erwinia amylovora*, a pathogen causing fire blight disease in apple and pear trees [23]. A **pyrazole molecule** that consisted of a substituted pyrazole, linked to the aspartate-b-carboxyl of the tripeptide L-alanyl-L-*homoserinyl*-L-aspartate, resulted in a deduced structure 3-[L-alanyl-L-*homoserinyl*-L-aspartyl-b-carboxy]-4-hydroxy-5-oxopyrazole (**5**). This compound was produced by *B. glumae* and was found to inhibit bacterial pathogens such as different species of *Erwinia*, *Pectobacterium*, *Pseudomonas*, and *Xanthomonas* [24]. **Pyrrolnitrin (6)**, 3-chloro-4-(2-nitro-3-chlorophenyl)pyrrole, is a microbial halometabolite (containing a halogen moiety) with a large antimicrobial significance in agricultural, pharmaceutical, and industrial implications [25]. This compound is produced by rhizospheric fluorescent and non-fluorescent pseudomonads, *Serratia* and *Burkholderia*. Pyrrolnitrin was first discovered in *Pseudomonas* (now *Burkholderia*) *pyrrocinia* in 1960 [26,27]; other species such as *Burkholderia cepacia* and *Burkholderia ambifaria* are able to synthesize it as well [28–31]. A number of phytopathogens are inhibited by **6**, e.g., Penicillium, Phytophthora, *Fusarium*, *Rhizoctonia*, *Colletotrichum*, and *Sclerotinia*, yeast such as *Candida*, *Hansenula*, and *Saccharomyces*, and bacteria such as *Bacillus* and *Streptomyces*. Interestingly, the production of pyrrolnitrin was induced when chloramphenicol was added to the culture medium of *B. ambifaria* AMMD^{T} [31]. **Phenazines** are a large group of nitrogen-containing heterocycles with diverse chemical structures and pharmacological activity such as antimicrobial, antiparasitic, neuroprotective, insecticidal, anti-inflammatory, and anticancer [32]. There are more than 100 phenazine derivatives produced by bacteria and archaea. *Burkholderia cepacia* 5.5B produces the phenazine 4,9-dihydroxyphenazine-1,6-dicarboxylic acid dimethyl ester (**7**), which inhibits *Rhizoctonia solani* [33]. The production of phenazines in *B. lata* was strongly affected by the growth conditions, the best production being observed in culture grown in King's B medium [34]. Moreover, the involvement of phenazine in the formation of biofilm by *B. lata* was analyzed using a phenazine-overproducing strain, a phenazine-deficient mutant, and the wild type of the strain. The results showed that both the wild type and the overproducing strain formed thicker biofilms and attached more quickly than the mutant, suggesting a role of phenazine in biofilm formation by *B. lata* and, therefore, a role in the pathogenicity of this member of Bcc. *Burkholderia glumae* 411gr-6 was found to synthesize **phencomycin** (**8**) (a phenazine with two substituents, a carboxyl and a carbomethoxy group) and two new derivatives, 4-hydroxyphencomycin (**9**) and 5,10-dihydro-4,9-dihydroxyphencomycin methyl ester (**10**) [35]. The three compounds inhibit several plant pathogenic fungi, yeasts, and bacteria. *Burkholderia* sp. HQB-1, closely related to *Burkholderia stagnalis*, produces PCA, **phenazine-1-carboxilic acid (11)**, which has been proposed to protect banana against *Fusarium* oxysporum wilt. Compound **11** produced by strain HQB-1 also inhibits the genera *Colletotrichum*, Botrytis, and *Curvularia* [36]. **Indole** compounds and derivatives are N-containing heterocycles; among this kind of compounds the VOC indole (**12**) produced by *Burkholderia cenocepacia* ETR-B22 inhibited the fungi *Alternaria, Aspergillus, Bipolaris, Bacillus, Fusarium, Helminthosporium, Mycosphaerella, Magnaporthe, Phyllosticta*, and *Rhizoctonia* [37]. The **pityriacitrin (13)**, a b-carboline alkaloid with an indole ring attached with a carbonyl group on C-1 position and the derivative **pityriacitrin B (14)** isolated and identified in *Burkholderia* sp. NBF227, was tested for cytotoxic activity against cancer cell lines, but chemically synthesized derivatives from the previous compounds were more effective [38]. Other synthesized pityriacitrin derivatives from *Burkholderia* sp. NBF227 were investigated for antifungal activity [39]. The fungicidal activity was tested with four taxonomically different plant pathogens (oomycetes, ascomycetes, deuteromycetes, and basidiomycetes), showing that pityriacitrin displayed broad-spectrum antifungal activity and protected pepper leaves and grapefruits against infection by *P. capsici* and *B. cinerea*, respectively. Some N-containing heterocycles are shown in Figure 1.

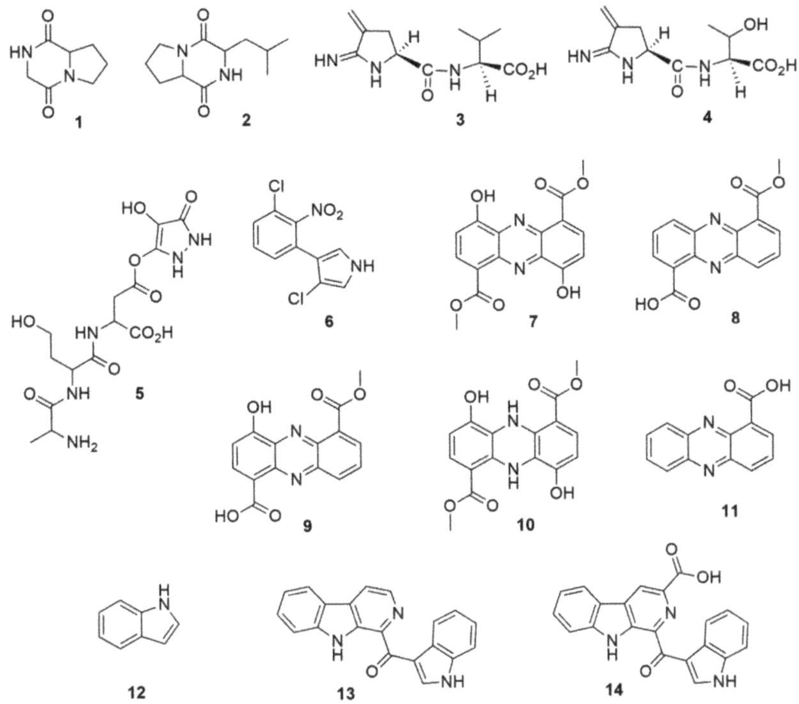

Figure 1. N-containing heterocycles. (**1**) pyrrolo (1,2-a) pyrazine-1,4-dione, hexahydro. (**2**) pyrrolo (1,2-a) pyrazine-1,4-dione, hexahydro-3(2-methyl-propyl). (**3**) 2-imino-3-methylene-5-L(carboxy-L-valyl)-pyrrolidine. (**4**) 2-imino-3-methylene-5-L(carboxy-L-threoninyl)-pyrrolidine. (**5**) 3-[L-alanyl-L-*homoserinyl*-L-aspartyl-b-carboxy]-4-hydroxy-5-oxopyrazole. (**6**) pyrrolnitrin, 3-chloro-4-(2-nitro-3-chlorophenyl)pyrrole. (**7**) phenazine, 4,9-dihydroxyphenazine-1,6-dicarboxylic acid dimethyl ester. (**8**) phencomycin. (**9**) 4-hydroxyphencomycin. (**10**) 5,10-dihydro-4,9-dihydroxyphencomycin methyl ester. (**11**) phenazine-1-carboxilic acid. (**12**) índole. (**13**) pityriacitrin. (**14**) pityriacitrin B.

3. Volatile Organic Compounds

Besides the VOCs mentioned in the N-containing heterocycles section, *B. cenocepacia* ETR-B22 also synthetizes other VOCs that lack nitrogen in their ring structure (Figure 2). These compounds are the benzyl derivatives **methyl anthranilate (15)**, **methyl salicylate (16)**, **methyl benzoate (17)**, **benzyl propionate (18)**, **benzyl acetate (19)**, **3,5-Di-tert-butylphenol (20)**, **allyl benzyl ether (21)**, and **benzyl benzoate (22)** (Figure 2), which inhibit an important number of fungal plant pathogens [37]. The VOCs **dimethyl trisulfide (23)**, **nonanoic acid (24)**, **2-pentadecanone (25)**, and 3-**hexen-1-ol, benzoate, (Z)- (26)** produced by the strain ETR-B22 also have antifungal activity. *Burkholderia gladioli* strain BBB-01, isolated from rice shoots, emits the VOCs **dimethyl disulfide (27)** and **2,5-dimethylfuran (28)** with inhibitory activity against the phytopathogenic fungi *M. oryzae*, *Gibberella fujikuroi*, *Sarocladium oryzae*, *Phellinus noxius*, and *Colletotrichum fructicola* and human pathogen *C. albicans* [40].

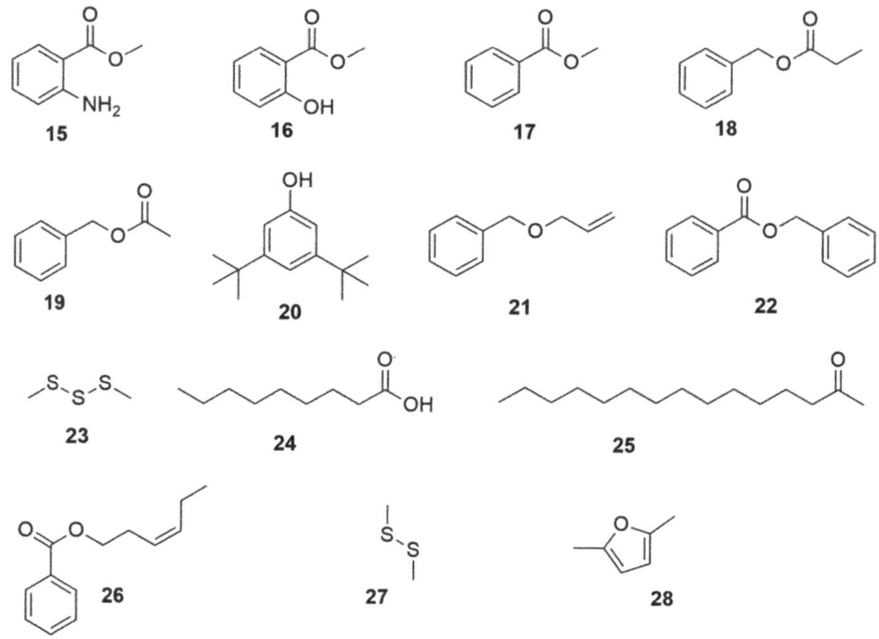

Figure 2. Volatile organic compounds. (**15**) methyl anthranilate. (**16**) methyl salicylate. (**17**) methyl benzoate. (**18**) benzyl propionate. (**19**) benzyl acetate. (**20**) 3,5-Di-tert-butylphenol. (**21**) allyl benzyl ether. (**22**) benzyl benzoate. (**23**) dimethyl trisulfide. (**24**) nonaoic acid. (**25**) 2-pentadecanone. (**26**) 3-hexen-1-ol, benzoate, (Z)-. (**27**) dimethyl disulfoxide. (**28**) 2,5-dimethylfuran.

4. Polyenes

Polyenes are poly-unsaturated organic compounds that contain at least three alternating double and single carbon–carbon bonds. Hunter and Manter [41] reported the isolation and purification of a compound with oxidizing and antibiotic properties from *B. cenocepacia* P525. The structure of this compound has not been reported but the preliminary chemical study showed that the compound could be a **polyene** with six conjugated double bonds and bacteriostatic activity against Enterobacter soli and Enterobacter aerogenes. *Burkholderia thailandensis* is a close relative of *B. pseudomallei* and therefore used as a model to study *B. pseudomallei* pathogenicity and biosynthetic pathways because *B. thailandensis* is not a pathogen. This species produces the polyene polyketide **thailandamide A (29)** (Figure 3) inhibiting notably bacteria such as *Bacillus subtilis*, *S. aureus*, and *Neisseria gonorrhoeae* [42]. Genetic analysis showed that **29** inhibits acetyl-CoA carboxylase (ACC), an essential enzyme responsible for the first step in fatty acid biosynthesis. Moreover, *B. thailandensis* synthetizes **thailandenes A (30), B (31)** and **C (32)** (Figure 3), which are linear formylated or acidic polyenes containing a combination of cis and trans double bonds [43]. Compounds **30** and **31** exhibited potent antimicrobial activity against *S. aureus* and *S. cerevisiae*. A polyketide (PK) **enacyloxin IIa (33)** (Figure 3) and its stereoisomer, designated **iso-enacyloxin IIa (34)**, produced by *B. ambifaria* AMMD^{T}, has activity against *Burkholderia multivorans*, *Burkholderia dolosa*, and *Acinetobacter baumannii* [44]. Expression analysis showed that enzymes-encoding genes for enacyloxin biosynthesis were among the most highly upregulated when strain AMMD^{T} was grown to stationary phase on glycerol. Moreover, enacyloxin targets protein biosynthesis by inhibition of the ribosomal elongation factor Tu [45]. *Burkholderia gladioli* pv. cocovenenans ATCC 33664^{T} produces **33** and **enacyloxin IIIa (35)**, and both were found to display equally potent activity against *Escherichia coli* and *P. aeruginosa* [46]. Moreover, traditionally used in food fermentations (tempe and sufu), Rhizopus microspores is accompanied by *B. gladioli* pv. cocovenenans. Thus, a coculture of both microorganisms showed that enacyloxins were found in high titers, with an increased production of the lethal toxin bongkrekic acid, showing the significance to food safety of this common microbial co-existence.

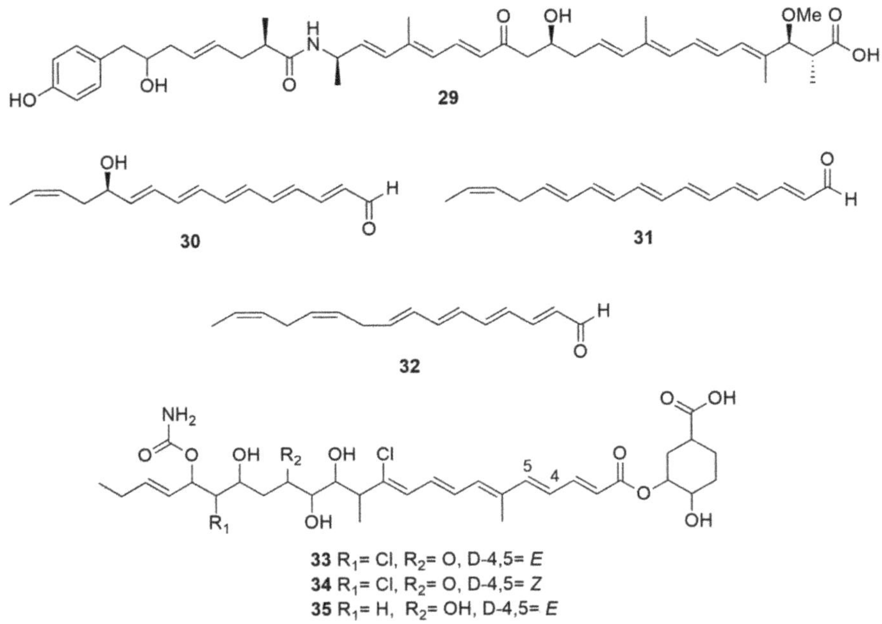

Figure 3. Polyenes. (**29**) thailandamide A. (**30**) thailandene A. (**31**) thailandene B. (**32**) thailandene C. (**33**) enacyloxin IIa. (**34**) iso-enacyloxin IIa. (**35**) enacyloxin IIIa.

5. Polyynes

Polyynes are organic compounds with alternating single and triple bonds, a series of consecutive alkynes. **Cepacin A (36)** and **cepacin B (37)** (Figure 4) are two acetylenic antibiotics produced by *B. cepacia* SC 11,783 with a strong activity against staphylococci [47] (Parker et al. 1984). *Burkholderia ambifaria* BCC0191 also synthetizes the metabolite cepacin A, which mediates protection of germinating crops against *Pythium* damping-off disease [48]. The activity was demonstrated when no biological control was observed with the inoculation of a cepacin mutant of strain BCC0191. *Burkholderia caryophylli*, a plant pathogen, produces the triple-bond compounds **caryoynecin A (38)**, **B (39)**, **and C (40)** (Figure 4). Although they are unstable, they can inhibit *E. coli*, *K. pneumoniae*, and *S. aureus* [49]. Caryoynecin analogues synthesized chemically were found more stable and demonstrated activity against *S. aureus*, *B. subtilis*, *Enterococcus faecalis*, *E. coli*, *Salmonella enteritidis*, *K. pneumoniae*, *Serratia marcescens*, *Proteus vulgaris*, *Shigella flexneri*, *Enterobacter cloacae*, *P. aeruginosa*, *T. mentagrophytes*, *Trichophyton interdigitale*, and *Trichophyton rubrum* [50]. *Burkholderia gladioli* also produces caryoynencin, which has activity against Purpureocillium lilacinum and has a role in the transition of the plant pathogen to an insect-defensive mutualism [51].

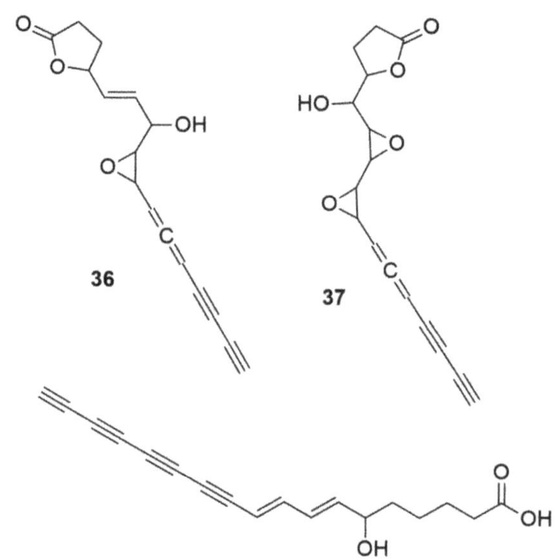

36 **37** **38**

39; 7Z, 7E
40; 7E, 7Z

Figure 4. Polyynes. (**36**) Cepacin. (**37**) Cepacin B. (**38**) Caryoynecin A. (**39**) Caryoynecin B. (**40**) Caryoynencin C.

6. Siderophores

Siderophores are low-molecular-weight organic compounds with high affinity to chelate iron (Fe). These compounds are produced by microorganisms and higher plants [52]. The typical siderophores ligands are cathecholate, *a*-hydrocycarboxylate, hydroxyphenyloxazolone, hydroxamate, *a*-aminocarboxylate, and a-hydroxyimidazole. Many bacterial siderophores are synthesized through **non-ribosomal peptide synthetases (NRPS)**. NRPS are a large family of biosynthetic enzymes that generate relevant natural compounds from amino acid precursors [53,54]. NRPSs are frequently categorized as type I and II [55]. Type I NRPSs are large modular complexes containing all the enzymes necessary to generate a peptide product in an assembly line fashion analogous to type I fatty acid synthases (FASs) and polyketide synthases (PKSs). Type II NRPS proteins are commonly standalone enzymes or didomains that coordinate to form unique amino acid derivatives. Unlike type II FAS and PKS, the type II NRPS proteins are linear, noniterative pathways that contain specialized tailoring enzymes and combine with other pathways to generate a final product [55]. **Pyochelin (41)** (Figure 5), a non-ribosomal peptide (NRP) purified from *P. aeruginosa* PAO1, was first found to display antibiotic activity against *S. aureus* and moderately against several species of *Xanthomonas* [56]. Pyochelin produced by "*Burkholderia paludis*", a non-validated species within the Bcc, inhibits three multidrug-resistant *E. faecalis* and four *S. aureus* strains but was not able to inhibit *Bacillus subtilis* ATCC 8188, *Bacillus cereus* ATCC 14579, *Aeromonas hydrophila* ATCC 49140, *E. coli* ATCC 25922, *Klebsiella pneumoniae* ATCC 10031, *Proteus mirabilis* ATCC 49140, *P. vulgaris* IMR, *P. aeruginosa* ATCC 10145 and ATCC BAA-47, *Salmonella Typhimurium* ATCC 14028, or *Shigella flexneri* ATCC 12022 [57]. This compound enhanced the production of intracellular reactive oxygen species (ROS), leading to cell death by disrupting the integrity of the bacterial membrane [58]. Pyochelin synthesized by *B. seminalis* TC3.4.2R3 inhibits *F. oxysporum*, which was demonstrated when a cepacin mutant was unable to inhibit the fungi [59]. **Cepabactin (42)** (Figure 5) is a 1-hydroxy-5-methoxy-6-methyl-2(1H)-pyridinone, a cyclic hydroxamate but also a heterocyclic analogue of catechol [60]. This compound produced by *B. cepacia* ATCC 25416^T has antimicrobial activity against *S. aureus*, *Staphylococcus epidermidis*, *Streptococcus faecalis*, *B. subtilis*, *Bacillus anthracis*, *E. coli*, *Salmonella Typhi*, *Salmonella Typhimurium*, *K. pneumoniae*, *P. vulgaris*, *P. mirabilis*, and *Proteus rettgeri* [60–62]. The production of **42** was present in only 12% of 65 *B. cepacia* strains, lower than other siderophores such as ornibactin (87%) or pyochelin (60%), showing that this siderophore is not largely produced in the species [63]. **Ornibactin (43)** (Figure 5), a NRP produced by most *Burkholderia* species [64], is a tetrapeptide siderophore with an l-ornithine-d-hydroxyaspartate-l-serine-l-ornithine backbone. A study with *Burkholderia* contaminans MS14, isolated from soil in Mississippi, USA, using transposon mutagenesis, resulted in two strains with insertional mutations in orbI gene (mutant MT577) and a luxR family transcriptional regulatory gene (mutant MT357) [65]. Both mutants lost bactericidal activity, relating the activity to siderophore ornibactin. This compound successfully inhibited *Xanthomonas citri* pv. malvacearum, *P. carotovorum* supsp. carotovorum, *Ralstonia solanacearum*, *P. syringae* pv. syringae, *E. amylovora*, *E. coli*, Clavibacter michiganensis subsp. Michiganensis, and *Bacillus* megaterium. The ornibactin mutant retained antifungal activity, showing that the antibacterial and antifungal action is independent, with the antifungal activity a result of occidiofungin. Similarly, ornibactin derivatives produced by *Burkholderia catarinensis* 89^T presented no activity against fungi [17]. Pyochelin and ornibactin are siderophores found in the genome of *Burkholderia orbicola* TAtl-371^T, and this bacterium produces siderophores in culture medium [66]. A test removing iron from the culture medium showed that the bacteria was able to inhibit *Paraburkholderia phenazinium* and *Candida glabrata*, suggesting the involvement of these siderophores in antagonism.

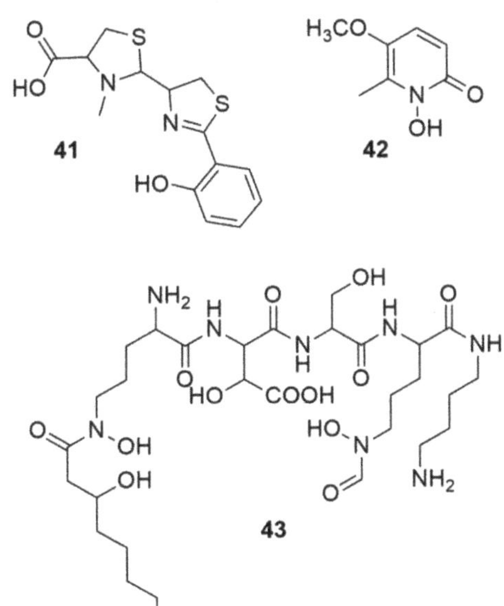

Figure 5. Siderophores. (**41**) Pyochelin. (**42**) Cepabactin. (**43**) Ornibactin.

7. Macrolides

Macrolides are various types of hydrophobic compounds containing a macrocyclic lactone ring and various side chains/groups [67]. *Burkholderia gladioli* BCC0238 isolated from a CF patient synthesize the PK, macrolide antibiotic **gladiolin (44)** (Figure 6), which has a strong activity against *M. tuberculosis* H37Rv and several other *M. tuberculosis* strains, *K. pneumoniae*, *A. baumannii*, *P. aeruginosa*, *E. clocae*, *Serratia plymuthica*, *"Ralstonia mannitolilytica"*, *B. multivorans*, *E. coli*, *Enterococcus faecium*, *S. aureus*, *B. subtilis*, and *C. albicans*, and was found to exhibit low toxicity toward an ovarian cancer cell line [68]. The mode of action of gladiolin is the inhibition of the RNA polymerase. Another PK macrolide **lagriene** (**45**) (Figure 6), produced by *B. gladioli* Lv-StA, has activity against *B. thuringiensis*, *M. vaccae*, vancomycin-resistant *E. faecalis* and *S. aureus* [51].

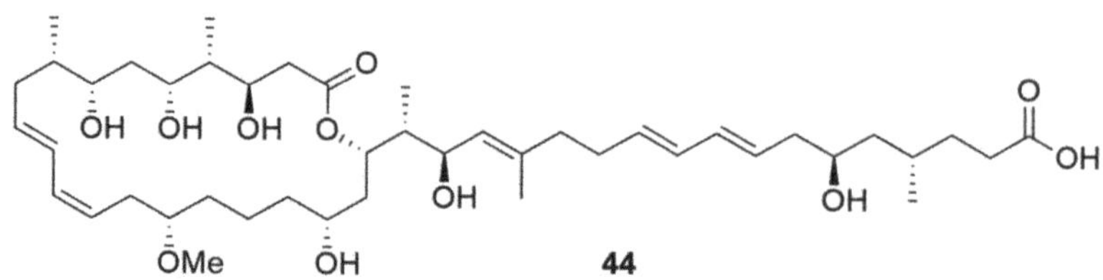

44

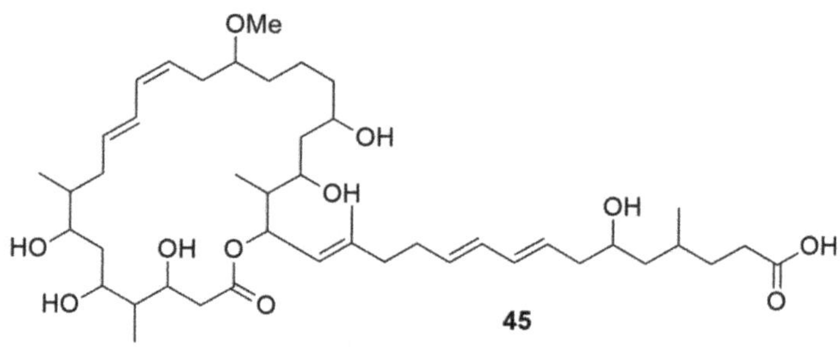

45

Figure 6. Macrolides. (**44**) Gladiolin. (**45**) Lagriene.

8. Bacteriocins

Bacteriocins are a varied class of bactericidal peptides or proteins produced by bacteria and archaea with bactericidal activity and specific immunity mechanisms toward strains closely related to the producer bacteria [69]. There are two central differences between bacteriocins and antibiotics: bacteriocins are ribosomally synthesized but antibiotics are not, and bacteriocins have a somewhat narrow killing spectrum while antibiotics have an extensive killing range. Bacteriocins vary in size, microbial target, mode of action, release, and immunity mechanism, and can be divided into two groups, the ones produced by Gram-negative bacteria and those by Gram-positive bacteria. Gram-negative bacteriocins are further classified according to their size into three main groups, namely colicins, phage-tail-like bacteriocins, and microcins [70]. Microcins are low-molecular-weight compounds grouped into class I (<5 kDa) or class II (5–10 kDa). Class I is now designated as ribosomally synthesized and post-translationally modified peptides (RiPP). *Burkholderia cenocepacia* BC0425 synthesizes the bacteriocin **tailocin**, a phage tail-like compound, named **BceTMilo** [71]. Unlike phages, tailocin injects through the cell membrane and disrupts the proton motive force [72]. Strains belonging to Bcc are sensitive to BceTMilo, and other non-Bcc such as *B. gladioli* and *B. glumae* are also sensitive to tailocin. Lectin-like bacteriocins (**LlpAs**) contain two monocot mannose-binding lectin (MMBL) domains, a module predominantly and abundantly found in lectins from monocot plants. *Burkholderia* strains can synthesize these bacteriocins. *B. cenocepacia* AU1054 (now *B. orbicola*) [11] produces an LlpA bacteriocin that inhibits *B. ambifaria*, *Burkholderia anthina*, *B. cenocepacia*, *B. contaminans* and *Burkholderia metallica* [73]. The homologue LlpA88 from *B. orbicola* TAtl-371^T inhibited the same species as strain AU1054 [66]. **Burkhocins M1** and **M2**, colicin M-like bacteriocins called ColM in *E. coli*, from *B. ambifaria* MEX-5 and AMMD^T were produced recombinantly, showing antagonistic activity against a number of Bcc strains [74]. Three strains from *Burkholderia ubonensis* inhibited *B. pseudomallei*; the antagonism from a representative strain (A21) was characterized, and a pepsin-sensitive moiety consistent with a bacteriocin-like compound was found, suggesting the antagonism is due to the production of a bacteriocin or bacteriocin-like inhibitory substance (**BLIS**) [75]. **Lasso peptides** are a structurally unique class of bioactive peptides characterized by a knotted arrangement where the C-terminus threads through an N-terminal macrolactam ring [76]. Lasso peptides are divided depending on the presence (class I) or absence (class II) of four conserved cysteine residues involved in the formation of two intramolecular disulfide bonds [77]. *Burkholderia thailandensis* produces the lasso peptide **capistruin**, a 19-amino-acid class II lasso peptide comprising an isopeptide bond between Gly1 and Asp9 resulting in a nine-residue macrolactam ring [78], which exhibits antimicrobial activity against *Burkholderia* (now *Paraburkholderia*) *caledonica*, *E. coli*, and *P. aeruginosa* [76]. **Ubonodin**, another lasso peptide produced by *B. ubonensis* MSMB2207, was heterologous expressed in *E. coli* BL21, displaying inhibition of *B. cepacia*, *B. multivorans* and *B. mallei*; it has a weak effect against *B. thailandensis* and had no effect on *B. gladioli* and *B. pseudomallei* [79]. This compound inhibits RNA polymerase in vitro and the narrow effect might allow therapeutic usage.

9. Quinolones

Quinolones were discovered as a by-product in the search for improved synthesis of the anti-malarial chloroquine; thus, they are fully synthetic molecules [80]. Today, it is known that molecules in the quinolone family are also present as natural products of plants and bacteria, although their potency has been tested only at the experimental level. The basic structure is a 3-carboxyquinolone and the first quinolone described was nalidixic acid. *Burkholderia thailandensis* contains a biosynthetic gene cluster (BGCs), which is a quorum-sensing-regulated *hmq* cluster that produces a diverse set of hydroxyalkylquinolines (HAQs). These compounds exist mainly in the 4(1H) quinolone type at neutral pH and are known as bioactive metabolites [81]. Two HAQ analogues, **HMNQ** (4-hydroxy-3-methyl-2-(2-nonenyl)-quinoline) (**46**) and **HQNO** (2-heptyl-4(1H)-quinolone N-oxide) (**47**) (Figure 7), synthesized by *B. thailandensis* E264^{T}, when challenged with antibiotics inhibit *B. subtilis* 168 but display weak activity against *E. coli* K12 [82]. It was found, as well, that both quinolones act synergistically to inhibit bacterial growth. Moreover, *B. thailandensis* produces **46** and rhamnolipids in outer membrane vesicles (OMV), which have antimicrobial and antibiofilm properties against methyl-resistant *S. aureus* [83]. Bacterial OMVs contain proteins, lipids, polysaccharides, and small molecules and serve numerous and versatile roles in intra- and interspecies interactions. *Burkholderia cepacia* RB425, isolated from lettuce root, makes the quinolone antibiotics **2-(2-heptenyl)-3-methyl-4-quinolinol** (**48**) and (**46**) (Figure 7) with high activity against fungal pathogen Verticillium dahlia, moderate inhibition of *Pyricularia oryzae* and *Cochliobolus myyabeanus*, and weak growth inhibition of *R. solani*, *F. oxysporum*, and *Gaeumannomyces graminis* [84]. A range of hydroxymethyl-alkylquinolines (**HMAQ**) produced by *B. cepacia* PC-II antagonizes *P. capsici*, which is responsible for Phytophthora blight in red peppers and many vegetables; in particular, **48** was the most potent against the oomycetes *P. capsica* and *Pythium* ultimum and the fungi *F. oxysporum* and *R. solanc* [85]. *Burkholderia* sp. QN15488 produces **burkholone (49)** (Figure 7), a (E)-3-methyl-2-(octenyl)-4-quinolone, this compound induces cell death in 32D/GR15 cells in IGF-I-containing medium [86]. Insulin-like growth factors (IGFs) play a key role in human cancer progression and IGF signals through the IGF-1 receptor are known to be significant for tumor cell growth and survival [87].

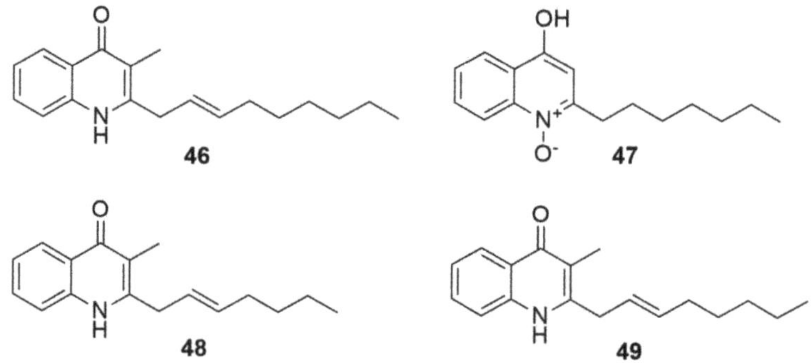

Figure 7. Quinolones. (**46**) HMNQ (4-hydroxy-3-methyl-2-(2-nonenyl)-quinoline). (**47**) HQNO (2-heptyl-4(1H)-quinolone N-oxide). (**48**) 2-(2-heptenyl)-3-methyl-4-quinolinol (C7Δ2). (**49**) Burkholone.

10. Other NPR-PK Compounds

Gladiofungin A (**50**) (Figure 8) is a novel antifungal PK that is highly unusual because it harbors a butanolide moiety. This compound is produced by the insect-associated bacteria *B. gladioli* HKI0739, which displays activity against *Penicillium notatum*, *Sprobolomyces salmonicolor*, and *P. lilacinum* [88]. The strains BCC0238 and BCC1622, belonging to *B. gladioli*, produce a PK antibiotic named **gladiostatin**, which has the same structure as (**50**) [89]. This molecule has promising activity against several cancer cell lines such as ovarian, pancreatic, and colon cancer, and inhibits tumor cell migration. Moreover, it was found to be inactive against a lung cell line, which indicates that it may exhibit some selectivity. Gladiostatin contains an unusual 2-acyl-4-hydroxy-3-methylbutenolide in addition to the glutarimide pharmacophore that also inhibits *S. cerevisiae*. **Glidobactins A** (**51**), **B** (**52**), and **C** (**53**) have a common cyclized tripeptide nucleus composed of L-threonine, 4(S)-amino-2(E)-pentoic acid, and erythro-4hydroxy-L-lysine but differ from each other in the unsaturated fatty acid moiety attached to the peptide [90]. These compounds were isolated from strain K481-B101, whose 16S rRNA sequence (Accession No. AM410613) analyzed in the EzBioCloud server (https://www.ezbiocloud.net/, accessed on 1 June 2022) was identified as *Schlegelella brevitalea*, a member of the Burkholderiales order and Comamonadaceae family. These glidobactins compounds have antifungal and antitumor activity [91]. Later, **cepafunings I**, **II** (**54**), and **III** (**55**) acylpeptides produced by *B. cepacia* CB-3 were described by Shoji et al. [92]. The mixture of cepafungins has moderate inhibitory activity against pathogenic yeast and fungi such as *C. albicans*, *Candida krusei*, *Aspergillus fumigatus*, *Microsporum canis*, and *T. mentagrophytes*. It showed no curative effect in mice infected with *C. albicans*, and there was no activity against bacteria, but it had a moderate effect on prolonging the survival period of mice in which murine lymphatic leukemia P388 cells were implanted [92]. The elucidation of cepafungin structures showed that compound I is identical to **51** (Figure 8) [93]. Later, Schellenberg et al. [94] named the group of cepafungins as glidobactins and, while studying the genes for the synthesis of **51** in S. brevitalea K481-B101, found homologous gene clusters in *B. pseudomallei* and *B. mallei*. The production but not the antimicrobial activity of **53** synthesized by *B. pseudomallei* was reported by Biggins et al. [95]. Moreover, a **53** variant was described as **deoxyglidobactin C**, which contains a lysine instead of a 4-hydroxylisine within the structure. **Occidiofungin A-D** (**56–59**) (Figure 8), synthesized by *B. contaminans* MS14, are glycopeptides with antifungal activity inhibiting a large spectrum of fungal pathogens, among them *Alternaria*, *Aspergillus*, *Fusarium*, *Geotrichum*, *Macrophomina*, *Microsporum*, *Penicillum*, *Pythium*, *Rhizoctonia*, *Trichophyton*, and several *Candida* species [96–98]. Occidiofungin disrupts fungal membrane morphology and induces apoptosis [96,99]. A recent study identified actin filaments as the primary cellular target of occidiofungin in fungi [98]. It also has antiparasitic activity, damaging the parasite Cryptosporidium parvum [100]. Occidiofungin is also produced by B. *pyrrocinia* Lyc2, having antifungal activity, attacking *Aspergillus*, *Cladosporium*, *Cochilobolus heterostro-*

phus, *Colletotrichum* acutatum, *Gaeumannomyces graminis*, *Geotrichum candidum*, *Glomerella cingulate* and *Thieloviopsis basicola* [101]. Moreover, occidofungin was tested in toxicological evaluations and it was found to have minimal toxicity in human fibroblasts and has potent anticancer activity [102]. **Cepacidine A_1 (60) and A_2 (61)** (Figure 8) are glycopeptides produced by *B. cepacia* AF 2001. They are highly similar, with molecular weights of 1199 and 1219 Da, respectively [103]. Cepacidine A_2 contains asparagine, and A_1 includes b-hydroxy aspargine, which combined have potent antifungal but no antibacterial activity [104]. The cepacidine mixture inhibits *C. albicans*, *C. glabrata*, *Cryptococcus neoformans*, *S. cerevisiae*, *A. niger*, *Microsporum* gypseum, *Epidermophyton floccosum*, *T. mentagrophyte*, *Trichophyton rubrum*, *F. oxysporum*, and *Rhizopus stolonifera*. Moreover, cepacidine A has immunosuppressive action involving in vitro inhibition of the proliferation of *murine lymphocytes* [105]. Cepacidine A has moderate anthelmintic in vitro but not in vivo activity [106]. **AFC-BC11** is a lipopeptide produced by *B. cepacia* BC11 [107]. This compound is involved in the biological control of *R. solani* damping-off in cotton. **Icosalide A1 (62)** (Figure 8) is an unusual two-tailed lipocyclopeptide antibiotic produced by *B. gladioli* HKI0739, which is active against entomopathogenic bacteria *B. thuringiensis* and *Paenibacillus larvae* and is involved in swarming inhibition [108]. *Burkholderia gladioli* BCC0238 was also found to synthesize (62), but its antimicrobial activity was not tested [109]. This compound was first reported from Aureobasidium and showed activity against *Streptococcus pyogenes* and *E. faecalis* [110]. *Burkholderia thailandensis* produces **bactobolins A-D (63–66)** (Figure 8), a group of polyketide-peptide molecules, some of which are potent antimicrobials [111]. The production of these compounds is temperature dependent with better results of production at 30 than 37 °C. The purification of the three most abundant bactobolins showed that **63** and **65** have strong activity against bacteria (*Bacillus* cereus, *B. subtilis*, *B. cenocepacia*, *Paraburkholderia kururiensis*, *Burkholderia vietnamiensis*, *Chromobacterium violaceum*, *E. coli*, *Flavobacterium johnsoniae*, *K. pneumoniae*, *Mycobacterium marinum*, *P. aeruginosa*, *Pseudomonas fluorescens*, *Ralstonia pickettii*, *S. Typhimurium*, *S. aureus*, and *Streptococcus pyogenes*) and fibroblasts. **Xylocandins A_1, A_2, B_1, B_2, C_1, C_2, D_1,** and **D_2** were isolated from *B. cepacia* ATCC 3927 [112]. Xylocandins are cyclic peptides containing glycine, serine, asparagine, β-hydroxytyrosine, and an unusual amino acid with the formula $C_{18}H_{37}NO_5$. The mixture of each compound showed that xylocandin A_1 and A_2 have a potent antifungal activity inhibiting several *Candida* species and dermatophytes such as *T. mentagrophytes*, *T. rubrum*, *Epidermophyton floccosum*, and *M. canis*, but does not inhibit Gram-negative and -positive bacteria, nor vaginitis in a rat model [113]. *Burkholderia cenocepacia* H111 was detected to produce a diazeniumdiolate metalophore compound called **fragin (67)** (Figure 8) [114]. When iron was added to the medium, the antifungal activity of strain H111 diminishes, suggesting that the metal chelation is the molecular basis for antifungal activity. Fragin enantiomers inhibit *F. solani*, *B. cereus*, *B. subtilis*, *B. thuringiensis*, *S. aureus*, and *S. cerevisiae*, but no Gram-negative bacteria such as *C. violaceum*, *E. coli*, *Klebsiella oxytoca*, and *P. syringae*. Besides fragin, strain H111 synthesizes a signal molecule called valdiazen, which shares a high degree of structural homology with fragin, but their function is different since valdiazen has no antimicrobial activity. Valdiazen is a diffusible signal that regulates both itself and fragin and the expression of more than 100 genes, representing a novel quorum-sensing signal. **Betulinans** are produced by *B. pseudomallei* K96243 [115]. During the screening of agonists for eukaryotic phosphodiesterase (PDE), the betulinan **BTH-II0204-207:A (68)** (Figure 8) compound produced by *B. pseudomallei* K96243 was found. PDE are divided into 11 families. PDE4 has been implicated in inflammation responses across multiple immune cell types; therefore, PDE4 inhibitors have been extensively investigated as potential therapeutic molecules for a number of inflammatory diseases. Bioactivity assays indicated that **68** is a PDE4 inhibitor. Microbial symbionts are often a source of chemicals that can contribute to host defense against antagonists. *Lagria* beetles live in symbiosis with multiple strains of *Burkholderia* that protect their offspring against pathogens. Among them, *B. gladioli* Lv-StB was found to produce the PK **lagriamide (69)** (Figure 8), which inhibits *A. niger* and *P. lilacinum* [116]. **Isosulfazecin (70)** (Figure 8) is a NRP b-lactam antibiotic produced by *Pseudomonas mesoacidophila* SB-72310 (now *B. ubonensis*) [117,118]. Compound **70** inhibits *S. Typhimurium* and moderately inhibits *E. coli, P. vulgaris, P. mirabilis, S. marcescens, E. faecalis*, and *B. subtilis. Burkholderia ubonensis* SB-72310 also produces **bulgecins**, glycopeptides that induce bulge formation in cooperation with b-lactams and enhance the lytic activity of b-lactam-antibiotics, but bulgecins show no antimicrobial activity [119]. *Burkholderia cepacia* CF-66 displays strong antifungal activity against *R. solani;* a compound named **CF66I** was purified and showed inhibition of *F. oxysporum, Fusarium sambucinum, Rosselinia necatrix, Aspergillus flavus, A. niger, Cochilobus carbonum, B. cinerea, Mucor hiemolis, Penicillum chrysogenum, Rhizopus oryzae, C. albicans, C. neoformens, Pichia membranae*, and *S. cerevisiae* but not *E. coli, B. subtilis*, or *S. aureus* [120]. *Pseudomonas aeruginosa*, causing nosocomial and wound infections, possess a signal molecule that integrates quorum sensing (QS) and stress response [121]. This integrated QS molecule (IQS) is identical to aeruginaldehyde from *P. fluorescens* [122]. IQS is effectively captured by the NRP siderophore malleobactin produced by *B. thailandensis*, which results in the formation of a rare nitrone bioconjugate called malleonitrone (**71**) (Figure 8) that is active against the IQS producer and therefore has significance from a pharmaceutical perspective [123]. **Spliceostatins** are spliceosome inhibitors, synthesized by a hybrid NRPS-PKS system of the trans-acyl transferase (AT) type, that show promising anticancer activity. *Burkholderia* sp. FERM BP-3421, identified by 16S sequence (KJ364655) as a member of the Bcc, produces the hemiketal spliceostatins, as well as analogs containing a terminal carboxylic acid [124]. Some spliceostatin analogues (**72–73**) and their semisynthetic analogues were evaluated in cell proliferation assays against a panel of solid tumor cell lines, showing potent cytotoxicity [125]. **Diketopiperazines (DKP)** are NRP-cyclized molecules comprising amino acid bounded by two peptide bonds [126]. *Burkholderia cepacia* CF-66 synthesizes the molecules diketopiperazines cyclo(Pro-Phe), cyclo(Pro-Tyr), cyclo(Ala-Val), cyclo(Pro-Leu), and cyclo(Pro-Val); all of these compounds are both D- and L-type [127]. These DKP showed a negative effect on the candidacidal activity of the culture supernatant extracts. *Burkholderia cepacia* CF-66 lacks the gen cepI that encodes for an acyl homoserine lactone (AHL), which is involved in QS; however, a study with *B. cenocepacia* J2315^{T} showed that new DKP molecules can inhibit CepI in vitro, impairing the ability of *B. cenocepacia* to produce proteases and siderophores and to form biofilms [128].

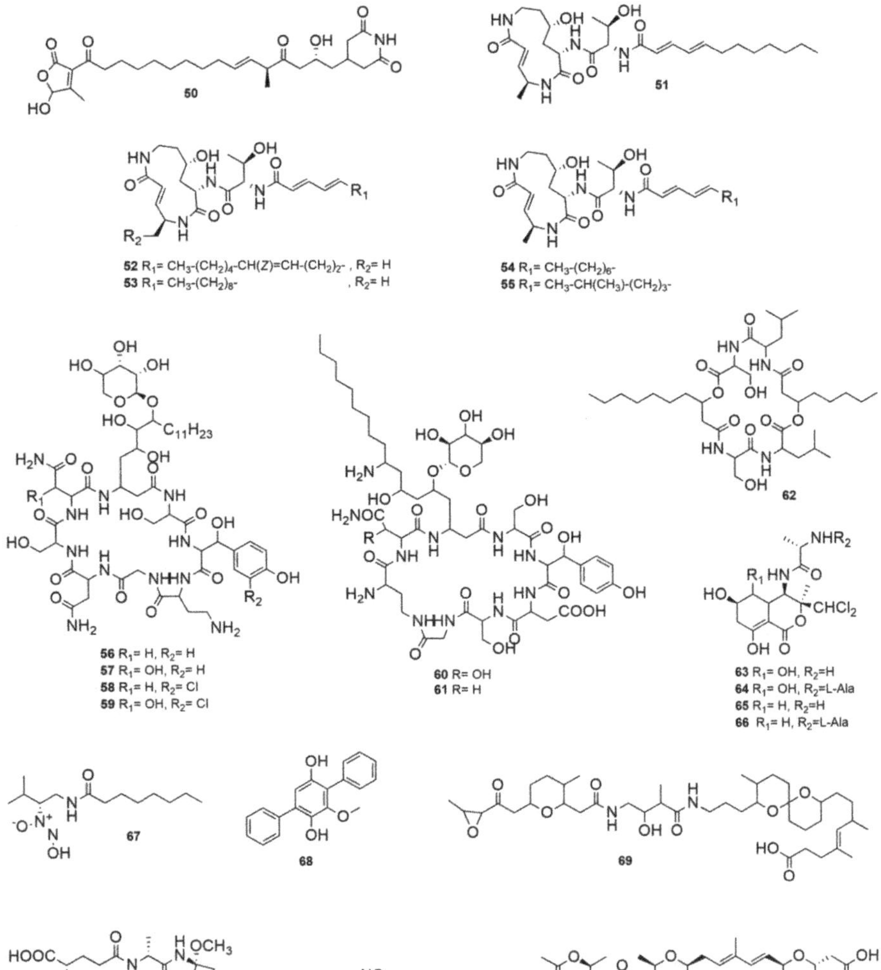

Figure 8. Other NPR-PK compounds. (**50**) Gladiofungin A. (**51**) Glidobactin A. (**52**) Glidobactin B. (**53**) Glidobactn C. (**54**) Cepafungin I, II (**55**) Cepafungin III. (**56–59**) Occidiofungins A–D. (**60**) Cepacidine A_1. (**61**) Cepacidine A_2. (**62**) Icosalide A1. (**63–66**) Bactobolins A–D. (**67**) Fragin. (**68**) BTH-II0204-207:A. (**69**) Lagriamide. (**70**) Isosulfazecin (iSZ). (**71**) Malleonitrone. (**72–73**) Spliceostatins.

11. Other Antimicrobial Compounds

Sinapigladioside (**74**) (Figure 9), an aromatic glycoside, contains an isothiocyanate moiety, a rare structural feature among bacterial metabolites [51]. This compound produced by *B. gladioli* displays antifungal activity against *P. lilacinum*, which is an egg entomopathogen, *A. fumigatus*, and *Penicillum notatum*. A not fully characterized molecule referred to as "**Compound 1**", an antimicrobial compound with an unknown structure but with an ion at m/z 391.2845 and produced by *B. orbicola* TAtl-371^T, was found to inhibit only Tatumella terrea SHS 2008^T [129]. **Cepaciamide A (75)** (Figure 9) is a (3R, 3′R, 2″R, 5″S, 6″R)-3-N-[3′-(2″-hydroxy-5″, 6″-methylenoctadecanoyl)-hexadecamido]-2-piperidinone isolated from *B. cepacia* D-202 [130]. This compound has toxicity activity against *B. cinerea*, which causes beet root rot in Japan. The bacterial type III secretion system (T3SS) acts as a complex multiunit nanomachine to translocate effector proteins across the bacterial membrane to deliver them directly into eukaryotic host cells [131]. *B. gladioli* NGJ1 produces a prophage tail-like protein (Bg_9562), which is a potential effector secreted by a T3SS and is essential for mycophagy in *R. solani* [132]. Bg_9562 protein showed antifungal activity against *S. cerevisiae*, *C. albicans*, *Alternaria* brassicae, *M. oryzae*, Venturia inaequalis, *F. oxysporum* 7063, *Alternaria* sp., Dedymella sp., Phytophthora sp., *Colletotrichum* sp., Ascochyta rabiei, and Neofusicoccum sp. Moreover, Bg_9562 protein showed no inhibition of *E. coli*, Pantoea ananatis, or B. glaidoli NGJ1. The compound **2-hydroxymethyl-chroman-4-one** (**76**), designated as **MSSP2** (Figure 9), is produced by *Burkholderia* sp. MSSP [133], whose 16S gene sequence (AY551271) indicates that it belongs to the Bcc. The compound MSSP2 displays an inhibitory effect against P. ultimum, *P. capsica*, and S. sclerotiorum. **Altericidins A**, **B**, and **C** produced by *B. cepacia* KB-1 can inhibit a wide range of fungi and yeasts, such as *Alteraria kikuchiana* and *Ustilago maydis*, but has no effect on bacteria [134]. **Rhamnolipids** (Rha) are glycolipidic biosurfactants consisting of rhamnose molecules linked through a β-glycosidic bond to 3-hydroxyfatty acids with various chain lengths produced by bacterial species with several functions such as antimicrobial activity [135]. The non-pathogenic *B. thailandensis* E264^T synthesize di-rhamnolipids C_{14}-C_{14} (**77**) and C_{12}-C_{14} (**78**) (Figure 9), which have antibacterial and antibiofilm activity against *Streptococcus sanguinis*, *Streptococcus oralis*, *Neisseria mucosa*, and *Actinomyces naeslundii* [136]. Other non-pathogenic *Burkholderia* synthesize Rha, such as *B. glumae* and *B. plantarii*; however, their antagonistic activity was not tested [137,138]. Additionally, *B. pseudomallei* produce Rha-Rha C_{14}-C_{14}, which showed cytotoxic and hemolytic activities [139,140].

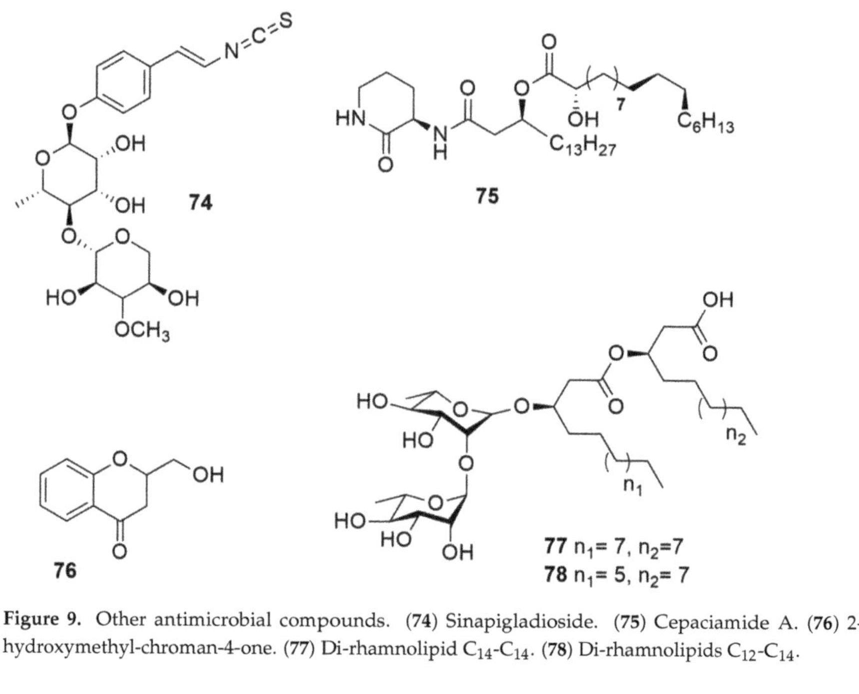

Figure 9. Other antimicrobial compounds. (**74**) Sinapigladioside. (**75**) Cepaciamide A. (**76**) 2-hydroxymethyl-chroman-4-one. (**77**) Di-rhamnolipid C_{14}-C_{14}. (**78**) Di-rhamnolipids C_{12}-C_{14}.

12. Compounds with Dual Effect

Many compounds are beneficial for humans since they are antifungal, antibacterial, or anticancer molecules. However, some of these compounds have a dual effect, both beneficial and toxic. For instance, **burkholdines** (Figure 10) are NRP-cyclic lipopeptides produced by *B. ambifaria* 2.2N with potent antifungal activity [141]. Many burkholdines have been described and the analysis of five representatives (**79–83**) showed antifungal activity on *S. cerevisiae*, *C. albicans*, and *A. niger*. However, they also exhibit hemolytic activity [142]. The latter results indicate that these compounds are important for *Burkholderia* virulence. **Tropolone** (**84**) (Figure 10) is a troponoid containing a seven-membered aromatic ring with various substitutions, produced by *B. plantarii* [143]. Tropolone shows broad-spectrum antimicrobial activity against bacteria and fungi, but it is the phytotoxin responsible for rice seedling blight [144,145]. **Cepalycin I** and **cepalycin II** were isolated from *B. cepacia* JN106 [146], but their structure was not reported. These compounds have both hemolytic and antifungal activity, inhibiting *S. cerevisiae*, *C. neoformans*, and *C. albicans*.

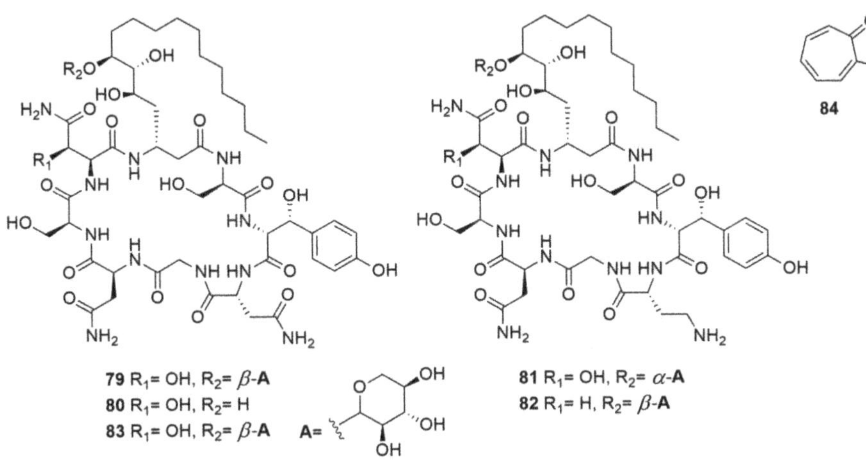

Figure 10. Compounds with dual effect. (**79–83**) Burkholdines. (**84**) Tropolone.

13. Metabolism as Control

Fusaric acid (**85**) (Figure 11) is a fungal metabolite produced by several *Fusarium* species, which is responsible for wilts and root rot diseases in a number of plants. *Burkholderia ambifaria* T16 can grow with **85** as a sole carbon, nitrogen, and energy source, and showed the ability to detoxify **85** in barley seedlings, suggesting that the strain might serve as a new source of metabolites or genes for the development of novel **85**-detoxification systems [147].

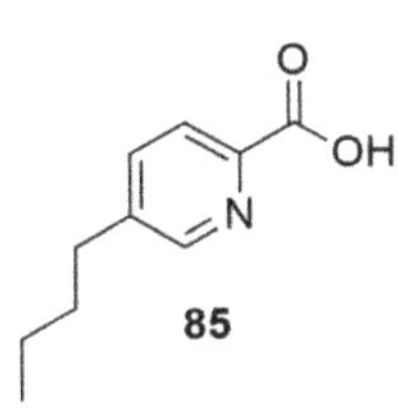

Figure 11. Metabolism as control. (**85**) Fusaric acid.

14. Data Mining

Genome mining is a promising tool in the search for new bioactive compounds produced by microorganisms. This strategy has been used in several bacterial genomes to analyze their potential as sources of new compounds with pharmacological potential. **Phenazines** are structurally diverse, but all share a conserved seven-gene operon, *phz-ABCDEFG*, termed the "core phenazine biosynthesis genes" [148]. A genome screening of *Burkholderia* genomes showed that phenazine gene clusters were identified in 20 strains belonging to *B. cepacia*, *Burkholderia lata*, *B. glumae*, B. singularis, *B. ubonensis*, and some *Burkholderia* sp. strains [34]. A genome mining of 64 *B. ambifaria* strains revealed an armory of known and unknown pathways within this species, among them the biosynthetic gene cluster to produce **cepacin**, which was the mode of action for the biopesticidal activity of *B. ambifaria* [48]. In this study [48], other compounds were found in *B. ambifaria* genomes, such as **pyrrolnitrin**, **burkholdines**, **hydroxyquinolines**, **bactobolins**, and **enacyloxina IIa**. Moreover, Mullins and Mahenthiralingam [149] analyzed 4000 genomes representing the genera of *Burkholderia* s.l.; among them, the *Burkholderia* species harbored more biosynthetic gene clusters and the more diverse clusters per species compared to the remaining genera from *Burkholderia* s.l. These clusters include genes involved in the production of **bacteriocins**, **phosphonates**, **lassopeptides**, **NRPS**, **betalactones**, **transAT-PKS**, and **terpenes**. **Chitinases** are glycosyl hydrolases that catalyze the hydrolytic degradation of chitin, one of the major constituents of cell walls of fungi. The genome analysis of *B. orbicola* TAtl-371^{T} showed the presence of the gene BCAL1722 that encodes a chitinase belonging to family 18 of the glycosyl hydrolases, as well as a gene that encodes a predicted chitinase [129]. The authors also found that several *B. cenocepacia* strains contain homologues; however, the activity of chitinase was not present in strain TAtl-371^{T}. Moreover, Rojas-Rojas et al. [129] also found that the genome of strain TAtl-371^{T} contains 30 genes reported for the biosynthesis of the bacteriocin **BceTMilo**. A genomic search for **LlpAs** in *Burkholderia* genomes and phylogenetic analysis showed two distinct clusters; one of them belongs to the *B. pseudomallei* group, including *Burkholderia oklahomensis*, *B. pseudomallei*, *B. mallei*, and *B. thailandensis* [73]. Another bacteriocin studied was ColM in *Burkholderia*. The colM-like bacteriocin gene was found mainly in Bcc and *B. oklahomensis* [74]. **Ubonodin** was found in the genome of 16 out of 306 *B. ubonensis* strains, which might be in relation to the intriguingly large size of 28 aa of the core peptides, longer than any previously characterized example [79]. HMAQ produced by the biosynthetic operon named *hmqABCDEFG* was searched for in the genome of Bcc strains [150]. The analysis showed that one-third of Bcc species carry a homolog of the *hmqABCDEFG*, and not all sequenced strains in each species possess this operon.

15. Conclusions

The ability of *Burkholderia* to produce antimicrobial compounds is remarkable, not just for the variety of molecules synthesized but also for the diversity of targets they attack, namely bacteria, fungi, cancer cells, tumor cells, or inflammatory processes. The compounds produced belong to a variety of chemical natures, such as N-containing heterocycles, volatile organic compounds, polyenes, polyynes, siderophores, macrolides, bacteriocins, quinolones, non-ribosomal peptides, polyketides, and other unclassified compounds such as sinapigladiosides, cepaciamide A, altericidins, and rhamnolipids, among others. Moreover, there are compounds that have both beneficial and toxic effects, such as burkholdines, tropolone, and others. The mining of genomes is another important method of finding new molecules. Certainly, *Burkholderia* is still a group of bacteria with as-yet unexplored compounds waiting to be discovered. Several papers about new *Burkholderia* strains are published daily, which may contain information about the new antimicrobial compounds they produce that have the potential to be used against multidrug-resistant microorganisms.

Supplementary Materials: The following supporting information can be downloaded at https://www.mdpi.com/article/10.3390/molecules28041646/s1: Table S1: Antimicrobial compounds produced by *Burkholderia* sensu stricto. References [151,152] are cited in the Supplementary Materials.

Author Contributions: Conceptualization, methodology, validation, formal analysis, investigation, data curation, writing—original draft preparation, L.M.M.-R., M.R.-C., A.S.-G., F.U.R.-R. and P.E.-d.l.S.; writing—review and editing, visualization, supervision, project administration, funding acquisition, P.E.-d.l.S. All authors have read and agreed to the published version of the manuscript.

Funding: This research was funded by the Secretaría de Investigación y Posgrado, Instituto Politécnico Nacional, grant numbers 2021-0392, 2022-0660, and 2023-0831.

Institutional Review Board Statement: Not applicable.

Informed Consent Statement: Not applicable.

Acknowledgments: P.E.S. thanks COFAA, EDI, and SNI. We thank Martha Thayer for proofreading this manuscript.

Conflicts of Interest: The authors declare no conflict of interest. The funders had no role in the design of the study, in the collection, analyses, or interpretation of data, in the writing of the manuscript, or in the decision to publish the results.

References

1. Sawana, A.; Adeolu, M.; Gupta, R.S. Molecular signatures and phylogenomic analysis of the genus *Burkholderia*: Proposal for division of this genus into the emended genus *Burkholderia* containing pathogenic organisms and a new genus *Paraburkholderia* gen. nov. harboring environmental species. *Front. Genet.* **2014**, *5*, 429. [CrossRef] [PubMed]
2. Dobritsa, A.P.; Samadpour, M. Transfer of eleven *Burkholderia* species to the genus *Paraburkholderia* and proposal of *Caballeronia* gen. nov., a new genus to accommodate twelve species of *Burkholderia* and *Paraburkholderia*. *Int. J. Syst. Evol. Microbiol.* **2016**, *66*, 2836–2846. [CrossRef] [PubMed]
3. Lopes-Santos, L.; Castro, D.B.A.; Ferreira-Tonin, M.; Correa, D.B.A.; Weir, B.S.; Park, D.; Mariscal-Ottoboni, L.M.; Rodrigues-Neto, J.; Lanza-Destefano, S.A. Reassessment of the taxonomic position of *Burkholderia andropogonis* and description of *Robbsia andropogonis* gen. nov., comb. nov. *Anton. Leeuw. Int. J. Gen.* **2017**, *110*, 727–736. [CrossRef] [PubMed]
4. Estrada-de los Santos, P.; Palmer, M.; Chavez-Ramírez, B.; Beukes, C.; Steenkamp, E.; Briscoe, L.; Khan, N.; Maluk, M.; Lafos, C.; Humm, E.; et al. Whole genome analyses suggests that *Burkholderia* sensu lato contains two additional novel genera (*Mycetohabitans* gen. nov. and *Trinickia* gen. nov.): Implications for the evolution of diazotrophy and nodulation in the *Burkholderiaceae*. *Genes* **2018**, *9*, 389. [CrossRef] [PubMed]
5. Lin, Q.H.; Lv, Y.Y.; Gao, Z.G.; Qiu, L.H. *Pararobbsia silviterrae* gen. nov., sp. nov., isolated from forest soil and reclassification of *Burkholderia alpina* as *Pararobbsia alpina* comb. nov. *Int. J. Syst. Evol. Microbiol.* **2020**, *70*, 1412–1420. [CrossRef]
6. Khakhum, N.; Tapia, D.; Torres, A.G. *Burkholderia mallei* and *glanders*. In *Defense Against Biological Attacks*; Springer: Manhattan, NY, USA, 2019; pp. 161–183. [CrossRef]
7. Gassiep, I.; Armostrong, M.; Norton, R. Human melioidosis. *Clin. Microbiol. Rev.* **2020**, *33*, e00006-19. [CrossRef]
8. Hall, C.M.; Baker, A.L.; Sahl, J.W.; Mayo, M.; Scholz, H.C.; Kaestli, M.; Schupp, J.; Martz, M.; Settles, E.W.; Busch, J.D.; et al. Expanding the *Burkholderia pseudomallei* complex with the addition of two novel species: *Burkholderia mayonis* sp. nov. and *Burkholderia savanae* sp. nov. *Appl. Environ. Microbiol.* **2022**, *11*, e0158321. [CrossRef]
9. Pereira-Andrade, J.; de Souza, H.G.; Carvalho-Ferreira, L.; Cnockaert, M.; De Canck, E.; Wieme, A.D.; Peeters, C.; Gross, E.; De Souza, J.T.; Santos-Marbach, P.A.; et al. *Burkholderia perseverans* sp. nov., a bacterium isolated from the Restinga ecosystem, a producer of volatile and diffusible compounds that inhibit plant pathogens. *Braz. J. Microbiol* **2021**, *52*, 2145–2152. [CrossRef]
10. Mahenthiralingam, E.; Urban, T.A.; Goldberg, J.B. The multifarious, multireplicon *Burkholderia cepacia* complex. *Nat. Rev. Microbiol.* **2005**, *3*, 144–156. [CrossRef]
11. Morales-Ruiz, L.M.; Rodriguez-Cisneros, M.; Kerber-Diaz, J.C.; Rojas-Rojas, F.U.; Ibarra, J.A.; Estrada-de los Santos, P. *Burkholderia orbicola* sp. nov., a novel species within the *Burkholderia cepacia* complex. *Arch. Microbiol.* **2022**, *204*, 178. [CrossRef]
12. Rose, H.; Baldwin, A.; Dowson, C.G.; Mahenthiralingam, E. Biocide susceptibility of the *Burkholderia cepacia* complex. *J. Ant. Chemother.* **2009**, *63*, 502–510. [CrossRef]
13. Coenye, T.; Vandamme, P. Diversity and significance of *Burkholderia* species occupying diverse ecological niches. *Environ. Microbiol.* **2003**, *5*, 719–729. [CrossRef]
14. Parke, J.L.; Gurian-Sherman, D. Diversity of the *Burkholderia cepacia* complex and implications for risk assessment of biological control strains. *Ann. Rev. Phytopathol.* **2001**, *39*, 225–258. [CrossRef]
15. Chain, P.S.G.; Denef, V.J.; Konstantinidis, K.T.; Vergez, L.M.; Agullo, L.; Reyes, V.L.; Hauser, L.; Cordova, M.; Gomez, L.; Gonzalez, M.; et al. *Burkholderia xenovorans* LB400 harbors a multi-replicon, 9.73-Mbp genome shaped for versatility. *Proc. Natl. Acad. Sci. USA* **2006**, *103*, 15280–15287. [CrossRef]
16. O'Sullivan, L.A.; Weightman, A.J.; Jones, T.H.; Marchbank, A.M.; Tiedje, J.M.; Mahenthiralingam, E. Identifying the genetic basis of ecologically and biotechnologically useful functions of the bacterium *Burkholderia vietnamiensis*. *Environ. Microbiol.* **2007**, *9*, 1017–1034. [CrossRef]

17. Bach, E.; Pereira-Passaglia, L.M.; Jiao, J.; Gross, H. *Burkholderia* in the genomic era: From taxonomy to the discovery of new antimicrobial secondary metabolites. *Rev. Microbiol.* **2022**, *48*, 121–160. [CrossRef]
18. Depoorter, E.; Bull, M.J.; Peeters, C.; Coenye, T.; Vandmme, P.; Mahenthiralingam, E. *Burkholderia*: An update on taxonomy and biotechnological potential as antibiotic producers. *Appl. Microbiol. Biotechnol.* **2016**, *100*, 5215–5229. [CrossRef]
19. Kerru, N.; Gummidi, L.; Maddila, S.; Gangu, K.K.; Jonnalagadda, S.B. A review on recent advances in nitrogen-containing molecules and their biological applications. *Molecules* **2020**, *25*, 1909. [CrossRef]
20. Valenca, C.A.S.; Barbosa, A.A.T.; Souto, E.B.; Caramao, E.B.; Jain, S. *Volatile nitrogenous* compounds from bacteria: Source of novel bioactive compounds. *Chem. Biodivers.* **2021**, *18*, e2100549. [CrossRef]
21. Prasad, J.; Pandey, P.; Anand, R.; Raghuwanshi, R. Drought Exposed *Burkholderia seminalis* JRBHU6 exhibits antimicrobial potential through pyrazine-1,4-dione derivatives targeting multiple bacterial and fungal proteins. *Front. Microbiol.* **2021**, *12*, 633036. [CrossRef]
22. Petri, G.L.; Spano, V.; Spatola, R.; Holl, R.; Raimondi, M.V.; Barraja, P.; Montalbano, A. Bioactive pyrrole-based compound with target selectivity. *Eur. J. Med. Chem.* **2020**, *208*, 112783. [CrossRef] [PubMed]
23. Mitchell, R.E.; Teh, K.L. Antibacterial iminopyrrolidines from *Burkholderia plantarii*, a bacterial pathogen of rice. *Org. Biomol. Chem.* **2005**, *3*, 3540. [CrossRef] [PubMed]
24. Mitchell, R.E.; Greenwood, D.R.; Sarajioni, V. An antibacterial pyrazole derivative from *Burkholderia glumae*, a bacterial pathogen of rice. *Phytochemistry* **2008**, *69*, 2704–2707. [CrossRef] [PubMed]
25. Pawar, S.; Chaudhari, A.; Prabha, R.; Shukla, R.; Singh, D.P. Microbial pyrrolnitrin: Natural metabolite with immense practical utility. *Biomolecules* **2019**, *9*, 443. [CrossRef] [PubMed]
26. Arima, K.; Imanaka, H.; Kousaka, M.; Fukuta, A.; Tamura, G. Pyrrolnitrin, a new antibiotic substance produced by *Pseudomonas*. *Agric. Biol. Chem.* **1964**, *28*, 575–576. [CrossRef]
27. Kilani, J.; Fillinger, S. Phenylpyrroles: 30 years, two molecules and (nearly) no resistance. *Front. Microbiol.* **2016**, *7*, 2014. [CrossRef]
28. El-Banna, N.; Winkelmann, G. Pyrrolnitrin from *Burkholderia cepacia*: Antibiotic activity against fungi and novel activities against streptomycetes. *J. Appl. Microbiol.* **1998**, *85*, 69–78. [CrossRef]
29. Sultan, M.Z.; Park, K.; Lee, S.Y.; Park, J.K.; Varughese, T.; Moon, S.S. Novel oxidized derivatives of antifungal pyrrolnitrin from the bacterium *Burkholderia cepacia* K87. *J. Antibiot.* **2008**, *61*, 420–425. [CrossRef]
30. Jung, B.K.; Hong, S.J.; Park, G.S.; Kim, M.C.; Shin, J.H. Isolation of *Burkholderia cepacia* JBK9 with plant growth-promoting activity while producing pyrrolnitrin antagonistic to plant fungal diseases. *Appl. Biol. Chem.* **2018**, *61*, 173–180. [CrossRef]
31. Webster, G.; Jones, C.; Mullins, A.J.; Mahenthiralingam, E. A rapid screening method for the detection of specialized metabolites from bacteria: Induction and suppression of metabolites from *Burkholderia* species. *J. Microbiol. Methods* **2020**, *178*, 106057. [CrossRef]
32. Yan, J.; Liu, W.; Cai, J.; Wang, Y.; Li, D.; Hua, H.; Cao, H. Advances in phenazines over the past decade: Review of their pharmacological activities, mechanisms of action, biosynthetic pathways, and synthetic strategies. *Mar. Drugs* **2021**, *19*, 610. [CrossRef]
33. Cartwright, D.K.; Chilton, W.S.; Benson, D.M. Pyrrolnitrin and phenazine production by *Pseudomonas cepacia*, strain 5.5B, a biocontrol agent of *Rhizoctonia solani*. *Appl. Microbiol. Biotechnol.* **1995**, *43*, 211–216. [CrossRef]
34. Hendry, S.; Steinke, S.; Wittstein, K.; Stadler, M.; Harmrolfs, K.; Adewunmi, Y.; Sahukhal, G.; Elasri, M.; Thomashow, L.; Weller, D.; et al. Functional analysis of phenazines biosynthesis genes in *Burkholderia* spp. *Appl. Environ. Microbiol.* **2021**, *87*, e02348-20. [CrossRef]
35. Han, J.W.; Kim, J.D.; Lee, J.M.; Ham, J.H.; Lee, D.; Kim, B.S. Structural elucidation and antimicrobial activity of new phencomycin derivatives isolated from *Burkholderia glumae* strain 411gr-6. *J. Antibiot.* **2014**, *67*, 721–723. [CrossRef]
36. Xu, Z.; Wang, M.; Du, J.; Huang, T.; Liu, J.; Dong, T.; Chen, Y. Isolation of *Burkholderia* sp. HQB-1, a promising biocontrol bacterium to protect banana against *Fusarium* wilt through phenazine-1-carboxylic acid secretion. *Front. Microbiol.* **2020**, *11*, 605152. [CrossRef]
37. Chen, J.H.; Xiang, W.; Cao, K.X.; Lu, X.; Yao, S.C.; Hung, D.; Huang, R.S.; Li, L.B. Characterization of volatile organic compounds emitted from endophytic *Burkholderia cenocepacia* ETR-B22 by SPME-GC-MS and their inhibitory activity against various plant fungal pathogens. *Molecules* **2020**, *25*, 3765. [CrossRef]
38. Xu, T.; Shi, L.; Zhang, Y.; Wang, K.; Yang, Z.; Ke, S. Synthesis and biological evaluation of marine alkaloid-oriented β-carboline analogues. *Eur. J. Med. Chem.* **2019**, *25*, 3765. [CrossRef]
39. Huang, D.; Zhang, Z.; Li, Y.; Liu, F.; Huang, W.; Min, Y.; Wang, K.; Yang, J.; Cao, C.; Gong, Y.; et al. Carboline derivatives based on natural pityriacitrin as potential antifungal agents. *Phytochemi. Lett.* **2022**, *48*, 100–105. [CrossRef]
40. Lin, Y.T.; Lee, C.C.; Leu, W.M.; Wu, J.J.; Huang, Y.C.; Meng, M. Fungicidal activity of volatile organic compounds emitted by *Burkholderia gladioli* strain BBB-01. *Molecules* **2021**, *26*, 745. [CrossRef]
41. Hunter, W.J.; Manter, D.K. Antimicrobial properties of an oxidizer produced by *Burkholderia cenocepacia* P525. *Curr. Microbiol.* **2014**, *68*, 610–614. [CrossRef]
42. Wu, Y.; Seyedsayamdost, M.R. The polyene natural product thailandamide A inhibits fatty acid biosynthesis in Gram-positive and Gram-negative bacteria. *Biochemistry* **2018**, *57*, 4247–4251. [CrossRef] [PubMed]

43. Park, J.D.; Moon, K.; Miller, C.; Rose, J.; Xu, F.; Ebmeier, C.C.; Jacobsen, J.R.; Mao, D.; Old, W.M.; DeShazer, D.; et al. Thailandenes, cryptic polyene natural products isolated from *Burkholderia thailandensis* using phenotype-guided transposon mutagenesis. *ACS Chem. Biol.* **2020**, *15*, 1195–1203. [CrossRef] [PubMed]
44. Mahenthiralingam, E.; Song, L.; Sass, A.; White, J.; Wilmot, C.; Marchbank, A.; Boaisha, O.; Paine, J.; Knight, D.; Challis, G.L. Enacyloxins are products of an unusual hybrid modular polyketide synthase encoded by a cryptic *Burkholderia ambifaria* genomic island. *Chem. Biol.* **2011**, *18*, 665–677. [CrossRef] [PubMed]
45. Parmeggiani, A.; Krab, I.M.; Watanabe, T.; Nielsen, R.C.; Dahlberg, C.; Nyborg, J.; Nissen, P. Enacyloxin IIa pinpoints a binding pocket of elongation factor Tu for development of novel antibiotics. *J. Biol. Chem.* **2006**, *281*, 2893–2900. [CrossRef]
46. Ross, C.; Opel, V.; Scherlach, K.; Hertweck, C. Biosynthesis of antifungal and antibacterial polyketides by *Burkholderia gladioli* in coculture with *Rhizopus microsporus*. *Mycoses* **2014**, *57*, 48–55. [CrossRef]
47. Parker, W.L.; Rathnum, M.L.; Seiner, V.; Trejo, W.H.; Principe, P.A.; Sykes, R.B. Cepacin A and cepacin B, two new antibiotics produced by *Pseudomonas cepacia*. *J. Antibiot.* **1984**, *37*, 431–440. [CrossRef]
48. Mullins, A.J.; Murray, J.A.H.; Bull, M.J.; Jenner, M.; Jones, C.; Webster, G.; Green, A.E.; Neill, D.R.; Connor, T.R.; Parkhill, J.; et al. Genome mining identifies cepacin as a plant-protective metabolite of the biopesticidal bacterium *Burkholderia ambifaria*. *Nat. Microbiol.* **2019**, *4*, 996–1005. [CrossRef]
49. Kusumi, T.; Ohtani, I.; Nishiyama, K.; Kakisawa, H. Caryoynencins, potent antibiotics from a plant pathogen. *Tetrahedron* **1987**, *28*, 3981–3984. [CrossRef]
50. Yamaguchi, M.; Park, H.J.; Ishizuka, S.; Omata, K.; Hirama, M. Chemistry and antimicrobial activity of caryoynencin analogs. *J. Med. Chem.* **1995**, *38*, 5015–5022. [CrossRef]
51. Florez, L.V.; Scherlach, K.; Gaube, P.; Ross, C.; Sitte, E.; Hermes, C.; Rodrigues, A.; Hertweck, C.; Kaltenpoth, M. Antibiotic-producing symbionts dynamically transition between plant pathogenicity and insect-defensive mutualism. *Nat. Commun.* **2017**, *8*, 15172. [CrossRef]
52. Hider, R.C.; Kong, X. Chemistry and biology of siderophores. *Nat. Prod. Rep.* **2010**, *27*, 637. [CrossRef]
53. Finking, R.; Marahiel, M.A. Biosynthesis of nonribosomal peptides. *Annu. Rev. Microbiol.* **2004**, *58*, 453–488. [CrossRef]
54. Hur, G.H.; Vickery, C.R.; Burkart, M.D. Explorations of catalytic domains in non-ribosomal peptide synthetase enzymology. *Nat. Prod. Rep.* **2012**, *29*, 1074–1098. [CrossRef]
55. Jaremko, M.J.; Davis, T.D.; Corpuz, J.C.; Burkart, M.D. Type II non-ribosomal peptide synthetase proteins: Structure, mechanism, and protein–protein interactions. *Nat. Prod. Rep.* **2020**, *37*, 355–379. [CrossRef]
56. Adler, C.; Corbalan, N.S.; Seyedsayamdost, M.R.; Pomares, M.F.; de Cristobal, R.E.; Clardy, J.; Kolter, R.; Vincent, P.A. Catecholate siderophores protect bacteria from pyochelin toxicity. *PLoS ONE* **2012**, *7*, e46754. [CrossRef]
57. Ong, K.S.; Aw, Y.K.; Lee, L.H.; Yule, C.M.; Cheow, Y.L.; Lee, S.M. *Burkholderia paludis* sp. nov., an antibiotic-siderophore producing novel *Burkholderia cepacia* complex species, isolated from Malaysian tropical peat swamp soil. *Front. Microbiol.* **2016**, *7*, 2046. [CrossRef]
58. Ong, K.S.; Cheow, Y.L.; Lee, S.M. The role of reactive oxygen species in the antimicrobial activity of pyochelin. *J. Adv. Res.* **2017**, *8*, 393–398. [CrossRef]
59. da Araujo, F.D.S.; Araujo, W.L.; Eberlin, M.N. Potential of *Burkholderia seminalis* TC3.4.2R3 as biocontrol agent against *Fusarium oxysporum* evaluated by mass spectrometry imaging. *J. Am. Soc. Mass Spectrom.* **2017**, *28*, 901–907. [CrossRef]
60. Meyer, J.M.; Hohnadel, D.; Halle, F. Cepabactin from *Pseudomonas cepacia*, a new type of siderophore. *J. Gen. Microbiol.* **1989**, *135*, 1479–1487. [CrossRef]
61. Itoh, J.; Miyadoh, S.; Takahasi, S.; Amano, S.; Ezaki, N.; Yamada, Y. Studies on antibiotics BN-227 and BN-227-F, new antibiotics. I. Taxonomy, isolation and characterization. *J. Antibiot.* **1979**, *32*, 1089–1095. [CrossRef]
62. Itoh, J.; Amano, S.; Ogawa, Y.; Kodama, Y.; Ezaki, N.; Yamada, Y. Studies on antibiotics BN-227 and BN-227-F, new antibiotics. II. Chemical structure of antibiotics BN-227 and BN-227-F. *J. Antibiot.* **1980**, *33*, 377–382. [CrossRef] [PubMed]
63. Darling, P.; Chan, M.; Cox, A.D.; Sokol, P. Siderophore production by cystic fibrosis isolates of *Burkholderia cepacia*. *Infec. Immun.* **1998**, *66*, 874–877. [CrossRef] [PubMed]
64. Thomas, M.S. Iron acquisition mechanisms of the *Burkholderia cepacia* complex. *BioMetals* **2007**, *20*, 431–452. [CrossRef] [PubMed]
65. Deng, P.; Foxfire, A.; Xu, J.; Baird, S.M.; Jia, J.; Delgado, K.H.; Shin, R.; Smith, L.; Lu, S.E. The siderophore product ornibactin is required for the bactericidal activity of *Burkholderia* contaminans MS14. *Appl. Environ. Microbiol.* **2017**, 83. [CrossRef]
66. Rojas-Rojas, F.U.; Salazar-Gomez, A.; Vargas-Diaz, M.E.; Vasquez-Murrieta, M.S.; Hirsch, A.M.; De Mot, R.; Ghequire, M.G.K.; Ibarra, J.A.; Estrada-de los Santos, P. Broad-spectrum antimicrobial activity by *Burkholderia cenocepacia* TAtl-371, a strain isolated from the tomato rhizosphere. *Microbiology* **2018**, *164*, 1072–1086. [CrossRef]
67. Lenz, K.D.; Klosterman, K.E.; Mukundan, H.; Kubicek-Sutherland, J.Z. Macrolides: From toxins to therapeutics. *Toxins* **2021**, *13*, 347. [CrossRef]
68. Song, L.; Jenner, M.; Masschelein, J.; Jones, C.; Bull, M.J.; Harris, S.R.; Harkoorn, R.C.; Vocat, A.; Romero-Canelon, I.; Coupland, P.; et al. Discovery and biosynthesis of gladiolin: A *Burkholderia gladioli* antibiotic with promising activity against Mycobacterium tuberculosis. *J. Am. Chem. Soc.* **2017**, *139*, 7974–7981. [CrossRef]
69. Riley, M.A. Bacteriocins, biology, ecology, and evolution. In *Encyclopedia of Microbiology*; Schaechter, M., Ed.; Elsevier: Oxford, UK, 2009; pp. 32–44.

70. Meade, E.; Slattery, M.A.; Garvey, M. Bacteriocins, potent antimicrobial peptides and the fight against multi drug resistant species: Resistance is futile? *Antibiotics* **2020**, *9*, 32. [CrossRef]
71. Yao, G.W.; Duarte, I.; Le, T.T.; Carmody, L.; LiPuma, J.J.; Young, R.; Gonzalez, C.F. A broad-host-range tailocin from *Burkholderia cenocepacia*. *Appl. Environ. Microbiol.* **2017**, *83*, e03414-16. [CrossRef]
72. Principe, A.; Fernandez, M.; Torasso, M.; Godino, A.; Fischer, S. Effectiveness of tailocins produced by *Pseudomonas fluorescens* SF4c in controlling the bacterial-spot disease in tomatoes caused by *Xanthomonas vesicatoria*. *Microbiol. Res.* **2018**, *212–213*, 94–102. [CrossRef]
73. Ghequire, M.G.K.; De Canck, E.; Wattiau, P.; Van Winge, I.; Loris, R.; Coenye, T.; De Mot, R. Antibacterial activity of a lectin-like *Burkholderia cenocepacia* protein. *MicrobiologyOpen* **2013**, *2*, 566–575. [CrossRef]
74. Ghequire, M.G.K.; De Mot, R. Distinct colicin M-like bacteriocin-immunity pairs in *Burkholderia*. *Sci. Rep.* **2015**, *5*, 17368. [CrossRef]
75. Marshall, K.; Shakya, S.; Greenhill, A.R.; Padilla, G.; Baker, A.; Warner, J.M. Antibiosis of *Burkholderia ubonensis* against *Burkholderia pseudomallei*, the causative agent for melioidosis. *J. Trop. Med. Public Health* **2010**, *41*, 904–912.
76. Knappe, T.A.; Linne, U.; Zirah, S.; Rebuffat, S.; Xie, X.; Marahiel, M.A. Isolation and structural characterization of capistruin, a lasso peptide predicted from the genome sequence of *Burkholderia thailandensis* E264. *Chem. Biol.* **2008**, *16*, 1290–1298. [CrossRef]
77. Rebufat, S.; Blond, A.; Destoumieux-Garzon, D.; Goulard, C.; Peduzzi, J. Microcin J25, from the macrocyclic to the lasso structure: Implications for biosynthetic, evolutionary, and biotechnological perspectives. *Curr. Protein Pept. Sci.* **2004**, *5*, 383–391. [CrossRef]
78. Knappe, T.A.; Linne, U.; Robbel, L.; Marahiel, M.A. Insights into the biosynthesis and stability of the lasso peptide capistruin. *Chem. Biol.* **2009**, *16*, 1290–1298. [CrossRef]
79. Cheung-Lee, W.L.; Parry, M.E.; Zong, C.; Jaramillo-Cartagena, A.; Darst, S.A.; Connell, N.D.; Rusoo, R.; Link, A.J. Discovery of ubonodin, an antimicrobial lasso peptide active against members of the *Burkholderia cepacia* complex. *ChemBioChem.* **2020**, *21*, 1335–1340. [CrossRef]
80. Millanao, A.; Mora, A.; Villagra, N.; Bucarey, S.; Hidalgo, A. Biological effects of quinolones: A family of broad-spectrum antimicrobial agents. *Molecules* **2021**, *26*, 7153. [CrossRef]
81. Heeb, S.; Fletcher, M.P.; Chhabra, S.R.; Diggle, S.P.; Williams, P.; Camara, M. Quinolones: From antibiotics to autoinducers. *FEMS Microbiol. Rev.* **2011**, *35*, 247–274. [CrossRef]
82. Wu, Y.; Seyedsayamdost, M.R. Synergy and target promiscuity drive structural divergence in bacterial alkylquinolone biosynthesis. *Cell Chem. Biol.* **2017**, *24*, 1437–1444. [CrossRef]
83. Wang, Y.; Hoffmann, J.P.; Chou, C.W.; Honer zu Bentrup, K.; Fuselier, J.A.; Bitoun, J.P.; Wimley, W.C.; Morici, L.A. *Burkholderia thailandensis* outer membrane vesicles exert antimicrobial activity against drug-resistant and competitor microbial species. *J. Microbiol.* **2020**, *58*, 550–562. [CrossRef] [PubMed]
84. Yoshihisa, H.; Sato, Z.; Hirayama, F.; Konno, K.; Shirahama, H.; Suzui, T. Production of antibiotics by *Pseudomonas cepacia* as an agent for biological control of soilborne plant pathogens. *Soil Biol. Biochem.* **1989**, *21*, 723–728. [CrossRef]
85. Saalim, M.; Villegas-Moreno, J.; Clark, B.R. Bacterial alkyl-4-quinolones: Discovery, structural diversity and biological properties. *Molecules* **2020**, *25*, 5689. [CrossRef] [PubMed]
86. Mori, T.; Yamashita, T.; Furihata, K.; Nagai, K.; Suzuki, K.I.; Hayakawa, Y.; Shin-ya, K. Burkholone, a new cytotoxic antibiotic against IGF-I dependent cells from *Burkholderia* sp. *J. Antibiot.* **2007**, *60*, 713–716. [CrossRef] [PubMed]
87. Baserga, R.; Hongo, A.; Rubini, M.; Prisco, M.; Valentinis, B. The IGF-I receptor in cell growth, transformation and apoptosis. *Biochim. Biophys. Acta* **1997**, *1332*, F105–F126. [CrossRef]
88. Niehs, S.P.; Kumpfmuller, J.; Dose, B.; Little, R.F.; Ishida, K.; Florez, L.V.; Kaltenpoth, M.; Hertweck, C. Insect-associated bacteria assemble the antifungal butanolide gladiofungin by non-canonical polyketide chain termination. *Angew. Chem. Int. Ed. Engl.* **2020**, *59*, 23122–23126. [CrossRef]
89. Nakou, I.T.; Jenner, M.; Dashti, Y.; Romero-Canelón, I.; Masschelein, J.; Mahenthiralingam, E.; Challis, G.L. Genomics-driven discovery of a novel glutarimide antibiotic from *Burkholderia gladioli* reveals an unusual polyketide synthase chain release mechanism. *Angew. Chem. Int. Ed. Engl.* **2020**, *59*, 23145–23153. [CrossRef]
90. Oka, M.; Yaginuma, K.; Numata, K.; Konishi, M.; Oki, T.; Kawaguchi, H. Glidobactins A, B and C, new antitumor antibiotics. II. Structure elucidation. *J. Antibiot.* **1988**, *41*, 1338–1350. [CrossRef]
91. Oka, M.; Nishiyama, Y.; Ohta, S.; Kamei, H.; Konishi, M.; Miyaki, T.; Oki, T.; Kawaguchi, H. Glidobactins A, B and C, new antitumor antibiotics. I. Production, isolation, chemical properties and biological activity. *J. Antibiot.* **1988**, *41*, 1331–1337. [CrossRef]
92. Shoji, J.; Hinoo, H.; Kato, T.; Hattori, T.; Hirooka, K.; Tawara, K.; Shiratori, O.; Terui, Y. Isolation of cepafungins I, II and III from *Pseudomonas* species. *J. Antibiot.* **1990**, *43*, 783–787. [CrossRef]
93. Terui, Y.; Nishikawa, J.; Hinoo, H.; Kato, T.; Shoji, J. Structures of cepafungins I, II and III. *J. Antibiot.* **1990**, *43*, 788–795. [CrossRef]
94. Schellenberg, B.; Bigles, L.; Dudler, R. Identification of genes involved in the biosynthesis of the cytotoxic compound glidobactin from a soil bacterium. *Environ. Microbiol.* **2007**, *9*, 1640–1650. [CrossRef]
95. Biggins, J.B.; Kang, H.S.; Ternei, M.A.; DeShazer, D.; Brady, S.F. The chemical arsenal of *Burkholderia pseudomallei* is essential for pathogenicity. *J. Am. Chem. Soc.* **2014**, *136*, 9484–9490. [CrossRef]
96. Lu, S.E.; Novak, J.; Austin, F.W.; Gu, G.; Ellis, D.; Kirk, M.; Wilson-Stanford, S.; Tonelli, M.; Smith, L. Occidiofungin, a unique antifungal glycopeptide produced by a strain of *Burkholderia contaminans*. *Biochemistry* **2009**, *48*, 8312–8321. [CrossRef]

97. Ellis, D.; Gosai, J.; Emrick, C.; Heintz, R.; Romans, L.; Gordon, D.; Lu, S.E.; Austin, F.; Smith, L. Occidiofungin's chemical stability and in vitro potency against *Candida* species. *Antimicrob. Agents Chemother.* **2012**, *56*, 765–769. [CrossRef]
98. Ravichandran, A.; Geng, M.; Hull, K.G.; Romo, D.; Lu, S.E.; Albee, A.; Nutter, C.; Gordon, D.M.; Ghannoum, M.A.; Lockless, S.W.; et al. Occidiofungin, and actin binding antifungal with in vivo efficacy in a vulvovaginal candidiasis infection. *bioRxiv* **2018**, 368720. [CrossRef]
99. Emrick, D.; Ravichandran, A.; Gosai, J.; Lu, S.; Gordon, D.M.; Smith, L. The antifungal occidiofungin triggers an apoptotic mechanism of cell death in yeast. *J. Nat. Prod.* **2013**, *76*, 829–838. [CrossRef]
100. Ma, J.; Guo, F.; Jin, Z.; Geng, M.; Ju, M.; Ravichandran, A.; Orugunty, R.; Smith, L.; Zhu, G.; Zhang, H. Novel antiparasitic activity of the antifungal lead occidiofungin. *Antimicrob. Agents Chemother.* **2020**, *64*, e00244-20. [CrossRef]
101. Wang, X.Q.; Liu, A.X.; Guerrero, A.; Liu, J.; Yu, X.Q.; Deng, P.; Ma, L.; Baird, S.M.; Smith, L.; Lu, S.E. Occidiofungin is an important component responsible for the antifungal activity of *Burkholderia pyrrocinia* strain Lyc2. *J. Appl. Microbiol.* **2016**, *120*, 607–618. [CrossRef]
102. Hing, S.L.; Ravichandran, A.; Escano, J.; Cooley, J.; Autin, F.; Lu, S.E.; Pruett, S.; Smith, L. Toxicological evaluation of occidiofungin against mice and human cancer cell lines. *Sci. Res.* **2014**, *5*, 1085–1093. [CrossRef]
103. Lim, Y.; Suh, J.W.; Kim, S.; Hyun, B.; Kim, C.; Lee, C. Cepacidine A, a novel antifungal antibiotic produced by *Pseudomonas cepacia*. II. Physico-chemical properties and structure elucidation. *J. Antibiot.* **1994**, *47*, 1406–1416. [CrossRef] [PubMed]
104. Lee, C.H.; Kim, S.; Hyun, B.; Suh, J.W.; Yon, C.; Kim, C.; Lim, Y.; Kim, C. Cepacidine A, a novel antifungal antibiotic produced by *Pseudomonas cepacia*. I. Taxonomy, production, isolation and biological activity. *J. Antibiot.* **1994**, *47*, 1402–1405. [CrossRef] [PubMed]
105. Lee, C.H.; Suh, J.W.; Cho, Y.H. Immunosuppressive activity of cepacidine A, a novel antifungal antibiotic produced by *Pseudomonas cepacia*. *J. Microbiol. Biotechnol.* **1999**, *9*, 672–674.
106. Lee, C.H.; Kempf, H.J.; Lim, Y.; Cho, Y.H. Biocontrol activity of *Pseudomonas cepacia* AF2001 and anthelmintic activity of its novel metabolite, cepacidine A. *J. Microbiol. Biotechnol.* **2000**, *10*, 568–571.
107. Kang, Y.; Carlson, R.; Tharpe, W.; Schell, M.A. Characterization of genes involved in biosynthesis of a novel antibiotic from *Burkholderia cepacia* BC11 and their role in biological control of *Rhizoctonia solani*. *Appl. Environ. Microbiol.* **1998**, *64*, 3939–3947. [CrossRef]
108. Dose, B.; Niehs, S.P.; Scherlach, K.; Florez, L.V.; Kaltenpoth, M.; Hertweck, C. Unexpected bacterial origin of the antibiotic icosalide: Two-tailed depsipeptide assembly in multifarious *Burkholderia symbionts*. *ACS Chem. Biol.* **2018**, *13*, 2414–2420. [CrossRef]
109. Jenner, M.; Jian, X.; Dashti, Y.; Masschelein, J.; Hobson, C.; Roberts, D.; Jones, C.; Harris, S.; Parkhill, J.; Raja, H.A.; et al. An unusual *Burkholderia gladioli* double chain-initiating nonribosomal peptide synthetase assembles "fungal" icosalide antibiotics. *Chem. Sci.* **2019**, *10*, 5489–5494. [CrossRef]
110. Boros, C.; Smith, C.J.; Vasina, Y.; Che, Y.; Dix, A.B.; Darveaux, B.; Pearce, C. Isolation and identification of the icosalides—Cyclic peptolides with selective antibiotic and cytotoxic activities. *J. Antibiot.* **2006**, *59*, 486–494. [CrossRef]
111. Chandler, J.R.; Truong, T.T.; Silva, P.M.; Seyedsayamdost, M.R.; Carr, G.; Radey, M.; Jacobs, M.A.; Sims, E.H.; Clardy, J.; Greenberg, E.P. Bactobolin resistance is conferred by mutations in the L2 ribosomal protein. *mBio* **2012**, *3*, e00499-12. [CrossRef]
112. Bisacchi, G.S.; Hockstein, D.R.; Koster, W.H.; Parker, W.L.; Rathnum, M.L.; Unger, S.E. Xylocandin: A new complex of antifungal peptides. II. Structural studies and chemical modifications. *J. Antibiot.* **1987**, *40*, 1520–1529. [CrossRef]
113. Meyers, E.; Bissachi, G.S.; Dean, L.; Liu, W.C.; Minassian, B.; Slusarchyk, D.S.; Sykes, R.B.; Tanaka, S.K.; Trejo, W. Xylocandin: A new complex of antifungal peptides. I. Taxonomy, isolation and biological activity. *J. Antibiot.* **1987**, *40*, 1515–1519. [CrossRef]
114. Jenul, C.; Sieber, S.; Daeppen, C.; Mathew, A.; Lardi, M.; Pessi, G.; Hoepfner, D.; Neuburger, M.; Linden, A.; Gademann, K.; et al. Biosynthesis of fragin is controlled by a novel quorum sensing signal. *Nat. Commun.* **2018**, *9*, 1297. [CrossRef]
115. Biggins, J.B.; Liu, X.; Feng, Z.; Brady, S.F. Metabolites from the induced expression of cryptic single operons found in the genome of *Burkholderia pseudomallei*. *J. Am. Chem. Soc.* **2011**, *133*, 1638–1641. [CrossRef]
116. Florez, L.V.; Scherlach, K.; Miller, I.J.; Rodrigues, A.; Kwan, J.C.; Hertweck, C.; Kaltenpoth, M. An antifungal polyketide associated with horizontally acquired genes supports symbiont-mediated defense in *Lagria villosa* beetles. *Nat. Commun.* **2018**, *9*, 2478. [CrossRef]
117. Imada, A.; Kitano, K.; Kintaka, K.; Muroi, M.; Asai, M. Sulfazecin and isosulfazecin, novel β-lactam antibiotics of bacterial origin. *Nature* **1981**, *289*, 590–591. [CrossRef]
118. Loveridge, E.J.; Jones, C.; Bull, M.J.; Moody, S.C.; Kahl, M.W.; Khan, Z.; Neilson, L.; Tomeva, M.; Adams, S.E.; Wood, A.C.; et al. Reclassification of the specialized metabolite producer *Pseudomonas mesoacidophila* ATCC 31433 as a member of the *Burkholderia cepacia* complex. *J. Bacteriol.* **2017**, *199*, e00125-17. [CrossRef]
119. Imada, A.; Kintaka, K.; Nakao, M.; Shinagawa, S. Bulgecin, a bacterial metabolite which in concert with β-lactam antibiotics causes bulge formation. *J. Antibiot.* **1982**, *35*, 1400–1403. [CrossRef]
120. Quan, C.S.; Zheng, W.; Liu, Q.; Ohta, Y.; Fan, S.D. Isolation and characterization of a novel *Burkholderia cepacia* with strong antifungal activity against *Rhizoctornia solani*. *Appl. Environ. Biotechnol.* **2006**, *72*, 1276–1284. [CrossRef]
121. Lee, J.; Wu, J.; Deng, Y.; Wang, J.; Wang, C.; Wang, J.; Chang, C.; Dong, Y.; Williams, P.; Zhang, L.H. A cell-cell communication signal integrates quorum sensing and stress response. *Nat. Chem. Biol.* **2013**, *9*, 339–343. [CrossRef]

122. Ye, L.; Cornelis, P.; Guillemyn, K.; Ballet, S.; Christophersen, C.; Hammerich, O. Structure revision of N-mercapto-4-formylcarbostyril produced by *Pseudomonas fluorescens* G308 to 2-(2-hydroxyphenyl)thiazole-4-carbaldehyde (aeruginaldehyde). *Nat. Prod. Commun.* **2014**, *9*, 789–794. [CrossRef]
123. Trottmann, F.; Franke, J.; Ishida, K.; Garcia-Altares, M.; Hertweck, C. A pair of bacterial siderophores releases and traps an intercellular signal molecule: An unusual case of natural nitrone bioconjugation. *Angew. Chem. Int. Ed.* **2019**, *58*, 200–2004. [CrossRef] [PubMed]
124. Eustaquio, A.S.; Janso, J.E.; Ratnayake, A.S.; O'Donnell, C.J.; Koehn, F.E. Spliceostatin hemiketal biosynthesis in *Burkholderia* spp. is catalyzed by an iron/-ketoglutarate-dependent dioxygenase. *Proc. Natl. Acad. Sci. USA* **2014**, *111*, E3376–E3385. [CrossRef] [PubMed]
125. He, H.; Ratnayake, A.S.; Janso, J.E.; Min, H.; Yang, H.Y.; Loganzo, F.; Shor, B.; O'Donnell, J.C.; Koehn, F.E. Cytotoxic spliceostatins from *Burkholderia* sp. and their semisynthetic analogues. *J. Nat. Prod.* **2014**, *77*, 1864–1870. [CrossRef] [PubMed]
126. Prasad, C. Bioactive cyclic dipeptides. *Peptides* **1995**, *16*, 151–164. [CrossRef]
127. Wang, J.H.; Quan, C.S.; Qi, X.H.; Li, X.; Fan, S.D. Determination of diketopiperazines of *Burkholderia cepacia* CF-66 by gas chromatography–mass spectrometry. *Anal. Bioanal. Chem.* **2010**, *396*, 1773–1779. [CrossRef]
128. Scoffone, V.C.; Chiarelli, L.R.; Makarov, V.; Brackman, G.; Israyilova, A.; Azzalin, A.; Forneris, F.; Riabova, O.; Savina, S.; Coenye, T.; et al. Discovery of new diketopiperazines inhibiting *Burkholderia cenocepacia* quorum sensing in vitro and in vivo. *Sci. Rep.* **2016**, *6*, 32487. [CrossRef]
129. Rojas-Rojas, F.U.; Sanchez-Lopez, D.; Tapia-Garcia, E.Y.; Arroyo-Herrera, I.; Maymon, M.; Humm, E.; Huntemann, M.; Clum, A.; Pillay, M.; Palaniappan, K.; et al. Draft genome of *Burkholderia cenocepacia* TAtl-371, a strain from the *Burkholderia cepacia* complex retains antagonism in different carbon and nitrogen sources. *Curr. Microbiol.* **2019**, *76*, 566–574. [CrossRef]
130. Jiao, Y.; Yoshihara, T.; Ishikuri, S.; Uchino, H.; Ichihara, A. Structural identification of cepaciamide A, a novel fungitoxic compound from *Pseudomonas cepacia* D-202. *Tetrahedron* **1996**, *37*, 1039–1042. [CrossRef]
131. Deng, W.; Marshall, N.C.; Rowland, J.L.; McCoy, J.M.; Worrall, L.J.; Santos, A.S.; Strynadka, N.C.J.; Finlay, B.B. Assembly, structure, function and regulation of type III secretion systems. *Nat. Rev. Microbiol.* **2017**, *15*, 323–337. [CrossRef]
132. Swain, D.M.; Yadav, S.K.; Tyagi, I.; Kumar, R.; Kumar, R.; Ghosh, S.; Das, J.; Jha, G. A prophage tail-like protein is deployed by *Burkholderia bacteria* to feed on fungi. *Nat. Commun.* **2017**, *8*, 404. [CrossRef]
133. Kang, J.G.; Shin, S.Y.; Kim, M.J.; Bajpai, V.; Maheshwari, D.K.; Kang, S.C. Isolation and anti-fungal activities of 2-hydroxymethyl-chroman-4-one produced by *Burkholderia* sp. *MSSP. J. Antibiot.* **2004**, *57*, 726–731. [CrossRef]
134. Kirinuki, T.; Ichiba, T.; Katamaya, K. General survey of action site of altericidins on metabolism of *Alternaria kikuchiana* and *Ustilago maydis*. *J. Pestic. Sci.* **1984**, *9*, 601–610. [CrossRef]
135. Abdel-Mawgoud, A.M.; Lepine, F.; Deziel, E. Rhamnolipids: Diversity of structures, microbial origins, and roles. *Appl. Microbiol. Biotechnol.* **2010**, *86*, 1323–1336. [CrossRef]
136. Elshikh, M.; Funston, S.; Chebbi, A.; Ahmed, S.; Marchant, R.; Banat, I.M. Rhamnolipids from non-pathogenic *Burkholderia thailandensis* E264: Physicochemical characterization, antimicrobial and antibiofilm efficacy against oral hygiene related pathogens. *N. Biotechnol.* **2017**, *36*, 26–36. [CrossRef]
137. Hormann, B.; Muller, M.M.; Syldatk, C.; Hausmann, R. Rhamnolipid production by *Burkholderia plantarii* DSM 9509^T. *Eur. J. Lipid Sci. Technol.* **2010**, *112*, 674–680. [CrossRef]
138. Costa, S.; Deziel, E.; Lepine, F. Characterization of rhamnolipid production by *Burkholderia glumae*. *Lett. Appl. Microbiol.* **2011**, *53*, 620–627. [CrossRef]
139. Funston, S.J.; Tsaousi, K.; Rudden, M.; Smyth, T.J.; Stevenson, P.S.; Marchant, R.; Banat, I.M. Characterizing rhamnolipid production in *Burkholderia thailandensis* E264, a non-pathogenic producer. *Appl. Microbiol. Biotechnol.* **2016**, *100*, 7945–7956. [CrossRef]
140. Haubler, S.; Nimtz, M.; Domke, T.; Wray, V.; Steinmetz, I. Purification and characterization of a cytotoxic exolipid of *Burkholderia pseudomallei*. *Infect. Immun.* **1998**, *66*, 1588–1593.
141. Tawfik, K.A.; Jeffs, P.; Bray, B.; Dubay, G.; Falkinham, J.O.; Mesbah, M.; Youssef, D.; Khalifa, S.; Schmidt, E.W. Burkholdines 1097 and 1229, potent antifungal peptides from *Burkholderia ambifaria* 2.2N. *Org. Lett.* **2010**, *12*, 664–666. [CrossRef]
142. Lin, Z.; Falkinham, J.O.; Tawfik, K.A.; Jeffs, P.; Bray, B.; Dubay, G.; Cox, J.E.; Schmidt, E.W. Burkholdines from *Burkholderia ambifaria*: Antifungal agents and possible virulence factors. *J. Nat. Prod.* **2012**, *75*, 1518–1523. [CrossRef]
143. Wang, M.; Tachibana, S.; Murai, Y.; Li, L.; Lau, S.Y.L.; Cao, M.; Zhu, G.; Hashimoto, M.; Hashidoko, Y. Indole-3-acetic acid produced by *Burkholderia* heleia acts as a phenylacetic acid antagonist to disrupt tropolone biosynthesis in *Burkholderia plantarii*. *Sci. Rep.* **2016**, *6*, 22596. [CrossRef] [PubMed]
144. Azegami, K.; Nishiyama, K.; Watanabe, Y.; Suzuki, T.; Yoshida, M.; Nose, K.; Toda, S. Tropolone as a root growth-inhibitor produced by a plant pathogenic *Pseudomonas* sp. causing seedling blight of rice. *Jpn. J. Phytopathol.* **1985**, *51*, 315–317. [CrossRef]
145. Wakimoto, S.; Hirayae, K.; Tsuchiya, K.; Kushima, Y.; Furuya, N.; Matsuyama, N. Production of antibiotics by plant pathogenic pseudomonads. *Ann. Phytopathol. Soc. Jpn.* **1986**, *52*, 835–842. [CrossRef]
146. Abe, M.; Nakazawa, T. Characterization of hemolytic and antifungal substance, cepalycin, from *Pseudomonas cepacia*. *Microbiol. Immunol.* **1994**, *38*, 1–9. [CrossRef] [PubMed]
147. Simonetti, E.; Roberts, I.N.; Montecchia, M.S.; Gutierrez-Boem, F.H.; Gomez, F.M.; Ruiz, J.A. A novel *Burkholderia ambifaria* strain able to degrade the mycotoxin fusaric acid and to inhibit *Fusarium* spp. growth. *Microbiol. Res.* **2018**, *206*, 50–59. [CrossRef]

148. Guttenberger, N.; Blankenfeldt, W.; Breinbauer, R. Recent developments in the isolation, biological function, biosynthesis, and synthesis of phenazine natural products. *Bioorg. Med. Chem.* **2017**, *25*, 6149–6166. [CrossRef]
149. Mullins, A.J.; Mahenthiralingam, E. The hidden genomic diversity, specialized metabolic capacity, and revised taxonomy of *Burkholderia* sensu lato. *Front. Microbiol.* **2021**, *12*, 726847. [CrossRef]
150. Coulon, P.M.L.; Groleau, M.C.; Deziel, E. Potential of the *Burkholderia cepacia* complex to produce 4-hydroxy-3-methyl-2-alkyquinolines. *Front. Cell. Infect. Microbiol.* **2019**, *9*, 33. [CrossRef]
151. Matthew, A.; Jenul, C.; Carlier, A.L.; Eberl, L. The role of siderophores in metal homeostasis of member of the genus *Burkholderia*. *Environ. Microbiol. Rep.* **2016**, *8*, 103–109. [CrossRef]
152. Li, X.; Quan, C.S.; Fan, S.D. Antifungal activity of a novel compound from *Burkholderia cepacia* against plant pathogenic fungi. *Lett. Appl. Microbiol.* **2007**, *45*, 508–514. [CrossRef]

Disclaimer/Publisher's Note: The statements, opinions and data contained in all publications are solely those of the individual author(s) and contributor(s) and not of MDPI and/or the editor(s). MDPI and/or the editor(s) disclaim responsibility for any injury to people or property resulting from any ideas, methods, instructions or products referred to in the content.

molecules

Article

Potential Inhibitors of Monkeypox Virus Revealed by Molecular Modeling Approach to Viral DNA Topoisomerase I

Xiaopeng Hu, Sanqi An *, Jiemei Chu, Bingyu Liang, Yanyan Liao, Junjun Jiang, Yao Lin, Li Ye * and Hao Liang *

Biosafety Level-3 Laboratory, Life Sciences Institute & Guangxi Collaborative Innovation Center for Biomedicine, Guangxi Medical University, Nanning 530021, China
* Correspondence: ansq@mail2.sysu.edu.cn (S.A.); yeli@gxmu.edu.cn (L.Y.); lianghao@gxmu.edu.cn (H.L.)

Abstract: The monkeypox outbreak has become a global public health emergency. The lack of valid and safe medicine is a crucial obstacle hindering the extermination of orthopoxvirus infections. The identification of potential inhibitors from natural products, including Traditional Chinese Medicine (TCM), by molecular modeling could expand the arsenal of antiviral chemotherapeutic agents. Monkeypox DNA topoisomerase I (TOP1) is a highly conserved viral DNA repair enzyme with a small size and low homology to human proteins. The protein model of viral DNA TOP1 was obtained by homology modeling. The reliability of the TOP1 model was validated by analyzing its Ramachandran plot and by determining the compatibility of the 3D model with its sequence using the Verify 3D and PROCHECK services. In order to identify potential inhibitors of TOP1, an integrated library of 4103 natural products was screened via Glide docking. Surface Plasmon Resonance (SPR) was further implemented to assay the complex binding affinity. Molecular dynamics simulations (100 ns) were combined with molecular mechanics Poisson–Boltzmann surface area (MM/PBSA) computations to reveal the binding mechanisms of the complex. As a result, three natural compounds were highlighted as potential inhibitors via docking-based virtual screening. Rosmarinic acid, myricitrin, quercitrin, and ofloxacin can bind TOP1 with KD values of 2.16 µM, 3.54 µM, 4.77 µM, and 5.46 µM, respectively, indicating a good inhibitory effect against MPXV. The MM/PBSA calculations revealed that rosmarinic acid had the lowest binding free energy at −16.18 kcal/mol. Myricitrin had a binding free energy of −13.87 kcal/mol, quercitrin had a binding free energy of −9.40 kcal/mol, and ofloxacin had a binding free energy of −9.64 kcal/mol. The outputs (RMSD/RMSF/Rg/SASA) also indicated that the systems were well-behaved towards the complex. The selected compounds formed several key hydrogen bonds with TOP1 residues (TYR274, LYS167, GLY132, LYS133, etc.) via the binding mode analysis. TYR274 was predicted to be a pivotal residue for compound interactions in the binding pocket of TOP1. The results of the enrichment analyses illustrated the potential pharmacological networks of rosmarinic acid. The molecular modeling approach may be acceptable for the identification and design of novel poxvirus inhibitors; however, further studies are warranted to evaluate their therapeutic potential.

Keywords: monkeypox virus; DNA topoisomerase I (TOP1); natural products; rosmarinic acid; molecular dynamics simulation; Surface Plasmon Resonance (SPR); enrichment analyses

check for updates

Citation: Hu, X.; An, S.; Chu, J.; Liang, B.; Liao, Y.; Jiang, J.; Lin, Y.; Ye, L.; Liang, H. Potential Inhibitors of Monkeypox Virus Revealed by Molecular Modeling Approach to Viral DNA Topoisomerase I. *Molecules* **2023**, *28*, 1444. https://doi.org/10.3390/molecules28031444

Academic Editors: Xun Song, Chenyang Li and Yifu Guan

Received: 6 December 2022
Revised: 13 January 2023
Accepted: 15 January 2023
Published: 2 February 2023

Copyright: © 2023 by the authors. Licensee MDPI, Basel, Switzerland. This article is an open access article distributed under the terms and conditions of the Creative Commons Attribution (CC BY) license (https://creativecommons.org/licenses/by/4.0/).

1. Introduction

Monkeypox virus (MPXV) is an emergent human pathogen, and the endemic was sporadically reported in Africa. However, it has posed as a new global outbreak since May 2022. MPXV is a double-stranded DNA virus, and it belongs to the Poxviridae orthopoxvirus family. The orthopoxviruses, including monkeypox virus, smallpox virus, vaccinia virus, cowpox virus, and buffalopox virus, infect various mammals and cause zoonosis [1]. Although MPXV is less lethal than smallpox, it is considered to be a dangerous virus, because it causes many challenges, including high prevalence, disability, and disfiguration [2]. In the past, there have been occasional outbreaks of the MPXV infection, which were mainly caused by people contacting infected rodents. Most infected cases experience fever, body aches, chills, and tiredness [3]. Patients with severe symptoms may develop a rash and cuts towards the face and hands that can spread the infection to other parts of the body. MPXV has a fatality rate of up to 10% and is thought to be more severe among children [1]. Previously, MPXV infection was mainly endemic in Western Africa [2]. However, with the number of cases outside Africa increasing rapidly in 2022, MPXV infection has led to high levels of concern. The World Health Organization (WHO) announced that the multi-country MPXV outbreak has become a global public health emergency at a conference of the International Health Regulations Emergency Committee (EC) on 23 July 2022.

More than 40 years have passed since the WHO declared the extinction of smallpox disease worldwide. The vaccination and immunization programs against smallpox have been halted since 1980. It is estimated that half of the general population is currently unvaccinated against smallpox [3]. Therefore, the general population is highly susceptible to outbreaks caused by the release of the smallpox virus. Moreover, the progressive move of zoonotic viruses from endemic reservoirs or the potential diffusion of variola virus due to bioterrorism behavior places animal poxviruses in the range of high epidemic risk pathogens. In this context, renewed interest in the discovery and exploitation of novel antiviral medicine can be applied to help mitigate the influence of future outbreaks [4].

The most effective choice for preventing orthopoxvirus infections is vaccination. However, the variability and mutation of viruses shorten the efficacious window of postexposure vaccination (from only 4 to 7 days) [5]. Moreover, complications upon vaccination, such as encephalitis, fetal vaccinia, progressive vaccinia, and eczema vaccinatum, often occur among individuals with acquired or congenital immunodeficiency [6]. Although vaccination remains critical to controlling outbreaks, the validity of antiviral medical treatment would be meaningful for individuals with vaccine contraindications and those who are infected [7].

Generally, the lack of valid and safe medicine hinders the eradication of orthopoxvirus infections. No specific treatment for MPXV has been demonstrated to date. The existing MPXV treatments mainly include nursing, symptomatic, and supportive treatment. Most existing drugs are nucleoside analogs, such as cidofovir, which inhibit virus replication. Other nucleoside analogs, such as adenozine N1-oxide (ANO) and its derivatives; N-methanocarbothymidine; or derivatives of aciklovir, peninclovir, and brivudin, have also been reported to be effective in inhibiting infection by limiting the release of progeny virions [8]. However, side effects and drug resistance limit the therapeutic potential of nucleoside analogs in response to new outbreaks of the orthopoxvirus.

Importantly, the transmission and mutation of MPXV are a high concern and priority. DNA topoisomerase I (TOP1) is a highly conserved DNA repair enzyme with a small size and low homology to human proteins, making it an attractive target for antiviral virtual screening [9]. Additionally, many enzymes are encapsidated in the orthopoxvirus. TOP1 is a crucial enzyme required for an early phase of viral transcription. Therefore, TOP1 is considered a potential antiviral target [10]. It functions by liberating the supercoiled and torsional tension towards the DNA duplex. It also introduces single-stranded breaks via transesterification at the specific site of 5′-[CT]CCTTp in the DNA duplex [9]. The scissile phosphodiesters are attacked through the catalytic tyrosine of the enzyme, providing the constitution of a DNA–(3′-phosphotyrosyl)–enzyme intermediate and the exclusion of a 5′-OH DNA strand [9,10].

Natural products including TCM have been an underexploited reservoir of novel compound candidates to combat a variety of illnesses. Numerous natural products have antiviral properties and have been tested [11]. Traditional Chinese Medicine (TCM) is rich in biological diversity and can be further developed to discover fresh compound candidates. Those therapeutic compounds are primarily prepared from heat-clearing and detoxifying herbs, which are rich in different kinds of alkaloids, terpenes, flavonoids, saponins, and carboxylic acids [11,12]. Computer-aided drug design (CADD) has become more advantageous as compared with the traditional approach of screening in reducing the dissipation of resources in terms of cost, time, and effort through markedly diminishing the quantity of compounds and filtering out hits for further HTS [12]. Here, the homology model of TOP1 of MPXV was obtained using SWISS-MODEL, and we identified potential natural product-derived MPXV inhibitory compounds via docking-based virtual screening. Molecular dynamics simulations (100 ns) were combined with molecular mechanics Poisson–Boltzmann surface area (MM/PBSA) computations to reveal the binding mechanisms of the complex. The natural product-derived compounds from Traditional Chinese Medicine (TCM) with limited side effects or complications will undoubtedly enlarge the arsenal of antiviral agents.

2. Results

2.1. Modeling the Structure of TOP1

We chose the protein sequence TOP1 (F1DIV3) from the UniProt database corresponding to the available experimental MPXV TOP1 model. The MPXV TOP1 sequences show only 28% were identical to the human P11387 when analyzed with the protein blast alignment tool (NCBI). Then, the amino acid sequences were submitted into SWISS-MODEL, which implemented a Basic Local Alignment Search Tool (BLAST) search to obtain reasonable templates that were identical to the target sequence. The TOP1 structures, from the variola virus or vaccinia virus (PDB ID: 3IGC, 2H7F, and 1A41) ranked top, with identical scores of 96.15%, 98.08%, 96.94%, respectively. Indeed, the sequence identity score between the template and the query exceeded 35%, indicating the model was created with high quality. Thus, the three structures were then chosen as the structural template to model MPXV TOP1 (Figure 1). Subsequently, MPXV TOP1 was verified as the best with the structure assessment (GMQE 0.93, QMEAND 0.85 ± 0.05). Therefore, MPXV TOP1 was successfully used for studies of structure-based molecular modeling.

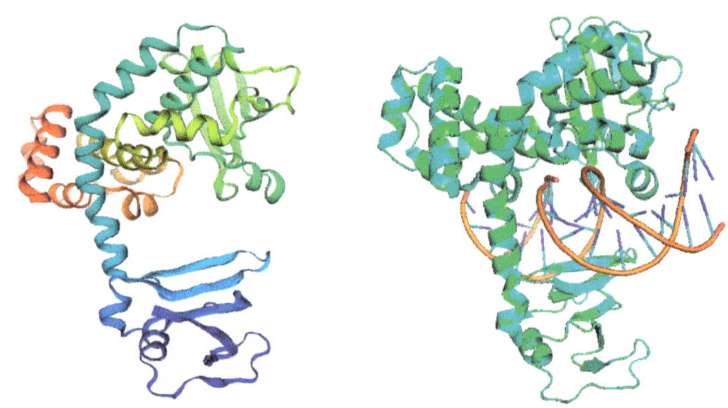

(**a**) (**b**)

Figure 1. (**a**) Cartoon views of predicted 3D structures of the TOP1 from the SWISS-MODEL; (**b**) the structural alignment between the modeled TOP1 and template TOP1 (PDB ID 3IGC) from the variola virus (green—modeled TOP1; cyan—template TOP1).

2.2. Quality Evaluation of MPXV TOP1 Models

The model was evaluated using SAVES v6.0. SWISS-MODEL had a good ERRAT. The Ramachandran plot was also procured utilizing PROCHECK, which assessed the stereochemistry of TOP1 by identifying residue-by-residue geometry and whole-structure geometry (Figure 2a) [12]. Due to the percentage of residues in the most favored (core), additionally allowed, generously allowed, and disallowed regions, the TOP1 structure was considered high quality [13,14]. The protein structure of TOP1 had 95.5%, 4.5%, 0.0%, and

0.0% towards the residues in the most favored, additionally allowed, generously allowed, and disallowed regions, respectively. Verify 3D was applied to determine the matched degree of the TOP1 model according to its amino acid sequence (Figure 2b) and a PROVE score of 95.082%. It was also forecasted by PROCHECK to have 0 faults and three passes, indicating that neither misfolded nor erroneous regions were identified.

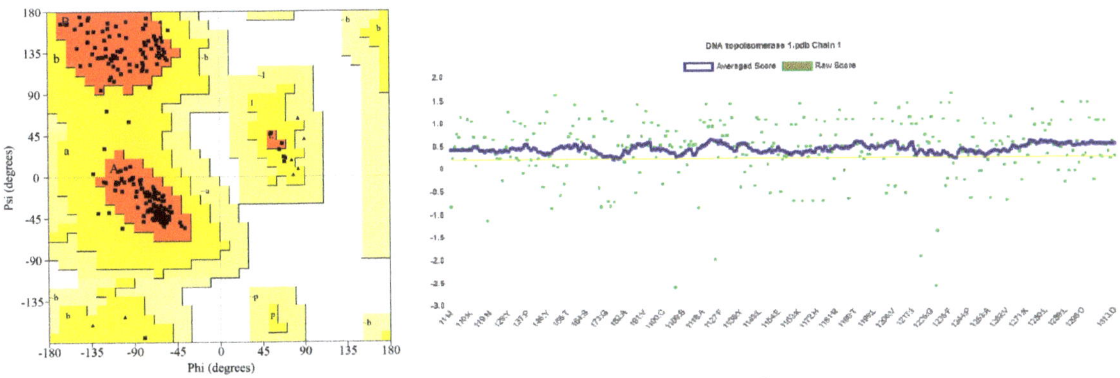

(a) (b)

Figure 2. (**a**) Ramachandran plot towards TOP1 model; (**b**) results of the Verify_3D analysis of the TOP1 model.

2.3. Docking-Based Virtual Screening

The integrated library from the Traditional Chinese Medicine Library (L8300) and the Natural Product Library (L1400) and the classical inhibitor ofloxacin docked with the energy minimized TOP1 at a grid pocket size of 28.32 × −10.89 × 29.89 Å. Centered at (10, 12, 14), Å was set as the binding domain of TOP1. Ofloxacin is one kind of fluoroquinolone. It acts on DNA gyrase and topoisomerase IV, enzymes which, similar to human topoisomerase, prevent the excessive supercoiling of DNA during replication or transcription. By inhibiting their function, the drug inhibits normal cell division. It was demonstrated that ofloxacin and levofloxacin inhibited the viral topoisomerase activity of the vaccinia virus but not of the herpes simplex virus and influenza virus [15]. A previous report also showed that fluoroquinolone enrofloxacin inhibits DNA relaxation by Vaccinia topoisomerase I [16]. Therefore, we chose ofloxacin as a positive inhibitor compared with other natural compounds, looking at the amount of 4103 compounds that were successfully sifted against TOP1. A stringent docking score standard of −7 kcal/mol was set to choose the compounds after the virtual screening. A total of three compounds were highlighted with docking scores of less than −7 kcal/mol. A negative approach was utilized to rank the export in the order of the diminishing binding affinity. As a result, the higher the negativity, the more reasonably the candidate performed as a feasible supreme compound [17].

Rosmarinic acid was proven to have the lowest binding energy to TOP1, with a docking score of −8.207 kcal/mol. Myricitrin, quercitrin, and ofloxacin also demonstrated low docking scores of −7.599 kcal/mol, −7.322 kcal/mol, and −6.046 kcal/mol, respectively. The other compounds demonstrated a weaker binding affinity to TOP1 than the classical inhibitor ofloxacin and are excluded in the following Table 1. As a classical inhibitor, ofloxacin was demonstrated to have a docking score of −6.246 kcal/mol to TOP1 in this study. Ofloxacin with ~2 mM was reported to induce ~30% inhibition of TOP I, which was purified from vaccinia [16]. This implies that compounds with a lower docking score than ofloxacin can feasibly demonstrate significant prohibitive activities towards MPXV.

Table 1. The chemical features and docking scores between TOP1 and selected compounds.

Molecule Name	Molecule Structure	Molecular Weight	Docking Score (kcal/mol)
Rosmarinic acid	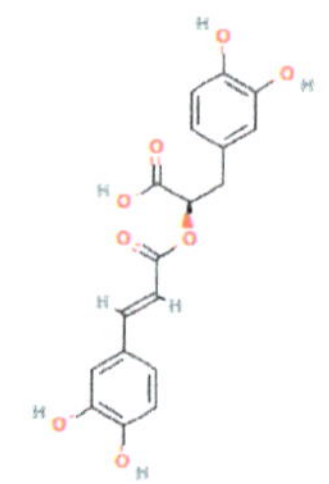	360.32	−8.207
Myricitrin	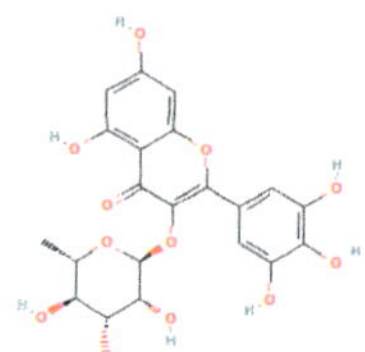	464.38	−7.599
Quercitrin	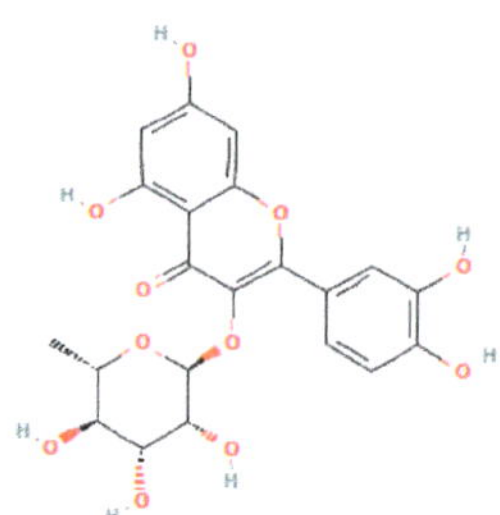	448.38	−7.322
Ofloxacin	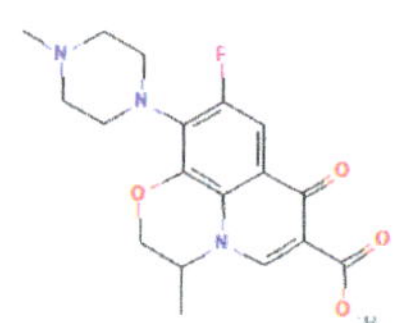	361.40	−6.046

2.4. Characterization of TOP1–Ligand Interactions

Rosmarinic acid was docked into the TOP1 pocket and shaped into two hydrogen bonds with TYR274 of lengths 2.67 and 2.71 Å. Other hydrogen bonds were also found, including LYS167, GLY132, LYS133, and ASP168. It also interacted with ARG223, LYS220, ARG130, PHE131, TYR136, and THR142 via hydrophobic bonds (Figure 3A, Table 2). The interaction between myricitrin and TOP1 was created through one hydrogen bond with TYR274 (2.94 Å), as well as with TYR209, LYS167, GLY132, and LYS133. It also formed hydrophobic contacts with ARG218, ILE219, LYS220, ARG223, ARG130, PHE131, TYR136, and THR142 (Figure 3B, Table 2). Quercitrin, which docked into the TOP1 pocket, interacted with TYR209, TYR274, LYS167, LYS133, and GLY132 via hydrogen bonding and ILE219, LYS220, ARG223, ARG130, PHE131, TYR136, and THR142 via hydrophobic bonding (Figure 3C, Table 2). Quercitrin's situation is quite close to that of myricitrin due to the similar structural types of the flavonoids. The interactions between the selected compounds and the residues may explain their high MPXV suppression. Ofloxacin targeted TOP1 via hydrogen bonds with TYR136 with bond lengths of 3.40 Å (Figure 3D, Table 2). Only one hydrogen bond was observed between ofloxacin and TOP1. This may be the critical reason for its weaker affinity compared with the above natural products. Ofloxacin also formed hydrophobic contacts with ARG67, ARG130, LYS133, TYR136, THR142, LYS167, ASP168, ILE219, and LYS220.

Table 2. The MM_PBSA and hydrogen/hydrophobic bonds between the selected compounds and TOP1.

Compounds	MM_PBSA (kcal/mol)	Hydrogen Bond	Hydrophobic Bond
Rosmarinic acid	−16.18	TYR274, LYS167, GLY132, LYS133, ASP168	ARG223, LYS220 ARG130, PHE131 TYR136, THR142 ASN140, LYS169
Myricitrin	−13.87	TYR209, TYR274, LYS167, GLY132, LYS133	ARG218, ILE219 LYS220, ARG223 ARG130, PHE131 TYR136, THR142
Quercitrin	−9.40	TYR209, TYR274, LYS167, LYS133, GLY132	ILE219,LYS220, ARG223,ARG130, PHE131, TYR136, THR142
Ofloxacin	−9.64	PHE131, GLY132, ASN140	ARG67, ARG130, LYS133, TYR136, THR142, LYS167, ASP168, ILE219, LYS220

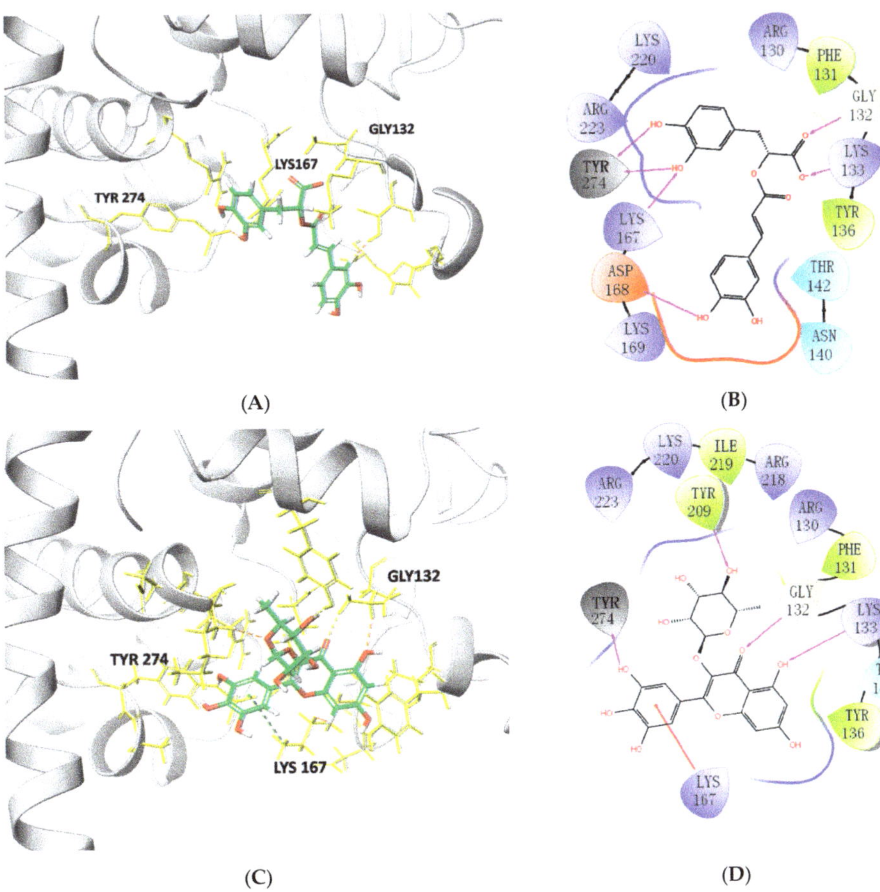

Figure 3. *Cont.*

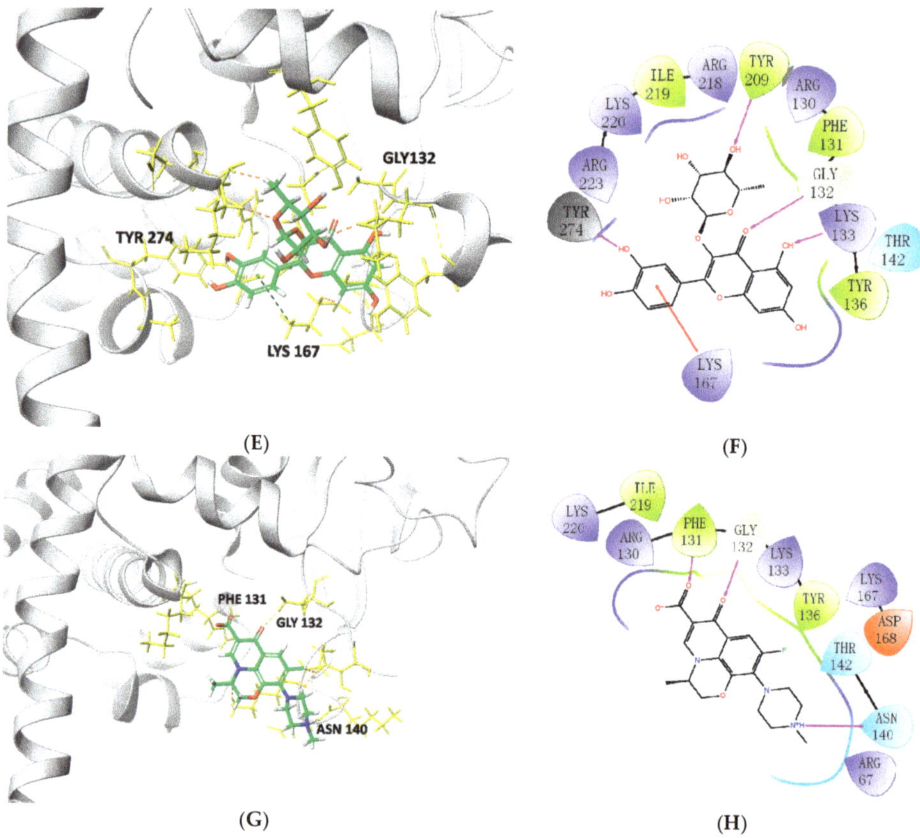

Figure 3. Cartoon on behalf of TOP1 targeted with: (**A**,**B**) rosmarinic acid; (**C**,**D**) myricitrin; (**E**,**F**) quercitrin; (**G**,**H**) ofloxacin. The 3D combining site was revealed as a cartoon representation, as well as the compounds, shown as sticks. The 2D one was further visualized and signed in the indicated atoms with the interaction force, including the hydrogen bond and hydrophobic contact.

2.5. Molecular Dynamics Simulations

The 100 ns MD simulations were implemented successfully on the compounds–TOP1 complexes to evaluate whether the MD simulations converged and whether they were stable. The root mean square deviation (RMSD), the root mean square fluctuation (RMSF), the radius of gyration (Rg), and the solvent-accessible surface area (SASA) were analyzed for unbound protein and protein–ligand complexes (Figure 4) [18]. The outputs indicated well-behaved systems.

TOP1 advanced to a mean of 5.7 Å until about 2 ns and then retained its balance during a period of 100 ns simulation. The RMSD plot for the complex presented the same tendency as the unbound TOP1 (Figure 4A). Interestingly, the TOP1–rosmarinic acid complex performed minimum fluctuations, maintaining an average 5.5 Å value during the period. Both the myricitrin and quercitrin complexes presented less fluctuations than ofloxacin. The TOP1–ofloxacin complex was proven to have the highest RMSD values, with an average of 5.8 Å, suggesting the most fluctuations.

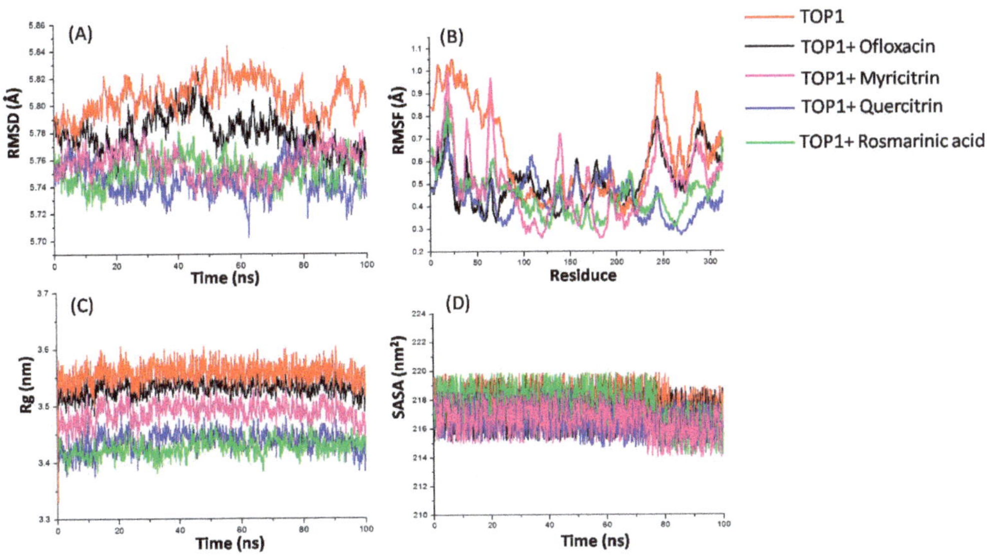

Figure 4. Root mean square deviation (RMSD), root mean square fluctuation (RMSF), radius of gyration (Rg), and solvent-accessible surface area (SASA) graphs of the TOP1–ligand complexes produced over a 100 ns molecular dynamics simulation. (**A**) RMSD versus time plot of TOP1–ligand complexes; (**B**) analysis of the RMSF trajectories of residues of TOP1–ligand complexes; (**C**) Rg versus time plot of TOP1–ligand complexes; (**D**) SASA versus time plot of TOP1–ligand complexes; as for the unbound protein, TOP1, TOP1+Ofloxacin, TOP1+Rosmarinic acid, TOP1+Myricitrin, and TOP1+Quercitrin are represented as red, black, purple, blue, and green, respectively.

RMSF discloses the elasticity of diverse sites of protein–ligand complexes [19,20]. The stability profile analysis was utilized to assess residues dedicated to the complex structural undulation. Higher RMSF values mean greater fluctuations in residues. The RMSF plots revealed that both rosmarinic acid and myricitrin caused some fluctuations in similar regions of TOP1 (Figure 4B). Fluctuations were perceived at regions from residue indices 1–75, 100–175, and 225–300. A high fluctuation was observed between residues 1–75 and 225–300, hinting they could be involved in compound binding. The outcome is consistent with the binding mode analysis above.

Rg was used to depict the characteristic of protein compactness and the folding mechanism [19]. The conformational behavior of the TOP1, TOP1+Ofloxacin, TOP1+Rosmarinic acid, TOP1+Myricitrin, and TOP1+Quercitrin systems was calculated with average Rg values of 3.44 nm, 3.48 nm, 3.43 nm, 3.53 nm, and 3.55 nm, respectively (Figure 4C). Rg values show a minor decrease when the selected compounds are bound with TOP1. The protein–ligand complexes remained stable with a stable equilibrium of Rg for 100 ns MD. A lower Rg value was observed while rosmarinic acid interacted with TOP1, indicating a higher compactness of TOP1. Therefore, the complex folding was less frequent.

SASA is a conducive parameter used to study the conformational dynamics of a protein in the surrounding solvent [19]. The average SASA values for TOP1, TOP1+Ofloxacin, TOP1+Rosmarinic acid, TOP1+Myricitrin, and TOP1+Quercitrin were calculated as 217.01 nm^2, 217.84 nm^2, 216.70 nm^2, 217.61 nm^2, and 216.5 nm^2, respectively (Figure 4D). A lower decrease in the SASA of TOP1 during targeting with rosmarinic acid occurred, possibly due to some external residues being concealed by the protein pocket, indicating that a potential hydrophobic force was formed within the complex.

The MM/PBSA calculations revealed that rosmarinic acid had the lowest binding free energy at −16.18 kcal/mol. It was also observed that myricitrin has a binding free energy of −13.87 kcal/mol, but quercitrin had −9.40 kcal/mol, and ofloxacin had −9.64 kcal/mol

(Table 2). The MM/PBSA method was also utilized to explore free binding energies using per-residue decomposition. The useful perception of meaningful interactions of crucial residues in free energy contributions was performed with decomposition. The TOP1–ligand spectra interactions presented a significant number of critical residues (PHE131, GLY132, LYS133, THR142, LYS167, LYS169, and TYR274), which contributes favorably to the binding of TOP1+Rosmarinic acid in comparison to TOP1+Ofloxacin (GLY132, LYS133, ASN140, THR142, LYS167, TYR209, and ILE219) (Figure 5A,D). Residues conferring a binding free energy greater than 1.0 kcal/mol or less than −1.0 kcal/mol were identified as critical residues for the interaction. It can further be observed from Figure 5A that TYR274 (−6.00 kcal/mol), THR142 (−2.60 kcal/mol), LYS167 (−2.02 kcal/mol), LYS133 (−1.02 kcal/mol), PHE131 (−1.00 kcal/mol), and LYS169 (−1.01 kcal/mol) contribute significantly to the binding between rosmarinic acid and TOP1, looking at the computations of MM/PBSA per-residue decomposition. Regarding the TOP1+Rosmarinic acid complex, PHE131, GLY132, LYS133, THR142, LYS167, LYS169, and TYR274 were observed to contribute individual energies beyond the −1.0 kcal/mol threshold, respectively. As for the TOP1+Myricitrin complex (Figure 5B), it was observed that PHE131, GLY132, THR142, LYS167, ILE219, LYS220, and TYR274 contributed individual energies above the −1.0 kcal/mol threshold. For the TOP1+Quercitrin complex (Figure 5C), LYS133, THR142, ILE219, and TYR274 were found to contribute individual energies above the −1.0 kcal/mol threshold. For the TOP1+Ofloxacin complex, GLY132, LYS133, ASN140, THR142, LYS167, TYR209, and ILE219 contributed individual energies above −1.0 kcal/mol, respectively.

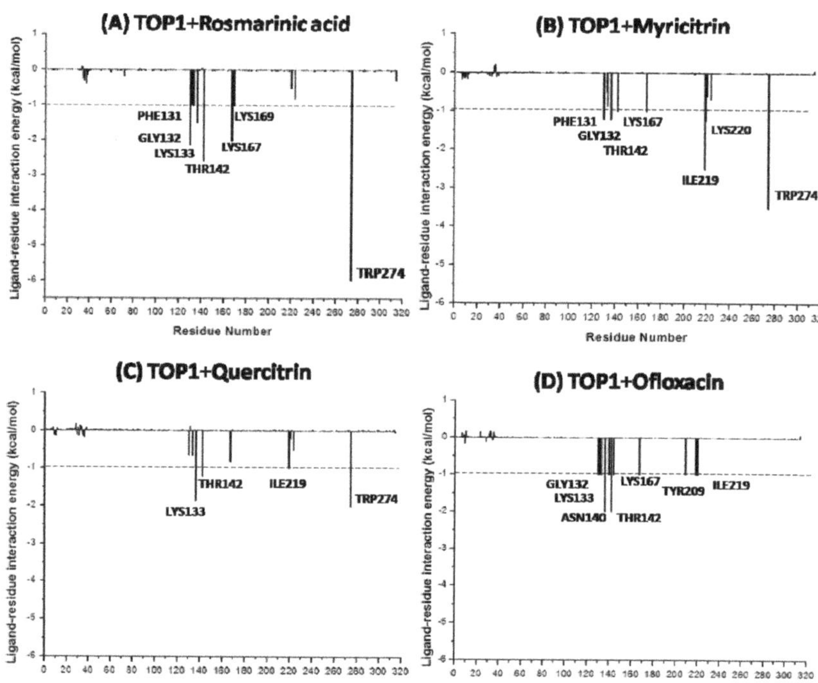

Figure 5. Decomposition of MM/PBSA into contributions from individual residues for (**A**) TOP1+Rosmarinic acid; (**B**) TOP1+Myricitrin; (**C**) TOP1+Quercitrin; (**D**) TOP1+Ofloxacin.

2.6. SPR Results

SPR experiments were applied to calculate the binding affinity between compounds and MPXV TOP1. Here, the binding affinity was measured with the equilibrium dissociation constant (KD). A wide range of half-diluted concentrations (from 6.25 to 100 μM) was employed to fit the KD values of four compounds (rosmarinic acid, myricitrin, quercitrin, and ofloxacin). A single-site model was used to fit compound half-diluted concentrations versus the response (RU). The KD of rosmarinic acid, myricitrin, quercitrin, and ofloxacin were also calculated as 2.16 μM, 3.54 μM, 4.77 μM, and 5.46 μM, respectively. The representative sensorgram of binding RU versus time is depicted in Figure 6.

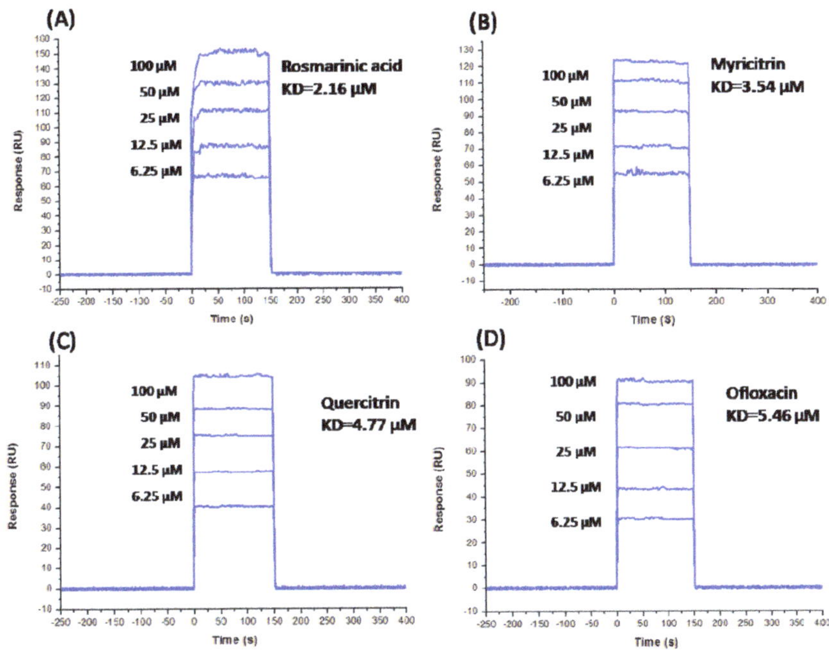

Figure 6. The direct binding affinity of selected compounds with vaccinia virus TOP1 was identified by SPR. (**A**) TOP1+Rosmarinic acid; (**B**) TOP1+Myricitrin; (**C**) TOP1+Quercitrin; (**D**) TOP1+Ofloxacin.

2.7. Analysis of Feasible Targets toward Rosmarinic Acid through GO and KEGG Enrichment

More than 100 target genes were yielded for rosmarinic acid using Swiss Target Prediction. Three enrichment branches were formed from the GO annotation output: molecular function (MF), cellular component (CC), and biological process (BP) (Figure 7A). Carbonate dehydratase and hydro-lyase activity were shown to be outstanding in rosmarinic acid-related MF. For CC, the rosmarinic acid-predicted target mostly took part in the membrane raft and extracellular space. The one-carbon metabolic process was thought to closely interplay with rosmarinic acid-predicted targets towards BP.

The KEGG pathway showed that nitrogen metabolism may be an important pathway towards rosmarinic acid targets (Figure 7B). These pathways are closely linked with innate and adaptive immune cells in the skin, indicating that the utilization of rosmarinic acid may allow for the suppression of inflammation during MPXV infection [21].

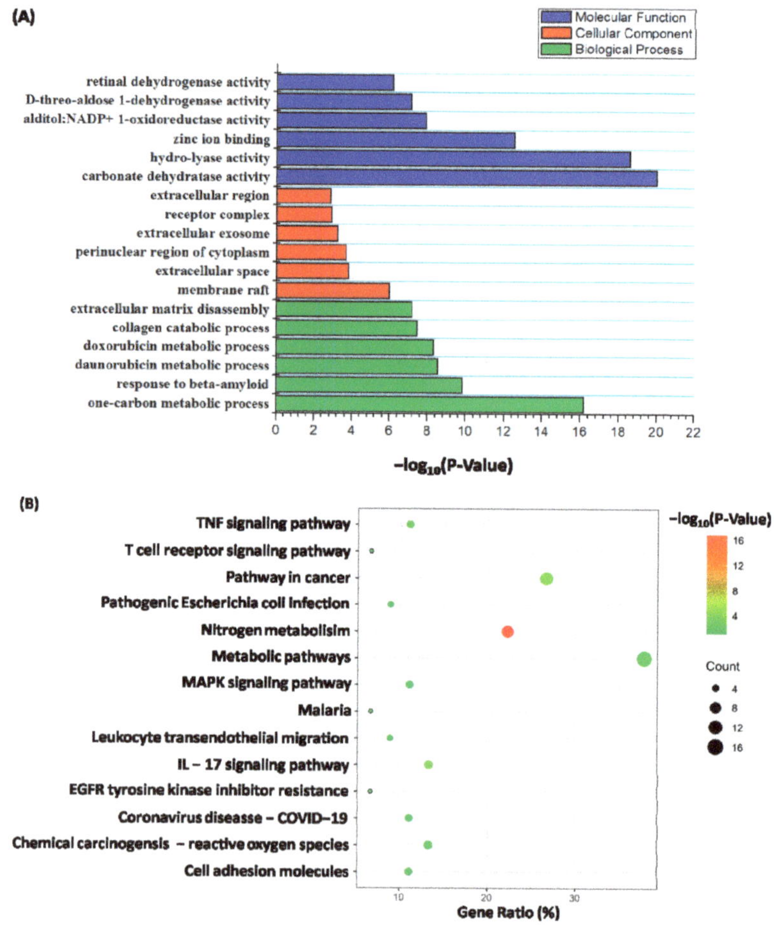

Figure 7. Analysis of the feasible targets toward rosmarinic acid through GO and KEGG enrichment. (**A**) The forecasted targets mostly take part in carbonate dehydratase activity in terms of the molecular function; as for the cellular component, the predicted targets mainly existed in cytosol; one-carbon metabolic is considered the primary progress during the biological process. (**B**) The circle diameter represents the number of rosmarinic acid-related genes. The deeper shield of orange shows the greater disparity. Forecasted genes towards rosmarinic acid (CA12, CA1, CA5B, CA2, CA4, CA7, CA6, CA9, CA14, and CA13) were allocated to the signaling pathway of nitrogen metabolism, with notable differences.

3. Discussion

Natural products, including TCM, have shown their potential to be repurposed as effective MPXV therapeutics. Here, we explored possible MPXV inhibitors by targeting TOP1. TOP1 is a topoisomerase with 314 amino acids, which recognizes and trans-esterifies at specific DNA sequences, such as 5′-(T/C) CCTT↓, where 30 phosphates of the incised strand bind with TYR274 towards the enzyme [22]. The amino acid at position 274 is the key amino acid towards the TOP1-binding pocket. Many enzymes are encapsidated in the orthopoxvirus. Topoisomerase I (TOP1) is a crucial enzyme required for the early phase of viral transcription. Therefore, TOP1 is considered a potential antiviral target [10]. However, the lack of a TOP1 tertiary structure in the recent outbreak of MPXV seriously narrowed the rational design of inhibitors. Therefore, a homology model was generated using the high-quality structural templates available from the smallpox virus, which shares 98.41% sequence identity with MPXV.

The novel TOP1 model identified three potential bioactive compounds comprising rosmarinic acid, myricitrin, and quercitrin with binding scores of −8.207 kcal/mol, −7.599 kcal/mol, and −7.322 kcal/mol, respectively. TYR274 was predicted to be a critical residue for those compounds anchored in the pocket of DNA TOP1 (Table 2, Figure 5). The selected compounds were demonstrated to have higher binding affinities than the clinical molecule (ofloxacin), with a docking score of only -6.046 kcal/mol. Furthermore, the potential inhibition of TOP1 by the compounds was corroborated through MD simulations, including MM/PBSA. Rosmarinic acid, myricitrin, and quercitrin demonstrated better affinity against the TOP1 of MPXV than ofloxacin, although ofloxacin was a typical inhibitor with defined targets in the vaccinia virus. To be more specific, a significant number of key residues (PHE131, GLY132, LYS133, THR142, LYS167, ASP168, LYS169, and TYR274) contribute favorably to the interaction with rosmarinic acid in comparison to ofloxacin (PHE131, GLY132, and ASN140). TYR274 was the unique residue that conferred individual energy that ranked top in the TOP1–compounds complexes. Thus, TYR274 means a lot for ligand binding with TOP1, which warrants studies for further validation and to identify its role. In this study, natural products from the TCM molecule library, which were identified to have desired binding affinities, deserve further experimental validation.

To explore whether the selected compounds could directly interact with TOP1, SPR experiments were administrated. The vaccinia TOP1 was chosen as the target protein in this study, because it presented high homology with monkeypox TOP1 (99%). A low equilibrium dissociation constant (KD) of rosmarinic acid, myricitrin, quercitrin, and ofloxacin was also calculated as 2.16 μM, 3.54 μM, 4.77 μM, and 5.46 μM, respectively. Figure 6 presents the binding affinity of TOP1 with rosmarinic acid, myricitrin, quercitrin, and ofloxacin, which increased with the level of concentration, and rosmarinic acid's affinity to TOP1 ranked as having the optimum sensitivity compared to the other compounds (Figure 5). Taken together, the above outcome proves that TOP1 may directly target rosmarinic acid, myricitrin, quercitrin, and ofloxacin.

Recently, Merecz-Sadowska et al. proved that extracts from Leonotis nepetifolia, including rosmarinic acid, had a strong impact on topoisomerase I activity, leading to cytotoxic potential against human melanoma cells [23]. Soluble rosmarinic acid was originally isolated from rosmarinus officinalis and used as an active ingredient and index component from herbs and nutraceuticals in TCM. It is widely distributed in a variety of plants of Labiaceae, Sycamaceae, Pomeraceae, Tiliaceae, and Umbelliferaceae. A special caffeoyl moiety confers rosmarinic acid with antiviral properties. Previous reports revealed that rosmarinic acid inhibits several viruses, such as HPV, VSV Ebola-pseudotyped, and vaccinia viruses [24]. In addition, heparan sulfate proteoglycans-mediating cellular attachments are required for viruses that strongly interfere with the caffeic chelates [24]. The result from the targeted prediction of rosmarinic acid during the cellular component also supported the assumption that the predicted target mainly participates in the membrane raft and extracellular space (Figure 6A).

Many flavonoids from natural products used both as medicine and food have strong in vitro and antiviral efficacy [25]. Both myricitrin and quercetin are dietary flavonoids with few adverse effects, which present potential anti-MPXV properties in our study. Multiple computational and experimental studies revealed that flavonoids, especially flavonols and their derivatives, are effective viral inhibitors, as well as possessing significant anti-inflammatory activities [26,27].

The bioinformatics analysis was utilized to briefly discuss the feasibility of cellular biological processes and pathways towards rosmarinic acid. Carbonic anhydrases ranked highly as the most likely target towards rosmarinic acid (Figure 6A). A previous study showed that carbonic anhydrases mainly regulate the pH and osmotic balance. Suri et al. further demonstrated that carbonic anhydrase II plays a critical role in Th2-relayed and toll-like receptor 3-caused pathways in inflammatory skin circumstances [28]. The functional annotation of GO and the enrichment analysis forecasted that the signaling pathways may participate in the anti-inflammatory response with rosmarinic acid (Figure 6B). Interestingly, nitric oxide (NO) was highlighted as a crucial molecule in series signaling pathways, including the vascular, metabolic, immune, and antiviral pathways. The deleterious physiological situation caused by viral infection may be partially reversed through the restoration of normative NO levels via combined interventions, such as pharmaceutical, dietary, or complex behavioral interventions [29]. Therefore, possible MPXV inhibitors, such as rosmarinic acid, may be exploited by targeting TOP1 or restoring NO-associated inflammatory responses. Nevertheless, further studies are required to clarify the role of rosmarinic acid in mediating nitrogen metabolism during MPXV infection. Overall, we aim to investigate whether the anti-MPXV activity of rosmarinic acid not only depends on the inhibition of viral functional protein such as TOP1 but also the alleviation of host inflammation caused by NO.

Multiple reports showed that both myricitrin and quercitrin are potential natural antioxidant agents [30,31]. Recent research further revealed that rosmarinic acid, coupled with flavonoids, conjugates with antioxidant and anti-inflammatory properties [32]. The application of those natural antioxidant agents is expected to inhibit viral replication, as well as alleviate the oxidative stress caused by viral hyperinflammation. TCM contains promiscuous phytonutrients of rosmarinic acid, myricitrin, and quercitrin. However, there is still a gap in the knowledge of its bioavailability towards rosmarinic acid, myricitrin, and quercitrin and their clinical applications. More studies emphasizing compound metabolites are still required. Although several vaccines and drugs against MPXV are being highlighted for their efficacy, exploring the repurposing of natural products from TCM may provide alternatives against MPXV. In summary, grouping some of these phytonutrients into the right combination in the form of a food supplement not only boosts the immune system to prevent viral spread but also further suppresses hyperinflammation, providing both prophylactic and therapeutic support against MPXV [33].

4. Materials and Methods

4.1. Homology Modeling

The TOP1 sequences were obtained from the NCBI (https://www.ncbi.nlm.nih.gov/protein/YP_010377093.1?report=genbank&log$=prottop&blast_rank=5&RID=J7W751W701R/ (accessed on 19 December 2022)) with the accession code YP_010377093.1 or were obtained from UniProt with ID F1DIV3. The homology model of TOP1 was obtained using SWISS-MODEL (SWISS-MODEL, https://swissmodel.expasy.org/ (accessed on 19 December 2022)) [34]. The TOP1 amino acid sequences were submitted into SWISS-MODEL, which implemented the Basic Local Alignment Search Tool (BLAST) search to obtain reasonable templates that were identical to the target sequence. The TOP1 structures from the variola virus or vaccinia virus (PDB ID: 3IGC, 2H7F, and 1A41) ranked top, with identity scores of 96.15, 98.08, and 96.94, respectively. The three structures were then chosen as the structural template to model the MPXV TOP1 [35].

4.2. Structural Validation

SAVES v6.0 (UCLA-DOE LAB, (Los Angeles, CA, USA)) was applied to assess the quality of the producing models (https://saves.mbi.ucla.edu/ (accessed on 22 December 2022)). The reliability of the model was validated by analyzing its Ramachandran plot, as well as the measurement of the 3D model's compatibility by the Verify 3D and PROCHECK services (https://saves.mbi.ucla.edu/results?job=1051156&p=procheck/ (accessed on 22 December 2022)) [11,35].

4.3. Prediction of Binding Sites

MOE 2015.10 (CCG, (Ottawa, CAN)) was applied to investigate possible binding sites of the TOP1 model protein, and the docking grid file centered on the cocrystal ligand at the active site was generated by the Receptor Grid Generation module of Schrödinger 2021 [36].

A total of 13 binding sites were predicted using MOE, and Hyd, which ranked top, was selected for further study. The selected Site Finder found 50, 1.96, 72, and 132 in terms of the sizes for PLB, Hyd, and Side, respectively. The site that ranked top was selected for further receptor grid generation. The site residues are listed as follows: ARG67; GLN69; VAL77; ARG80; ASN81; LYS83; ARG84; ARG86; ILE87; ARG90; PHE127; PHE128; ILE129; ARG130; PHE131; GLY132; LYS133; TYR136; ASN140; THR142; VAL143; GLY144; PHE164; GLY166; LYS167; ASP168; LYS169; HIS172; PHE174; TYR209; ILE212; ARG218; ILE219; LYS220; ARG223; THR224; VAL263; GLY264; HIS265; ARG272; ALA273; TYR274;

4.4. Preparation of Protein Targets and Ligand Libraries

The compounds were acquired from the Traditional Chinese Medicine Library and the Natural Product Library (https://www.selleck.cn/screening/natural-product-library.html/ (accessed on 23 December 2022)). Schrödinger Maestro was utilized to generate the necessary format of the ligands or receptor grid. A total of 4103 natural micromolecules in the format of 2D spatial data files (sdf) were uploaded to the LigPrep module process. The possible ionization states of the small molecules were generated in the specified PH range (PH = 7.0 $\pm$ 2.0) under the condition of an OPLS4 force field; then, all possible combinations of the chiral atoms were generated. Each ligand could generate up to 32 different conformations and generate one low-energy ring conformation. Eventually, the small molecules were transformed into 3D structures (maestro). In addition, the compound libraries were sifted according to Lipinski's rule of five. Ofloxacin, which presented a strong inhibition of the vaccinia TOP1 poxvirus, was also included in this study [37].

The structure minimization step was done in the preparation of protein targets via the Protein Preparation Wizard module in Schrödinger 2021, including the addition of hydrogen atoms and residue sidechains, structure optimization, and energy minimization. Water was removed from the modeled TOP1, while other impurity molecules were removed from the model structure. The protonation state of the residues in the protein was generated under a specific PH condition (PH = 7.0). Finally, the system was optimized under the OPLS4 force field to prepare a more reasonable protein structure.

4.5. Receptor Grid Generation and Virtual Screening

Schrödinger was applied for the process of virtual screening. The Receptor Grid Generation module was used to generate the docking grid file centered on the binding site. The coordinate of the binding site was parameterized based on the MOE-binding site prediction. The virtual screening was based on three kinds of different precision screenings and MM/GBSA [36]: (1) high-throughput virtual screening (HTVS), (2) standard precision screening (SP), and (3) extra-precision screening (XP). Firstly, HTVS filtered 50% unfitted compounds; the remaining 50% matched compounds were further employed for SP. Secondly, the top 15% fitted compounds of SP with better docking scores were performed for XP screening. Eventually, XP highlighted the remaining 10% of the compounds with the desired docking score. Meanwhile, MM/GBSA was utilized to calculate the binding affinity between the compound and TOP1, respectively.

4.6. Characterisation of Binding Mechanism

The binding between TOP1 and the compounds was evaluated and analyzed via Glide docking (Maestro).

4.7. MD Simulations of Protein–Ligand Complexes and Proteins

The mechanism of the protein–ligand interactions was clarified through energy minimization in a precise solvent with AMBER18. The forcefields of ff14SB and GAFF were used for the protein and the small molecules, respectively. To be specific, the complex of TOP1–ligands was dissolved virtually in a rectilinear box of TIP3P3 water molecules buffering 6 Å from the macromolecular system, with a total amount of 17,990 immersed water molecules used for solvation. Energy minimization was implemented in the following two consecutive steps with the default parameters. Canonical ensemble (NVT) and isothermal-isobaric ensemble (NPT) were used for equilibrium system, and a MD simulation of 100 ns was performed at a normal temperature and pressure. The RMSD, RMSF, Rg, and SASA were employed to test the complicated stability after the project was completed [38]. The computation of the complexes was implemented using MM/GBSA, where the binding energy was calculated, and the individual energy contributions of the residues were found [38].

4.8. Surface Plasmon Resonance (SPR) Analysis

Four selected compounds, including rosmarinic acid, myricitrin, quercitrin, and ofloxacin, as well as the vaccinia DNA Topoisomerase I, were procured from Selleck (Shanghai, China) and Beyotime (Beijing, China), respectively. The SPR experiments were implemented on the system of the ProteOn XPR36TM SPR instrument (Bio-Rad, Hercules, CA, USA) [39]. Firstly, the vaccinia TOP1 was immobilized with standard amine coupling on the EDC/NHS and activated the GLH biosensor chip surface (Bio-Rad). A TOP1 solution of 1 mg/mL PBST (5 mM, 7.4 pH) was diluted to 30 μg/mL (pH 4.5). Then, TOP1 (5 μL/min for 400 s) was charged and covalently fixed. The final immobilization level for TOP1 was around 18,000 RU. The selected compounds were provided in PBS with 0.005% Tween-20 (pH 7.4) and injected at 20 mL/min for 150 s at concentrations of 6.25–100 μM (doubling dilution). The related phases with 5 concentrations towards 150 s were simultaneously injected at a flow rate of 30 mL/min, and then, the phases of dissociation were injected at 25 °C for 250 s. The chip surface was regenerated with 30 s pulses of running buffer after compound injection. The collection of data was reference-subtracted via ProtedOn ManagerTM 2.0. OriginPro 8 software was used for the data analysis.

4.9. Forecasting for Mechanism of Action towards Rosmarinic Acid

The structure of rosmarinic acid was submitted to Swiss Target Prediction (http://www.swisstargetprediction.ch/ (accessed on 3 December 2022)) to screen the potential target gene [40]. The functional annotation of the Database for Annotation, Visualization and Integrated Discovery (DAVID) v6.8 (Laboratory of Immunopathogenesis and Bioinformatics, (Maryland, USA)) was implemented to annotate the target gene, and the Official Gene Symbol was selected as the identifier in DAVID v6.8. Both Gene Ontology (GO) and the Kyoto Encyclopedia of Genes and Genomes (KEGG) were used to analyze each target gene [41]. The enrichment bubble plot of the KEGG pathway was formed by R program v3.5.0.

5. Conclusions

In this study, natural products from TCM molecule library were screened via Glide docking (Maestro). Firstly, the acceptable MPXV TOP1 structure was modeled according to the 3D structure of variola virus TOP1 via SWISS-MODEL due to the lack of an available structure in the Protein Data Bank. Subsequently, the virtual screening was based on three kinds of different precision screenings (HTVS, SP, and XP) and MM/GBSA. Three compounds (rosmarinic acid, myricitrin, and quercitrin) presenting with good docking scores and binding affinity were highlighted and purchased for further SPR measurements.

It can be assumed that three potential antiviral compounds, comprising rosmarinic acid, myricitrin, and quercitrin, were discovered through docking-based virtual screening. The following study clarified TYR274 as a crucial residue for compound binding affinities. MD simulations, including MM/PBSA, corroborated the potential inhibition of TOP1 by the compounds. SPR was applied to evaluate the binding affinity of the vaccinia virus TOP1 in vivo. Further experiments are needed to determine whether the compounds have inhibitory activity against MPXV TOP1. Additionally, most of the data provided in this paper were based on a computer-aided drug design. Therefore, the clinical application requires further evaluation of the possible inhibitors through experimental confirmation both in vitro and in vivo.

Author Contributions: Conceptualization, L.Y. and H.L.; methodology, S.A.; software, X.H.; validation, J.C.; formal analysis, B.L.; investigation, J.J.; resources, X.H.; data curation, Y.L. (Yanyan Liao) and Y.L. (Yao Lin); writing—original draft preparation, X.H.; writing—review and editing, S.A.; visualization, B.L.; supervision, H.L.; project administration, L.Y.; funding acquisition, S.A. All authors have read and agreed to the published version of the manuscript.

Funding: This research was funded by National Natural Science Foundation of China, grant number 82160389; Guangxi Science and Technology Plan, grant number GuiKe AB19245038; National Key R&D Program of China, grant number 2022YFC2305001C.

Institutional Review Board Statement: Not applicable.

Informed Consent Statement: Not applicable.

Data Availability Statement: Not applicable.

Conflicts of Interest: The authors declare no conflict of interest.

Sample Availability: Samples of the compounds (Rosmarinic acid, myricitrin, quercitrin, and ofloxacin) are available from the authors.

References

1. Alakunle, E.; Moens, U.; Nchinda, G.; Okeke, M. Monkeypox Virus in Nigeria: Infection Biology, Epidemiology, and Evolution. *Viruses* **2020**, *12*, 1257. [CrossRef]
2. Reynolds, M.G.; Doty, J.B.; Mccollum, A.M.; Olson, V.A.; Nakazawa, Y. Monkeypox re-emergence in Africa: A call to expand the concept and practice of One Health. *Expert Rev. Anti-infective Ther.* **2019**, *17*, 129–139. [CrossRef] [PubMed]
3. Thèves, C.; Crubézy, E.; Biagini, P. History of Smallpox and Its Spread in Human Populations. *Microbiol. Spectr.* **2016**, *4*, 161–172. [CrossRef] [PubMed]
4. Zhu, X.; Hu, Z.; Yu, T.; Hu, H.; Zhao, Y.; Li, C.; Zhu, Q.; Wang, M.; Zhai, P.; He, L.; et al. The Antiviral Effects of Jasminin via Endogenous TNF-α and the Underlying TNF-α-Inducing Action. *Molecules* **2022**, *27*, 1598. [CrossRef] [PubMed]
5. Nguyen, P.Y.; Ajisegiri, W.S.; Costantino, V.; Chughtai, A.A.; MacIntyre, C.R. Reemergence of Human Monkeypox and Declining Population Immunity in the Context of Urbanization, Nigeria, 2017–2020. *Emerg. Infect. Dis.* **2021**, *27*, 1007–1014. [CrossRef]
6. Reeves, P.M.; Smith, S.K.; Olson, V.A.; Thorne, S.H.; Bornmann, W.; Damon, I.K.; Kalman, D. Variola and monkeypox viruses utilize conserved mechanisms of virion motility and release that depend on abl and SRC family tyrosine kinases. *J. Virol.* **2011**, *85*, 21–31. [CrossRef]
7. Bray, M. Pathogenesis and potential antiviral therapy of complications of smallpox vaccination. *Antivir. Res.* **2003**, *58*, 101–114. [CrossRef]
8. Baker, R.; Bray, M.; Huggins, J.W. Potential antiviral therapeutics for smallpox, monkeypox and other orthopoxvirus infections. *Antivir. Res.* **2003**, *57*, 13–23. [CrossRef]
9. Yakovleva, L.; Shuman, S. Chemical Mutagenesis of Vaccinia DNA Topoisomerase Lysine 167 Provides Insights to the Catalysis of DNA Transesterification. *Biochemistry* **2013**, *52*, 984–991. [CrossRef]
10. Sekiguchi, J.; Stivers, J.T.; Mildvan, A.S.; Shuman, S. Mechanism of Inhibition of Vaccinia DNA Topoisomerase by Novobiocin and Coumermycin. *J. Biol. Chem.* **1996**, *271*, 2313–2322. [CrossRef]
11. Broni, E.; Kwofie, S.K.; Asiedu, S.O.; Miller, W.A.; Wilson, M.D. A Molecular Modeling Approach to Identify Potential Antileishmanial Compounds Against the Cell Division Cycle (cdc)-2-Related Kinase 12 (CRK12) Receptor of *Leishmania donovani*. *Biomolecules* **2021**, *11*, 458. [CrossRef]
12. Sarkar, B.; Kulharia, M.; Mantha, A.K. Understanding human thiol dioxygenase enzymes: Structure to function, and biology to pathology. *Int. J. Exp. Pathol.* **2017**, *98*, 52–66. [CrossRef] [PubMed]
13. Hoda, S.; Gupta, L.; Shankar, J.; Gupta, A.K.; Vijayaraghavan, P. *cis*-9-Hexadecenal, a Natural Compound Targeting Cell Wall Organization, Critical Growth Factor, and Virulence of *Aspergillus fumigatus*. *ACS Omega* **2020**, *5*, 10077–10088. [CrossRef]
14. Tam, B.; Sinha, S.; Wang, S.M. Combining Ramachandran plot and molecular dynamics simulation for structural-based variant classification: Using TP53 variants as model. *Comput Struct Biotechnol J.* **2020**, *18*, 4033–4039. [CrossRef]
15. Dalhoff, A. Antiviral, Antifungal, and Antiparasitic Activities of Fluoroquinolones Optimized for Treatment of Bacterial Infections: A Puzzling Paradox or a Logical Consequence of Their Mode of Action? *Eur. J. Clin. Microbiol. Infect. Dis.* **2015**, *34*, 661–668. [CrossRef] [PubMed]
16. Kamau, E.; Grove, A. Fluoroquinolone-dependent DNA Supercoiling by Vaccinia Topoisomerase I. *J. Mol. Biol.* **2004**, *342*, 479–487. [CrossRef] [PubMed]
17. Saikia, S.; Bordoloi, M. Molecular docking: Challenges, advances and its use in drug discovery perspective. *Curr. Drug Targets* **2019**, *20*, 501–521. [CrossRef] [PubMed]
18. Akaji, K.; Konno, H. Design and Evaluation of Anti-SARS-Coronavirus Agents Based on Molecular Interactions with the Viral Protease. *Molecules* **2020**, *25*, 3920. [CrossRef]

19. Mohammad, T.; Shamsi, A.; Anwar, S.; Umair, M.; Hussain, A.; Rehman, M.T.; AlAjmi, M.F.; Islam, A.; Hassan, M.I. Identification of high-affinity inhibitors of SARS-CoV-2 main protease: Towards the development of effective COVID-19 therapy. *Virus Res.* **2020**, *288*, 198102. [CrossRef] [PubMed]
20. Parashar, A.; Shukla, A.; Sharma, A.; Behl, T.; Goswami, D.; Mehta, V. Reckoning γ-Glutamyl-S-allylcysteine as a potential main protease (mpro) inhibitor of novel SARS-CoV-2 virus identified using docking and molecular dynamics simulation. *Drug Dev. Ind. Pharm.* **2021**, *47*, 699–710. [CrossRef]
21. Hickman, H.D.; Reynoso, G.V.; Ngudiankama, B.F.; Rubin, E.J.; Magadán, J.G.; Cush, S.S.; Gibbs, J.; Molon, B.; Bronte, V.; Bennink, J.R.; et al. Anatomically Restricted Synergistic Antiviral Activities of Innate and Adaptive Immune Cells in the Skin. *Cell Host Microbe* **2013**, *13*, 155–168. [CrossRef]
22. Reed, B.; Yakovleva, L.; Shuman, S.; Ghose, R. Characterization of DNA Binding by the Isolated N-Terminal Domain of Vaccinia Virus DNA Topoisomerase IB. *Biochemistry* **2017**, *56*, 3307–3317. [CrossRef] [PubMed]
23. Merecz-Sadowska, A.; Sitarek, P.; Śliwiński, T.; Zajdel, K.; Malinowska, K.; Zielińska-Bliźniewska, H.; Kucharska, E.; Zajdel, R. In Vitro and In Silico Studies on *Leonotis nepetifolia* (L.) R. Br. Root Extract against Cancer Cells. *Curr. Pharm. Biotechnol.* **2022**, *23*, 1383–1395. [CrossRef]
24. Langland, J.; Jacobs, B.; Wagner, C.E.; Ruiz, G.; Cahill, T.M. Antiviral activity of metal chelates of caffeic acid and similar compounds towards herpes simplex, VSV-Ebola pseudotyped and vaccinia viruses. *Antivir. Res.* **2018**, *160*, 143–150. [CrossRef]
25. Jannat, K.; Paul, A.K.; Bondhon, T.A.; Hasan, A.; Nawaz, M.; Jahan, R.; Mahboob, T.; Nissapatorn, V.; Wilairatana, P.; Pereira, M.D.L.; et al. Nanotechnology Applications of Flavonoids for Viral Diseases. *Pharmaceutics* **2021**, *13*, 1895. [CrossRef]
26. Mouffouk, C.; Mouffouk, S.; Mouffouk, S.; Hambaba, L.; Haba, H. Flavonols as potential antiviral drugs targeting SARS-CoV-2 proteases (3CL(pro) and PL(pro)), spike protein, RNA-dependent RNA polymerase (RdRp) and angiotensin-converting enzyme II receptor (ACE2). *Eur. J. Pharmacol.* **2021**, *891*, 173759. [CrossRef]
27. Hu, X.; Cai, X.; Song, X.; Li, C.; Zhao, J.; Luo, W.; Zhang, Q.; Ekumi, I.O.; He, Z. Possible SARS-coronavirus 2 inhibitor revealed by simulated molecular docking to viral main protease and host toll-like receptor. *Futur. Virol.* **2020**, *15*, 359–368. [CrossRef]
28. Suri, B.; Verma, N.; Schmidtchen, A. Toll-like Receptor 3 Agonist, Polyinosinic-polycytidylic Acid, Upregulates Carbonic Anhydrase II in Human Keratinocytes. *Acta Derm. Venereol.* **2018**, *98*, 762–765. [CrossRef]
29. Åkerström, S.; Mousavi-Jazi, M.; Klingström, J.; Leijon, M.; Lundkvist, A.; Mirazimi, A. Nitric Oxide Inhibits the Replication Cycle of Severe Acute Respiratory Syndrome Coronavirus. *J. Virol.* **2005**, *79*, 1966–1969. [CrossRef]
30. Mangmool, S.; Kunpukpong, I.; Kitphati, W.; Anantachoke, N. Antioxidant and Anticholinesterase Activities of Extracts and Phytochemicals of *Syzygium antisepticum* Leaves. *Molecules* **2021**, *26*, 3295. [CrossRef] [PubMed]
31. Ramos, A.S.; Mar, J.M.; da Silva, L.S.; Acho, L.D.; da Silva, B.J.P.; Lima, E.S.; Campelo, P.H.; Sanches, E.A.; Bezerra, J.A.; Chaves, F.C.M.; et al. Pedra-ume caá fruit: An Amazon cherry rich in phenolic compounds with antiglycant and antioxidant properties. *Food Res. Int.* **2019**, *123*, 674–683. [CrossRef] [PubMed]
32. Huerta-Madronal, M.; Caro-Leon, J.; Espinosa-Cano, E.; Aguilar, M.R.; Vazquez-Lasa, B. Chitosan–Rosmarinic acid conjugates with antioxidant, anti-inflammatory and photoprotective properties. *Carbohydr. Polym.* **2021**, *273*, 118619. [CrossRef]
33. Mrityunjaya, M.; Pavithra, V.; Neelam, R.; Janhavi, P.; Halami, P.M.; Ravindra, P.V. Immune-Boosting, Antioxidant and Anti-inflammatory Food Supplements Targeting Pathogenesis of COVID-. *Front. Immunol.* **2020**, *11*, 570122. [CrossRef]
34. Waterhouse, A.; Bertoni, M.; Bienert, S.; Studer, G.; Tauriello, G.; Gumienny, R.; Heer, F.T.; De Beer, T.A.P.; Rempfer, C.; Bordoli, L.; et al. SWISS-MODEL: Homology modelling of protein structures and complexes. *Nucleic Acids Res.* **2018**, *46*, W296–W303. [CrossRef]
35. Bowie, J.U.; Lüthy, R.; Eisenberg, D. A Method to Identify Protein Sequences That Fold into a Known Three-Dimensional Structure. *Science* **1991**, *253*, 164–170. [CrossRef]
36. Friesner, R.A.; Murphy, R.B.; Repasky, M.P.; Frye, L.L.; Greenwood, J.R.; Halgren, T.A.; Sanschagrin, P.C.; Mainz, D.T. Extra Precision Glide: Docking and Scoring Incorporating a Model of Hydrophobic Enclosure for Protein−Ligand Complexes. *J. Med. Chem.* **2006**, *49*, 6177–6196. [CrossRef]
37. Ikeda, S.; Yazawa, M.; Nishimura, C. Antiviral activity and inhibition of topoisomerase by ofloxacin, a new quinolone derivative. *Antiviral Res.* **1987**, *8*(3), 103–113. [CrossRef] [PubMed]
38. Sk, M.F.; Roy, R.; Jonniya, N.A.; Poddar, S.; Kar, P. Elucidating biophysical basis of binding of inhibitors to SARS-CoV-2 main protease by using molecular dynamics simulations and free energy calculations. *J. Biomol. Struct. Dyn.* **2020**, *38*, 1–13. [CrossRef]
39. Zhang, H.; Liu, W.; Liu, Z.; Ju, Y.; Xu, M.; Zhang, Y.; Wu, X.; Gu, Q.; Wang, Z.; Xu, J. Discovery of indoleamine 2,3-dioxygenase inhibitors using machine learning based virtual screening. *MedChemComm* **2018**, *9*, 937–945. [CrossRef] [PubMed]
40. Gfeller, D.; Grosdidier, A.; Wirth, M.; Daina, A.; Michielin, O.; Zoete, V. SwissTargetPrediction: A web server for target prediction of bioactive small molecules. *Nucleic Acids Res.* **2014**, *42*, W32–W38. [CrossRef] [PubMed]
41. Huang, D.W.; Sherman, B.T.; Tan, Q.; Kir, J.; Liu, D.; Bryant, D.; Guo, Y.; Stephens, R.; Baseler, M.W.; Lane, H.C.; et al. DAVID Bioinformatics Resources: Expanded annotation database and novel algorithms to better extract biology from large gene lists. *Nucleic Acids Res.* **2007**, *35*, W169–W175. [CrossRef] [PubMed]

Disclaimer/Publisher's Note: The statements, opinions and data contained in all publications are solely those of the individual author(s) and contributor(s) and not of MDPI and/or the editor(s). MDPI and/or the editor(s) disclaim responsibility for any injury to people or property resulting from any ideas, methods, instructions or products referred to in the content.

MDPI
St. Alban-Anlage 66
4052 Basel
Switzerland
www.mdpi.com

Molecules Editorial Office
E-mail: molecules@mdpi.com
www.mdpi.com/journal/molecules

Disclaimer/Publisher's Note: The statements, opinions and data contained in all publications are solely those of the individual author(s) and contributor(s) and not of MDPI and/or the editor(s). MDPI and/or the editor(s) disclaim responsibility for any injury to people or property resulting from any ideas, methods, instructions or products referred to in the content.

www.ingramcontent.com/pod-product-compliance
Lightning Source LLC
Chambersburg PA
CBHW061625210926
51850CB00009B/2441
* 9 7 8 3 7 2 5 8 0 7 9 0 1 *